D0258077

The Next World War

The Next World War

James Adams

HUTCHINSON
LONDON

1 3 5 7 9 10 8 6 4 2

This edition first published in 1998 by
Hutchinson

Random House (UK) Limited
20 Vauxhall Bridge Road, London, SW1V 2SA

Random House Australia (Pty) Limited
20 Alfred Street, Milsons Point, Sydney,
New South Wales 2061, Australia

Random House New Zealand Limited
18 Poland Road, Glenfield,
Auckland 10, New Zealand

Random House South Africa (Pty) Ltd
Endulini, 5A Jubilee Road, Parktown 2193, South Africa

A CIP record for this book is available
from the British Library

Papers used by Random House UK Limited are natural,
recyclable products made from wood grown in sustainable
forests. The manufacturing processes conform to the
environmental regulations of the country of origin.

ISBN 0 091 80232 6

Printed and bound in Great Britain by
Mackays of Chatham PLC, Chatham, Kent

Acknowledgments

F I R S T , I must thank Nick Peters, a good friend without whom this book would not have been completed. He helped me immeasurably throughout the long and difficult process of researching and writing, and because he is a fine journalist and a wonderful colleague it all happened.

Allison Bozniak has worked with me for two years and I have learned to value her advice, trust her judgment and recognize how important it is to have an assistant who is loyal and tolerant of the ups and downs of the writing process. Allison found information where I believed none existed, created order out of chaos and generally helped make the process work.

Gannon Pitre did some early research that set hares running down the right track. He showed a real understanding of the perils and the promise of information warfare and helped set me on the correct course.

I am very grateful to my assistant, Victoria Brown, who uses the word "excellent" to describe her pleasure at achieving the impossible and her enthusiasm for carrying it out. She truly is the calm in the eye of the storm and I readily acknowledge her contribution to this book. She has my thanks.

A large number of people in several countries went out of their way to explain their views and to open doors. Many requested anonymity, which I will respect. Others I can thank publicly: Stewart Baker, David Bickford, Stephen Braithwaite, Mike Brown, Al Campen, Dom Cardonita, Captain Stan Coerr, Barry Collin, Doug Dearth, Dorothy Denning, Matt Devost, Ken Gabriel, Fred Giessler, Maj. Gen. Michael Hayden, Professor Chih-Ming Ho, Chris Hoenig, Glen Howard, Martin Libicki, Lt. Brenda Malone, Gen. Robert Marsh, Mike Vlahos, Erin Whitney-Smith and Ira Winkler.

Finally, my wife, Rene, has patiently put up with the interruptions in the lives of our family that this book has caused. For ten years she has lived through a succession of books and faithfully managed the crises and supported me throughout. This has been both the most challenging book and the most difficult to research and write, and I am truly grateful that I have a soul mate on whom I can depend completely.

TO SANDY

Contents

Introduction

KINKEAD'S is one of Washington, D.C.'s, most fashionable restaurants. A few blocks from the White House, it is frequented by President Clinton, White House staffers and other movers and shakers who like to see and be seen. I like it because I can always get the same quiet corner table far removed from prying eyes or the eavesdroppers that are a constant nuisance in a city where people consider everyone else's business to be their own.

I had invited the Washington station chief of MI6 for one of our regular lunches and we had been enjoying some rockfish and an espresso when he began describing the problems confronting the intelligence community with the growth of organized crime and the expected developments in cryptography. The world he described was indeed frightening: gangs who cared little for national boundaries, who had enormous wealth that far exceeded the MI6's own budget and the ability to communicate in unbreakable codes over the Internet.

"It's all part of information warfare and it's so much what we are all about these days," he explained.

I had heard the term before but knew little about it. In the course of researching *The New Spies*, a book that examined the role of intelligence after the Cold War, I had heard occasional mutterings about the possibilities of a new form of warfare to meet the challenges of the postcommunist world. But it meant little to me then and at that lunch with the MI6 officer, I simply nodded wisely, as if I understood perfectly what the man was talking about.

Intrigued, I began to investigate the term "information warfare" to try to discover what it meant. I quickly learned that it meant different things to different people and that within the defense, political and intelligence establishments there was not only disagreement about its meaning but about its purpose. As I delved deeper, I discovered that information warfare was already a reality. In military bases, research centers and defense manufacturing complexes, new weapons and methods of waging war were being developed. In secret programs across the United States, billions of dollars had been quietly allocated to this promising new area of research.

This effort had resulted in a whole new generation of weapons and a new breed of soldiers, sailors and airmen ready to use them in the new kinds of wars that were being quietly planned inside the Pentagon. This is the world of the cyberknight, as some information warriors call themselves; of new weapons such as microwave cannons, plasma guns and fire ants. It is a place where dominance of the battle space is created not by bullets but by bits and bytes.

As the research unfolded and I began to travel across America and then to Europe and Russia, it became clear to me that the world was on the brink of a fundamental revolution in warfare. Not only had the end of the Cold War meant the demise of the old superpower rivalries but it also marked the end of our comparatively stable dependence on nuclear deterrence to keep the peace. Our understanding of war itself had changed dramatically. While the risk of a massive confrontation on the central plains of Europe between the massed armies of the Warsaw Pact and NATO had evaporated, new conflicts have come to fill the vacuum.

Today, we are at war on several fronts. The fights against terrorism, organized crime, economic espionage and weapons proliferation are permanent conflicts that are likely to confront us through the next century. At the same time, some of the old tensions remain, with ethnic conflicts in Bosnia and Somalia reminding the revolutionaries and the traditionalists that wars will continue however many chips and computers may populate the world.

What I also learned was that a new generation of visionaries has sprung up, enabled by the computer revolution. They predict a very different world that will emerge from the current uncertainty. It will be a place where wars of every type will be fought not by soldiers confronting soldiers but by new warriors engaging in the infosphere, the virtual world where commerce, conversation and connectivity will all occur. In those wars, new weapons will be needed and real power will accrue to the nation or group that understands the use of this new technology.

In this new world, the soldier will be the young geek in uniform who can insert a virus into Teheran's electricity supply to plunge the city into darkness. His civilian equivalent will be able to read every e-mail, crash any office computer anywhere in the world, invade networks and destroy systems, all from thousands of miles away. The bank account of the drug baron or organized criminal located in an apparently safe offshore haven will be an open book.

The soldier will also be the man or woman equipped with a uniform powered by body heat that automatically adjusts to the environment and that relays location and vital signs back to base. That soldier will have on his head a helmet that allows him to see in all conditions, to locate incoming fire and return it with deadly accuracy, and an eyepiece that will provide his location, the location of the enemy and the locations of others in his patrol. He will have in his backpack "ants" powered by new microchips that will be able to see, smell and hear or even explode on command. He will be equipped with tiny airplanes no larger than a small notebook that will fly ahead and show him the terrain and the enemy.

The enemy will be different, too. No longer will it be the simple terrorist armed with an AK-47 or the Semtex bomb (although he will still be around); the new threat will be groups who will bond in cyber-space and attack using the new weapons of war: viruses, bugs, worms and logic bombs.

Although little of the perils and possibilities of this new form of warfare has reached the public or the political leadership, I discovered that across America thousands of men and women have been quietly working to make what sounds like science fiction into a reality. And such is the pace of the information revolution that many of the tactics and weapons that sound like fantasy are already in place.

When I began the research for this book, I found at first that every door was shut. The intelligence communities refused to talk. Nearly every defense program on the subject was "black" (top secret), so it was hard to penetrate there, too. The dangers have been discussed only in hushed tones in the sanctuaries of the Pentagon or CIA headquarters. There has been almost no public discussion and no opportunity for an informed debate. Yet, information warfare is something that will affect every man, woman and child on this planet.

After many years of covering the covert world as a journalist and author, I found I knew enough people in strange places to begin to get a glimpse of what lay behind the curtain of secrecy. The result was visits to secret information warfare centers across America, to laboratories where twenty-first-century weapons are already being made, to think tanks where the new wars are being discussed and modeled. I took an extraordinary trip to Moscow where the Russian defense and intelligence community shared with me their real fears about the future of information warfare. They see it as threatening and likely to cause a new arms race between the major powers. Researching this book was the most exciting, fascinating and frightening journey of my life, rich in vision and color and filled with people of enthusiasm and insight. At the same time I realized that we are beginning a journey into a world that is fraught with danger.

It was disturbing also to learn that the countries that have the most effective information warfare capabilities are also the most vulner-able to attack. Uniquely in the history of the world, a single individual armed with just a computer and a modem can literally hold America to ransom.

A glimpse of that vulnerability was provided on February 5, 1997, when George Tenet, then the proposed Director of Central Intelligence, during his confirmation hearings took the Senate Intelligence Committee through the usual litany of threats and potential crises that confronted the United States.

"First is the continuing transformation of Russia and the evolution of China," he said. "Second are those states—North Korea, Iran and Iraq —whose hostile policies can undermine regional stability. Third are very important transnational issues—terrorism, proliferation, international drug trafficking and international organized crime. Fourth are those regional hot spots—such as the Middle East, the south Asian subcontinent, Bosnia and the Aegean—which carry a high potential for conflict. Fifth are states and regions buffeted by human misery and large-scale suffering, states involved in or unable to cope with ethnic and civil conflict, forced migration, refugees and the potential for large-scale deaths from disease and starvation."

There was nothing particularly new in any of this. It was a speech that could have been made by most CIA directors at any time over the previous twenty years. Even the names would have been the same. What set Tenet's remarks apart was the few sentences at the end of his lengthy prepared statement.

"There's a new threat I've put in this transnational threat area and that is security to information systems in the United States. The tremendous growth in communications technology is shrinking distances and weakening the barriers to the flow of information. This technology also presents us with an important transnational challenge—protecting our information systems. Recognizing this problem, we are assessing countries that have such potential, including those which appear to have instituted formal information warfare programs."

Despite the apparent recognition that information warfare is the next revolutionary technology, there remain fundamental disagreements among individuals, groups and nations about the significance of the Information Age in general and information warfare in particular. Some of the traditionalists in the military see IW as simply part of an evolutionary process that began with the longbow and has continued through gunpowder and repeating rifles, guided missiles and stealth technology. For those on the evolutionary wing, IW is both a promise and a threat. IW may allow for better use of existing forces but if it develops too far, it may threaten the very existence of those forces.

The real revolutionaries believe that for countries like the United States, IW offers the possibility of fighting and winning wars without the commitment of troops on the ground—something that is considered heresy to military historians, who argue that troops will always be needed to take and hold ground. The revolutionaries argue that even the very definition of "ground" is changing as the world migrates from earth to cyberspace.

This debate is ongoing and will likely not be resolved for some years. In preparing this book, I have chosen to use a broader definition of both "information" and "warfare" to accommodate both sides of the argument. Some experts argue that our old definitions of both words are redundant as we migrate into the infosphere.

A conventional military view would suggest that information is simply a message or set of messages that flows from commanders to their troops and back again. In the context of information warfare, disrupting the flow can have a critical impact on the course of a war. But information is also a medium or glue that holds systems together. For example, a modern Abrams tank has fifty microprocessors, many of which talk to each other and are dependent on each other to work effectively. Disrupt their ability to communicate—destroy the medium—and the system becomes ineffective.

Finally, information can also be the actions taken to obtain intelligence from an opponent's information flows or databases. For example, the CIA might use computers to go inside the database of a criminal gang in Moscow to obtain the organization's financial records. The information gathered from that operation could be exploited to help combat global organized crime.

Information warfare therefore seems to break down into three distinct pieces: perception management where information is the message, systems destruction where information is the medium, and information exploitation where information is an opponent's resource to be targeted. To adequately address all three of these categories, I have cast a wide net to embrace the intelligence community, the Pentagon, and law enforcement, which all use weapons as diverse as viruses, sponge balls, and precision-guided ammunitions to achieve information dominance in the new conflicts.

Making full use of today's information revolution implies not only adopting new technologies but also rethinking the very bases of military organization, doctrine, and strategy. All this requires reformulation in order to fulfill Clausewitz's exhortation that 'knowledge must become capability' in the information age. The information revolution is not simply technological in nature; it has powerful conceptual and organizational dimensions as well. The new meanings of power and information . . . favor the argument that wars and other conflicts in the information age will revolve as much around organizational as technological factors.

The struggle confronting modern societies is how to incorporate the opportunities presented by information warfare while holding on to the foundations that have made societies and cultures function effectively. The scale of this challenge is enormous and the stakes just as large, for as the information revolution gathers pace, so information warfare in all its aspects can threaten us all. Just how this struggle will be resolved is unclear and will probably not become clear for some years. This book is a first attempt to frame the debate, to freeze a moment in time on the information roller coaster so that we may understand where we have come from, where we are, and where we are heading.

Part 1

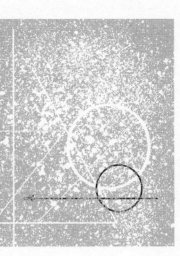

ONE

War in the Infosphere

IT had taken him months to get close enough, but now, Allah be praised, his duty was done. The shaped charges had been concealed, the detonators hooked to a microreceiver that used the tall metal tower above as an antenna. Using the tower as a means toward its own destruction was an irony that gave him pleasure. He checked again.

The desert night was cold. There was no moon, and the brilliant stars shed insufficient light for him to be seen. Security had been rigid at the site from the very first, but in the way of these people it allowed no room for individual initiative on the part of the guards. Thus it was open to exploitation by a motivated and cunning adversary. He smiled to himself at the ease with which he had been able to fool the guard with his fake ID. It would be that particular soldier's last mistake.

Once more, using a tiny pen flashlight held in his teeth, he checked his connections. He refused to allow his natural nervousness to make him hurry. The connections were good. He switched on the receiver.

Out in the desert, an animal sent out a lonely cry. He stood up, looking out toward the beckoning darkness. Stuffing his tools beneath his clothes, he headed back, past the guard, to the compound. He had no illusions about what lay ahead. Once the explosion occurred and his absence was noted, he was bound to be the prime suspect. If they found him he would no doubt be tortured and killed. The thought did not trouble him. He had participated in a historic venture, and his life was of little value compared with the aim he was trying to achieve.

Allah be praised.

He had easily found work in the oil field. Production was being accelerated as reserves elsewhere in China dried up. The region was inhospitable but he was used to that. In fact the oil company relied on people like him; tough, seemingly immune to the extremes of weather, without need of even the simplest luxuries. He had made no effort to conceal his faith, but at the same time neither had he flaunted it. He knew the locals were nervous of religious fanaticism, but that anxiety was far outweighed by their need for workers capable of consistent performance.

In recent years, the Beijing government had settled hundreds of thousands of ethnic Chinese into this predominantly Muslim region, trying to neutralize the growing revolutionary influence from across the border. Even so, it would take several years yet for these implants to become properly acclimatized to the rigors of the desert life, which is why the oil company was still obliged to rely on local workers, who provided excellent cover for an agent like himself.

He had picked up the Semtex from the safe house in the desolate suburbs of Kashi, the region's main city one hundred miles east of the border with Tajikistan. Workers were given seventy-two hours R&R there every two weeks, flown in by their employers, as were his brothers in arms at sites all over the region. All of them found their way to the tumbledown house behind the old, barely used mosque. Minor but persistent persecution had made life difficult for true believers, but worship continued, and it was not unusual to see followers of Islam in the area. Despite the danger inherent in a centralized point of supply, the convenience had been deemed worth the risk.

Thus from Kashi went a stream of explosives and ancillary equipment, smuggled in the linings of suitcases, under flowing robes, even in turbans, unto the moment when a mighty blow for Allah would be struck.

As the saboteur melted into the desert, his job complete, that moment was just hours away.

From the celestial perspective of the KH-15 spy satellite, it looked like a grand fireworks display. Across the vast reaches of the Taklimakan Shamo in Xinjiang province (also known as Chinese Turkistan), a series of brilliant, simultaneous explosions punctuated the drab landscape.

The mundane buzz of chatter that provided the routine backdrop to activity in the monitoring section at CIA fell away into silence. As the initial pinpricks of light blossomed into clouds of black and red, the duty officer called for close-ups on the explosions.

As he did so, he set in train the alert sequence, a kind of automatic telephone tree by which key Agency officials were rousted from their beds in the event of an emergency. The system worked well. Drivers, security details, canteen staff, all were instantly brought up to speed to administer the necessary logistic backup.

In a nondescript house in suburban Washington, the Director of Central Intelligence pulled on his suit jacket and reached for the front door. As he stepped onto his front path, his bodyguards fell into step beside him and his black official car pulled quietly to a stop by the curbside.

• • •

The reality of the White House situation room, the place where America's wars are won and lost, was far removed from the technological wizardry on which America lives or dies.

An apparently small, drab, windowless room in the White House basement, it seems to completely lack any electronic gadgets. But, a push of a button reveals a secret world of maps, surveillance relays and video conferencing.

The atmosphere in the room was charged with anticipation, yet laced with a nagging worry that the unfolding crisis was somehow beyond the United States's abilities to control. For two decades now, the Islamic Arc of Crisis had been at the forefront of global flashpoints. Scenarios for developing conflict in the region that stretched from the Caspian Sea to western China had been mapped, drawn, plotted and endlessly debated. Encompassing the Islamic republics of the former Soviet Union, known in shorthand as the Stans, as well as Iran, Afghanistan and Pakistan, it was the land of promise for whichever country sought to lead a new Islamic empire.

Two nations were fighting for that crown, in a secret, undeclared war in which holy books were as much part of the battlefield as guns and money. Iran and Saudi Arabia had been stirring up the region's religious and political ambitions, hoping to capitalize on whatever ferment resulted. Rebels in Chechnya had benefited from the military largess of both suitors. Religious leaders in Tajikistan, Uzbekistan and Turkmenistan gratefully accepted Saudi help in rebuilding mosques and stocking schools with religious textbooks.

Critical to the Saudi strategy was support for the Taliban leadership in Afghanistan. The radical Taliban had seized power in the wake of the Soviet disaster in Afghanistan, and had proved eager to spread its revolutionary influence throughout the region. The Taliban could not care less that in accepting arms and money from Riyadh they were acting as Saudi surrogates.

As the Saudis devoted more and more energy to staking their claim to Islamic leadership—or doing whatever it took to deny Iran that opportunity—so they put more and more distance between themselves and the Americans, anxious not to be open to accusations of being too close to the Great Satan. Wahhabism, Saudi Arabia's own brand of Islamic fundamentalism, had achieved almost irresistible momentum in its attempt to propel the country from the ways of the corrupt and factionalized monarchy toward a future that would be dominated by the mullahs. American influence in the region was on the wane.

Iran was playing a brilliant double game. Despite turning a more friendly, market-oriented face toward Western Europe, and developing

strong trade links with Russia, it was at the same time losing none of its
fervor for the Islamic revolution that fueled its national life. Its success
in employing both strategies simultaneously had been demonstrated in
1997, when despite being found guilty by a German court of sponsoring
the assassinations of Iranian dissidents in Germany, it continued to
enjoy the friendship of Moscow and of several West European states.
America at times seemed the only major power that kept a consistently
hostile attitude toward Teheran. Memories of 1979 burned deep in the
American consciousness.

The first priority was to establish who had set off the blasts. The National Security Adviser desperately needed some kind of answer to lay
before the President at their hastily convened meeting that would start
in twenty minutes.

The second urgent question was, What would Beijing do in response?

Immediate suspicion fell on the Tibetans, who for decades had suffered terribly at the hands of China's brutal occupation and repression
of their country. Analysts from the CIA suggested the Tibetans did not
have the sophistication to carry out such an attack.

The next possibility was that the oil fields were being put out of
action by a Middle Eastern power that hoped to derail the Chinese oil
operations and in so doing win back some lost business. China's oil
exploration in the desert region of its western province was made necessary by the country's burgeoning demand for energy. Even in the late
1990s, China's energy supply system was crude, and fell far short of
what its rapidly expanding industrial base required. At any one time, 20
percent of factories stood idle because of power shortages. Coal was
the country's principal energy source, with nuclear and hydroelectric
generation a major priority. But the country also needed oil, up to 1.3
million barrels a day in 2000, with over 68 percent of that coming from
imports. The country's main oil fields in Daqing, in northeastern China,
were increasingly less productive, so the push was on to increase domestic production and reduce reliance on imports.

The Tarim Basin in the Taklimakan Shamo was one of the richest
fields in the world, but the hostile local environment and its distance
from the sea made exploration, production and transportation extremely
difficult. The benefits were worth the investment. Reducing oil imports
also reduced China's foreign exchange imbalance, and foreshadowed the
day when the country could become a regional energy exporter.

Suspicion thus fell on the Saudis. Their economy was in serious
trouble. The oil boom of the 1970s and early 1980s had produced
billions in revenues for the nation. Instead of hedging against the day
when the revenues would take a downturn, the money had been spent
lavishly on public works, education and the creation of a huge national

bureaucracy that offered employment to virtually anyone who wanted it. When revenues did indeed start to slow down, the country found itself incapable of sustaining such a grand national lifestyle. There were inevitably economic casualties, all of which played into the hands of the burgeoning radical opposition.

If it were the Saudis, it was likely they had used as their agents the Taliban. The Afghan rebels had benefited handsomely from Saudi financial and logistic support in their fight to rid Afghanistan of the last vestiges of Russian influence, and had learned the dark arts of sabotage and terrorism during the course of the Soviet occupation.

The other natural suspect was Iran. Thoughts that the Teheran government might not have carried out such an act out of respect for a long-standing trade relationship with China were swiftly quashed. The trade between the two countries was based strictly on self-interest, on both their parts. China desperately needed Iran's oil, Iran needed China's help in creating an arsenal of nuclear missiles and chemical weapons, augmented by sales from Russia and North Korea. For some time Iran had had virtually all it needed to successfully fight a sizable regional war at any level from conventional through to nuclear and chemical. In creating this stockpile of weapons of mass destruction Iran had never lost sight of the dream that one day it would be the center of a new Islamic empire. That made the western province of China a real target of opportunity; indeed one of urgency, too. As the Chinese resettlement program started to dilute the Muslim population of Xinjiang, Iran saw a clear duty in rescuing the region for Islam.

As the moment arrived for the National Security Adviser to meet the President, both Saudi Arabia and Iran remained coequal suspects.

Sitting at his desk in the Oval Office, the President focused on what China's next step would be.

Since becoming a manufacturing powerhouse over the previous decade China had built a sizable trade surplus with the United States and other nations of the Pacific Rim. International concern for human rights, which had so dominated world affairs in the 1970s, evaporated in the years that followed as the power of the marketplace proved more potent than the power of the human spirit. Ignoring some minor handwringing from Congress, successive administrations had put national economic self-interest over outrage at the repression of political expression in China, the imprisonment of dissidents and the use of prison and slave labor to produce the goods American consumers craved. At times it seemed Washington was more embarrassed by the Tiananmen Square massacre of 1989 than Beijing, such was the certitude with which the Chinese leadership viewed their purpose and the moral ambiguity with which the Americans viewed theirs. That the Chinese had managed to turn their vast backward nation into the economic titan of Asia without compromising the central authoritarian control of the communist party was testament to the party leadership's resilience, resourcefulness and ruthlessness.

The sophistication of the Chinese economy was roughly on a par with what the Japanese had achieved by the early 1960s. Marvelously adept, as the Japanese had been, at copying Western designs, something to which American software and entertainment companies could ruefully attest, they had entered phase two of their economic revolution. They were moving from being a cheap assembly plant for American multinational companies to creating their own products. China now eclipsed Japan as the predominant regional economic power, something which terrified the Japanese. It was, after all, fear of growing Chinese power in the 1930s that led to Japan's invasion of China and the terrible depredations the Imperial Army visited upon the Chinese people. Japan realized that memories of that invasion would only spur any Chinese ambitions that included Japan's eclipse or even subjugation.

The Japanese also labored under the supreme disadvantage of their complete reliance on imported energy. It was a vulnerability that informed the country's entire strategic and diplomatic thinking. It was also a vulnerability that a foe like China could exploit in a time of crisis.

The President had long struggled to come to terms with the Chinese mind-set. He could understand national pride. Pride in one's country was, after all, part of being an American; it was natural to see it manifested in other nations. What eluded his understanding was the sense of national destiny that drove China's ambitions. The Chinese felt it completely natural to dominate lesser nations; it was their country's rightful role, and now that their economic and military power had grown sufficiently strong, it was only a matter of time or circumstance before they fulfilled their destiny. The Western concept of peaceful cohabitation was as alien to them as the concept of destiny was to the President.

Carrying out a terrorist attack in China was no easy task. A police state where torture of suspects was routine and families were encouraged to inform on one another, even a normally secure terrorist cell structure proved vulnerable to Chinese penetration. Over previous years, as attacks on political centers and other symbols of Chinese rule had intensified, hundreds of Muslims had been arrested and remained in labor camps where they made cheap plastic toys for export to America. Those attacks were annoying but they were tolerable. The latest incident was different and signaled a new escalation in the war. In only a matter of hours a sweep by Chinese counterintelligence agents had produced a culprit: Iran.

Now that an external force had dared to launch an unprovoked attack on their western oil field, what would they do next? The possibilities ranged from specific retaliation to using the incident as an excuse to launch a long-fermented plan for regional domination.

As the President listened to a depressingly long list of options, all of them seemingly intractable, an aide brought fresh intelligence that had been passed on by the British intelligence service. No matter that American intelligence services had by far the most sophisticated technological apparatus for gathering information, they had a notable weakness when it came to the oldest form of intelligence gathering of all—human beings. The British, however, had a long-established tradition of human intelligence, particularly in the Middle East, a legacy of their imperial past. British agents roamed the souks and bazaars, and penetrated high councils of state across the region.

Now came word from a British asset in Teheran that the Iranian leadership was in a state of high excitement and that there was much rejoicing over the disaster wrought on the hated Chinese. The National Security Agency, which managed America's eavesdropping around the world, confirmed significant increases in signals traffic between Teheran and military units ranged along the nation's borders with Iraq, Afghanistan, the former Soviet republics and the Persian Gulf.

That part of the puzzle in place, the President continued to ponder what would happen next.

He did not have to wait long, but the next event was of such startling unpredictability that it served only to deepen his dilemma.

In the late 1990s, the crucial issue that dominated Middle Eastern affairs was the construction of the Caspian Oil Pipeline. This would carry oil from the fabulously oil-endowed Azerbaijan on the landlocked Caspian Sea to ports where the oil could be refined and shipped. The choice facing the Azerbaijan government in Baku and the multinational oil companies building the pipeline was whether it should run through Turkey to the Mediterranean port of Ceyhan or through Russia to the Black Sea port of Novorossiysk. The enterprise lay at the root of regional conflict for years, such as in Chechnya where the separatist uprising had to be quelled by Moscow. To cede control of the area would sever Russia's land links with Azerbaijan, and thus end Russia's interest in the pipeline and the fabulous transit revenues that would accrue. Georgia, to the west of Azerbaijan and with its own Black Sea coastline, had become a hotbed of mafia activity as criminals saw the bounty that would flow to the region from the oil bonanza. Throughout, Iran played a careful waiting game, knowing that an independent Azerbaijan would eventually opt for freedom from its former ruler in Moscow and security from another potential ruler in the West.

The Iranians desperately wanted the pipeline to traverse their country. It would seal their economic and political ascendancy. Few in the oil business were persuaded of the sincerity of Iran's newly found reasonableness toward the market economies of the West. Western companies had severe doubts about the strategic wisdom of placing in Iranian hands such a precious resource as the pipeline, so they structured a deal that benefited Iran immensely for as long as oil flowed freely through the pipeline to its terminus at Bandar-e Abbas on the Strait of Hormuz. Iran would be hurt more than anyone by holding the pipeline hostage.

The building of the pipeline gave an enormous boost to Iran's regional ambitions and proved a severe setback to those of Saudi Arabia.

So when an intelligence flash from the CIA reported that the Caspian Oil Pipeline had been breached in three places inside Iran, it was not hard to imagine who had been responsible.

Explosions had ripped apart large sections of the pipeline in areas of Iran that were least accessible to construction equipment. Oil was gushing, and fires spewed black smoke heavenward. While it would take relatively little time to turn off the flow upstream from the breaches, it would take weeks to repair them, and it would take even longer for Iran to recover from the humiliation.

Analysis suggested it would have taken a great deal of planning to organize such a raid, which implied that the explosives had been planted earlier, waiting for a moment when maximum political and economic damage would be caused. In the wake of the Iranian attack on China, the pipeline assault would sow confusion in Teheran.

Hard on the heels of the pipeline attack came the awaited response from China, and it took the form of what U.S. national security analysts had come to describe as the Malaccan Option.

The Strait of Malacca is the narrow strip of water between Malaysia and Indonesia. Through here sail oil tankers on their way from the Persian Gulf to the oil-importing nations of Asia. While not as critical a choke point as the Suez Canal, the Strait nevertheless is an important factor in the shipment of oil to the Far East. The alternative requires sailing around Indonesia, almost to the northern coast of Australia.

A battle group of the Chinese navy had long been stationed in the area of the South China Sea known as the Spratly Islands, a haphazard collection of rocks that sixty-eight nations had claimed as their own but that only China had occupied by force. The oil deposits in the region were thought to be considerable, but so far China had made no effort to drill, preferring to expend available resources on the Tarim Basin deposits.

The battle group was now on the move, heading southwest from the Spratlys toward Singapore and the Strait of Malacca.

The President called for a readout on the battle group's power. The response was not reassuring. The recent buildup of an oceangoing navy by the Chinese had included diesel submarines, as well as surface ships with precision-guided vertical launch systems, cruise missiles and antiship mines. Naval air forces had in-flight refueling capability, giving the Chinese military a substantial reach.

As he thought about how to couch a message urging restraint on the Chinese leadership, another flash from both the Pentagon and the CIA reported that a Chinese cruise missile had hit and sunk *The Sword of Islam,* an Iranian-owned oil supertanker that had been carrying oil to Japan. The timing of the strike could not have been more perfect. The vessel was steaming south of the Malaysian port of Kukup where the strait narrowed and turned into the Strait of Singapore. The explosion and wreck, clearly visible from the shore and, more importantly, to the cameras of the international media, stood as a signal to any vessel thinking of following the same route. Just to ensure that the message was received loud and clear, the government in Beijing issued a statement warning that all oil shipments would be subject to attack unless permission for passage was received from the Chinese government and a fee paid. The statement regretted the necessity of the act but said that the attack on the western oil fields by aggressors in Iran had jeopardized the security of the People's Republic and that of its neighbors. This last was a gratuitous bit of bluster that no one was expected to take seriously. Clearly the Chinese had spotted a triple opportunity to exert their power across the entire Southeast Asian region, to retaliate against Iran and to threaten the Japanese, cloaking it all in the language of victimhood.

Almost immediately the President was alerted to an urgent request for an audience from the Japanese ambassador in Washington.

The President sat in the Oval Office deep in thought. The Japanese had informed him that within the last year they had perfected their own missile-deliverable nuclear stockpile and stood ready to deploy it in defense of their energy supply, which was tantamount to saying in defense of their national survival.

Messages had been sent by the United States to the capitals of China, Japan, Iran and Saudi Arabia urging restraint. The United Nations Security Council was meeting in emergency session in New York, but the President despaired of any progress there, what with China wielding the power of veto as a permanent member of the council.

The Joint Chiefs of Staff had made it clear to him that while America had a duty to seek an immediate end to hostilities before a war that would consume China, Japan, Iran, parts of the Stans, Saudi Arabia and who knew where else, it lacked the ability to do anything of conventional military significance without making the situation significantly worse. Military downsizing left the United States with insufficient power to step into a fight between two belligerents, let alone to step into a conflict as complex as the one now brewing. Even sending reinforcements to the U.S. military base in Korea seemed a pathetically inadequate response.

And there was no doubt in anyone's mind that the Chinese were prepared to launch one of their long-range nuclear missiles at the west coast of the United States if the U.S. Navy took aggressive action such as launching a strike against the battle group.

What America did have in this stalemate was surveillance capabilities beyond anyone's imagining. As soon as the crisis unfolded, U.S. planes from their newly established base in Ukraine and from South Korea scrambled to establish stare surveillance over the areas where the crisis was taking shape. Stare surveillance permits constant viewing of a region for as long as the planes carrying the long-range cameras are able to stay in position, which given in-flight refueling is a considerable period. It was a quantum jump over the keyhole satellites that traversed areas of the world as they traveled.

The President, the Chairman of the Joint Chiefs, the National Security Adviser and their aides could therefore watch the drama as it progressed, but short of launching first strikes against Beijing and Teheran, there was little they appeared capable of doing except watching.

It was then that the Chairman of the Joint Chiefs moved that the President consider WBOM—War by Other Means.

WBOM was a catch-all title for a grab bag of capabilities, including information warfare, or IW. (Despite attempts by the Pentagon to neutralize this term by changing it to IO, for Information Operations, the term IW had stuck.)

IW was as old as the hills and as new as the latest technology from Silicon Valley. From the Trojan horse of ancient times to the XX (the Twenty or Double Cross) Committee of British Intelligence that did so much to turn the tide against the Germans in World War II, IW stood for disinformation, deception, propaganda and, more recently, electronic subversion.

The information revolution had not bypassed countries like China or Iran. Both had managed to build national electronic infrastructures without subverting their traditional, rigidly controlled societies. This was something of a tribute to the power of authoritarian rule, and bucked analysis that said the flow of information would eventually be the undoing of authoritarian regimes. There was precedent for the argument. The fax machine had undermined Soviet rule in Poland by making information available to dissident groups and allowing coordination of anti-government activities. In China, too, the fax machine and the Internet had played important roles in coordinating student activities and conducting a free flow of information into and out of the country. But the rumble of tanks in Tiananmen Square and the deaths of several thousand people demonstrated that the flow of information could not compete with sheer force if those in authority were prepared to wield it.

But if Iran and China had only selectively joined in the information revolution, by laying fiber-optic cables to create a sophisticated internal communications system, and by contracting with Western and Russian satellite companies to give them state-of-the-art military capabilities such as GPS (Global Positioning System), they still had to rely on outside sources to supply the equipment. They did not yet have the technology to originate world-class electronic products.

In China's case, that nation had brought its technology in through its newly acquired port of Hong Kong, which under British rule had made itself an Asian center for high-tech commerce. When the British left in 1997, they had bequeathed the colony's new Chinese masters a secret legacy that was to have profound consequences in this crisis.

Using a complex series of cutouts (apparently legitimate companies that were, in fact, fronts for intelligence work) and corporate shells, British intelligence had many years before established the key technology-importing companies in Hong Kong. From anyone else's perspective, these companies were owned and operated by ethnic Chinese. In fact, they were surrogates for the British, who used them to plant Trojan horses in all the high-tech machinery China was importing to support its economic growth. The Trojan horses were the equivalent of espionage "sleepers," agents put in place by a foreign power and then activated many years later when there was a mission to fulfill. By adding lines of computer code to software and planting new chips into hardware, the British had given themselves, and their American allies, a lever to use against the Chinese.

The practice of subverting technology was not new. The Americans had done it to the Russians and others, including the Iranians, for years. They had also tried it on the Chinese, although anything that came directly from the United States was regarded by Beijing with suspicion and scrupulously checked. The British operation was altogether smoother for the elaborate cover under which it operated.

The order went out from the White House to the Joint Services Information Warfare Command (JSIWC) at Norfolk, Virginia, the body that brings all the competing and overlapping IW sections of the four main services and the intelligence agencies under one roof, to institute an escalating series of attacks on Iran and China.

The top general in the PLA (People's Liberation Army) was receiving a telephone briefing on the South China Sea operation from his naval counterpart when the phone went dead. Across Beijing, communications ceased as the phone system inexplicably went haywire. Calls were cut off, or they went to the wrong destination. The leadership in the

Politburo found itself isolated, unable to talk to its military commanders. Radio systems were rushed into operation, and for a few minutes contact was restored. Then the satellite uplinks that bounced the signals to the iridium satellite system and down to receivers on land and sea went dead.

The General Secretary of the communist party sat impassively in the middle of the maelstrom as chaos reigned all around him.

The phone in front of him rang. He picked it up gingerly. It was the President of the United States, offering personal greetings, and a fervent desire that the current crisis could be resolved peacefully.

Refusing to admit he was having a communications meltdown, the General Secretary returned the President's greetings and regretted that as his nation's security was under threat he had no choice but to exercise any and all options to defend his country. The fact that the President had managed to get through directly to the phone in front of him told him at once who was disrupting his communications.

The center of Teheran was a mess. Traffic lights had all gone dead as a major power blackout paralyzed the city. Roads were jammed, offices were silent and the streets were packed with people wondering what had happened.

Several thousand miles away, in an underground computer center in Norfolk, a small team of IW warriors understood perfectly what had happened. Using their own Trojan horses, planted long ago in the main-frame computers that ran Iran's national grid, they had sent a specialized computer virus, a worm, into the electricity system. A worm is an aggressive virus that can be targeted to consume and destroy vital lines of code. As the people of Teheran piled into the streets demanding answers, the electricity system they depended on for everything from cooking to national defense was dying.

In Riyadh, Saudi Arabia, the national defense force leadership was in a panic. The country's multibillion-dollar radar defense system had gone blank. The country was paranoid about air attack from Iraq or Iran and had bought the finest equipment oil money could buy to ring the country with early warning radar.

They had also unwittingly bought some extra little bits of hardware that on a signal from Washington had brought the entire system to a halt.

The country was defenseless against air attack.

The President waited to see what effect the IW strikes had had. Calls were flooding in from the Saudi embassy and from Riyadh itself. The Secretary of State had a very quiet word with the Saudi ambassador to suggest that the fault with the radar system was no doubt temporary and that American engineers were brainstorming ways to bring it back online. And, the Secretary added, perhaps the Riyadh government might consider ending terrorist operations against Iran?

Riyadh got the message.

Iran was in chaos. Civil engineers went into overdrive to restore power, but every avenue they explored was blocked by the balky central computer system. Backup generators in key military and civilian centers restored a modicum of control, but so pervasive had the worms been that it would take a long time to put everything right. Calls were going out to European and American hardware manufacturers seeking urgent advice. Through cutouts long in place, the Iranians were advised that engineers would be sent immediately to rectify the breakdown. Those engineers would also be intelligence agents whose job was to see that the system was restored, eventually, while leaving in position the Trojan horses that made a repeat performance possible.

The President was not too concerned about Iran and Saudi Arabia. It was China and its threat to Japan that was worrying him.

In the South China Sea, the Chinese naval battle group continued to steam south to the Strait of Malacca.

He put another call in to the Chinese General Secretary, noting that the operation was still ongoing. The response was as regretful, but firm, as before.

As well as buying computers to run civil and military communications, the Chinese had invested heavily in mainframes to run their power generation industry, controlling the grid that spread across the vast land of China.

Critical to the nation's power supply was not only oil, but coal, which the Chinese had in huge supply, and nuclear and hydroelectric generation.

The jewel in the hydroelectric crown was the Three Gorges Dam on the Yangtze, capable of providing 18GW (gigawatt) of power. The carefully controlled flow of water through the turbines also irrigated the agricultural land downstream. Balancing the flow to maximize energy efficiency and keep crops watered was handled by computers.

As the battle group steamed closer to the strait, engineers at the Three Gorges Dam flew into a panic as the water flow reduced to a trickle and then stopped. It was the dry season and water levels upstream from the dam were lower than usual and not being replenished quickly, so the danger of an overflow was not an immediate worry. But the generation of power was at a standstill, while downstream, farmers were beginning to wonder why their water supply had dried up. With mounting concern, they realized their paddies would soon bake dry in the hot sun and their rice crop would shrivel. Given that the Yangtze valley was one of the major rice-producing areas for the nation, the consequences of large-scale crop failure were too awful to contemplate.

Once again, the phone in front of the General Secretary rang. The

President offered his condolences on the unfolding agricultural crisis in the Yangtze valley, and said if China needed help, the people of the United States stood ready to do what they could.

Mystified, the General Secretary thanked the President, then hung up. With internal communications dead, he had no idea what the President had been talking about.

Miraculously, phone service between the Yangtze region and Beijing was restored, and the grim news of the breakdown at the dam was relayed to the Politburo.

The General Secretary quickly understood that he was not in as much control of his nation as he had thought. He ordered the Malacca-bound battle group home.

TWO

A Desert Myth

THE 1991 Gulf War was the last hurrah of the armed forces and generals who had trained on the legacy of the Second World War. War had changed its form with the arrival of standoff weapons and precision-guided munitions, the former designed to allow the military forces to fight a battle with long-range missiles and guns and the latter ensuring pinpoint accuracy. However, the tactics had remained essentially evolutionary and revolved around the application of mass on a battlefield. In the deserts of Saudi Arabia as the allied forces built toward their critical mass, the scene unfolded like a World War II movie. Mile after mile of tanks, armored personnel carriers and fuel trucks; serried ranks of tents marching in perfect order over the horizon and runways shimmering in the desert heat with aircraft poised for the inevitable air strikes. It was a stirring sight the likes of which the post–World War II generation had never seen before. Nor would they see its like again.

The war that was gradually taking shape in the desert was in many respects the antithesis of the Vietnam War that had scarred so many of the senior American officers who were charged with executing the battles to come. Unlike that dirty little war that had been such a failure for America, those officers were determined that this time there would be certainty and conviction in the prosecution of the conflict. Part of that certainty would be produced by the use of new weapons and systems that had never before been tested in battle. Principal among these was the microchip and its big brother, the computer. For the first time warriors had access to a bewildering array of equipment that delivered more information, intelligence and firepower than at any other time in military history.

This ability to apply force in new ways was a revelation not just to the Americans and their coalition partners but to the rest of the world as well. The lesson that every country drew from the war was that conflict from now on was going to be very different. The templates that had held good for forty-five years were clearly outmoded and new models would have to be made. The shape of those models—the kinds of soldiers, the type of equipment and the tactics that would be employed—would be the subject of considerable debate in the years ahead, a debate that is largely unresolved today. But as the coalition forces struggled to prepare for war in the dying days of 1990, the focus was on defeating Saddam Hussein's forces, which, the intelligence community was convinced, were the most powerful in the Middle East.

When Saddam poured his troops across the border into Kuwait on August 1, 1990, he had little idea that he was about to provoke the most profound change in modern military forces since the Germans unleashed their blitzkrieg campaign that launched World War II. At the outset, his intentions were clear and simple enough: to take the oil-rich kingdom of Kuwait and use its natural resources, which constituted 20 percent of the world's known oil reserves, to help pay off $90 billion in debts incurred as a result of the protracted war with Iran. Saddam calculated that Kuwait was of little concern to the major powers and that no nation in the region had either the firepower or the will to fight his aggression. He was almost right but as the world now knows, his confidence proved badly misplaced.

A formidable political effort by President George Bush, who had himself been encouraged by British Prime Minister Margaret Thatcher, led to the formation of a powerful coalition force. Exactly how that force was to be deployed and its precise composition owed much to the influence of General Colin Powell, the Chairman of the Joint Chiefs of Staff. Powell was one of a new generation of military leaders who had been hardened by their experience in the Vietnam War. For Powell, his tours in Vietnam had soured him on an out-of-touch military and a weak political leadership and he had formulated his own tough guidelines for the prosecution of war:

> Wars should be the politics of last resort. And when we go to war, we should have a purpose that our people understand and support; we should mobilize the country's resources to fulfill that mission and then go in to win. In Vietnam, we had entered into a half-hearted half war with much of the nation opposed or indifferent, while a small fraction carried the burden.

This view was honed in the Pentagon into what became known as the Powell Doctrine, which was interpreted to mean: Define your objective, bring massive force to bear and take on only those battles you can be sure of winning. Be sure to win political and public support before conflict starts.

To Powell's critics inside the Pentagon and overseas, this was a classic

American solution to the thorny problem of military intervention abroad. America would join the fray only if victory was assured, casualties were acceptably low (essential for continued political and public support) and mass could be brought to bear. This was war American-style and generally bore little reality to the pattern of late-twentieth-century conflict, which tended to be dirty, bloody and with uncertain outcomes. But understanding the Powell Doctrine is important not just for the prosecution of the Gulf War but for everything that came afterward, including the information revolution and America's ability to take advantage of it.

From the moment that the allied forces began to gather in the Gulf, Saddam prepared for a conflict that would never be fought. He constructed massive defenses similar to those built by the French at the Maginot Line between the world wars. In both countries, such static defenses were to prove inadequate against an enemy committed to maneuver warfare. Saddam seemed to understand only the trench warfare that was modeled on the carnage of World War I and his own recent experience against Iran. The allies were trained in mobility and well understood the effective use of firepower.

From the outset, the coalition forces had determined not to fight Saddam on his own terms. This meant using all the equipment available in the allied arsenal to ensure that from the moment the first air strike was launched, Saddam and his forces would be rendered deaf, dumb and blind. It was a daunting task that required the deployment of equipment that had previously been secret and that would be used in anger for the first time.

While innovation through accident has been commonplace throughout history, it was a peculiar accident of the weather that delivered to the Americans their first secret weapon of the Gulf War. An unexpectedly stiff westerly wind blowing onto the Pacific coast in the early 1980s led a decade later to the first action of the air war that eventually forced Saddam out of Kuwait.

During a routine war-fighting exercise called Hey Rube, U.S. Navy aircraft dropped rope chaff over the Pacific Ocean to interfere with a theoretical enemy's radar. Chaff was not a new concept; the Royal Air Force had dropped metal chaff, thin strips of flexible metal, to achieve the same goal against the Germans during World War II. The chaff dropped on this day was different. It was made of rope to which were bonded shards of glass impregnated with metal. It was also somewhat sensitive to the vagaries of the elements, which is where the unexpected strength of the westerly wind enters the picture. For instead of floating harmlessly to sea as intended, the chaff was blown toward land ninety miles away, ending up draped over electric power lines. The resulting sparks shorted out transformers and caused blackouts in parts of San

Diego. Initial displeasure over an accident that cost the Navy significant sums in compensation quickly turned to intrigue as the ramifications of what had happened were worked out, namely that the mass placement of disruptive material on an enemy's power supply system could cause power outages on a major scale. After all, if you could black out San Diego by accident, what damage might be done to an enemy city if you really put your mind to it?

The concept was refined over the remaining years of the Cold War. For instance, fibers were developed that would be sucked into the air intakes of Soviet ships. These were then refined further into a payload for a version of the Tomahawk cruise missile known as Kit 2. Its existence had remained a closely guarded secret for a very good reason. If the world knew that America had devised a fairly simple method of blacking out an enemy's power supply, it would not be long before an enemy tried to do the same to America. This is a phenomenon that runs consistently through the analysis of war fought by modern means: namely that the greater the advantage derived from information technology, often the greater is the potential vulnerability. In the event, the decision to deploy Kit 2s against Iraq was made despite concerns about what the exposure of this weapon might mean to America and what the effects of a widespread, and possibly long-term, blackout might have on the Iraqi civilian population. U.S. leaders felt that the military and psychological advantage to be derived from creating mass blackouts was worth the risk.

A list of twenty-eight electrical supply sites critical to Iraqi civilian and military life was drawn up. The operation was given a name, which happened to be the radio call sign of the Air Force general charged with planning electronic warfare (EW) against Iraq. They called it Pooh Bah's Party.

At the same time, the intelligence community examined what it could do to influence what everyone was increasingly certain would be a bitterly fought ground war. For much of the 1980s as the computer revolution began to take hold across the world, both British and American intelligence had been working on new methods of waging war from afar. In particular, America's CIA and the NSA (National Security Agency) and Britain's GCHQ (Government Communication Headquarters) and SIS (Secret Intelligence Service) had all experimented with new types of bugs and viruses that could be injected into an enemy's computer system. A prime target had been Russia (see Chapter 16) and there had been considerable success at planting viruses inside the Soviet military-industrial structure that could be activated in the event of war. The CIA had also been able to plant bugs inside computer systems to feed back via satellite information that had been leeched off hard drives in the Soviet Defense Ministry and elsewhere.

Immediately after the invasion of Kuwait, GCHQ and the NSA began a joint project to see if they could employ the same techniques against Saddam's forces. Over a two-month period intelligence was gathered about exactly the kind of computers that were used in Saddam's command and control networks. It turned out that Saddam had bought most of his computers in the West and that they could be successfully penetrated.

The challenge was how to deliver the virus so that it could do its work inside the system. A number of options were devised that included sending viruses from a remote location, doctoring floppy disks and even supplying computers that Saddam urgently needed and ensuring that they were suitably doctored. Finally, the decision was made to insert some hardware into a cargo of computer equipment that also included a truckload of tires destined for the Iraqi military. In December, CIA operatives working in Jordan infiltrated the viruses into the hardware and then tracked it across the border into Iraq. Once in place, the NSA/GCHQ experts believed that the virus would spread like a virulent cancer through the Iraqi command and control network, infecting and corrupting every system it came across.

Satellites recorded that the tires had been unloaded at an Iraqi air defense site, and the intelligence community assumed that the computer hardware, too, had reached its target. But, before the virus could knock network after network off line, the air war began. At the top of the target list was Saddam's command and control network and one of the first targets destroyed was the very building where the infected computer hardware had been so carefully inserted. One of the most successful intelligence operations of the Gulf War was buried beneath the rubble.

"It was a very frustrating experience for everyone involved," said one Pentagon official. "The intelligence people were very, very pissed at the Air Force. All that work for nothing."

CENTCOM, the U.S. Central Command in Florida, planned two distinct phases in the war: an air assault designed to weaken the enemy and degrade his ability to fight, followed by a ground offensive that would deliver the killer punch. The Kit 2s' mission to render Iraq blind and befuddled would clearly be an opening element of that air assault. But it lay five months in the future. Before then one of the most complex planning and communications systems in the history of warfare had to be put in place.

Advance elements of the U.S. Air Force (USAF) started to arrive in Saudi Arabia in the second week of August 1990, within days of Iraq's invasion of Kuwait. Contingency plans for fighting an air war in the Gulf were ready to be taken from the shelf and dusted off, just as there are contingency plans for combat in virtually all areas of the world. Yet it became swiftly apparent that there was a yawning gap between what the U.S. military thought would be needed to fight a war in the Gulf and what was actually required. The demands being placed on air operations planners were so great that the system came close to breakdown. For example, from the arrival of the first U.S. fighter aircraft on August 9, 1990, pilots had no means of communicating with TACC (Tactical Air Control Center) in Riyadh. In turn, TACC could not communicate with home base in the United States.

The burden of creating a network that patched together the pilots from the USAF and all allied air forces with their commanders in Riyadh, Dhahran and aboard naval vessels in the Gulf with links back to the Pentagon and the White House fell on the shoulders of the communications planners of the USAF Tactical Air Command at Langley Air Force Base in Virginia and the Ninth Air Force at Shaw Air Force Base in South Carolina. The system also had to be capable of creating and transmitting daily Air Tasking Orders (ATOs) covering air sorties all over Iraq, while at the same time coordinating weather forecasts, intelligence, airborne surveillance (AWACS, Airborne Warning and Control System, and J-Stars), and, eventually, Bomb Damage Assessments (BDAs).

When the planners set to their task, they had considerably less technology at their disposal than the required minimum as laid out in the doctrine that governed their mission. For instance, there was no satellite earth station capable of instantaneous voice and data transmission to the United States available locally, despite years of negotiations with the Saudis and others. The terrestrial system that the doctrine said should and would be in place to link Joint Headquarters with the various air, land and sea commands simply did not exist. Most of the tactical communications equipment required for that was in the hands of Air National Guard units that would never be deployed.

Thus began one of the most innovative, seat-of-the-pants operations ever undertaken. As more and more units were deployed in theater, the scientists, computer nerds and whiz kids had to patch together a voice and data network using whatever came to hand, while in the United States their support services scrambled to pull together and ship out the hardware they needed. In the end, over 200 sorties by C-141 Starlifter transport aircraft were required to carry this equipment to the Gulf, in the process virtually draining the USAF of tactical and strategic communications equipment. In the end, the communications effort involved 2,300 personnel, 12 combat communications squadrons, 7,000 radio frequencies, 1,000 miles of land links, 59 communications centers, over 29 million calls, all in support of the 350,000 air operations in Desert Shield and the 225,000 in Desert Storm. It is testament to the skill and imagination of the technicians involved that the system worked, because nothing on that size or scale had ever been envisioned, let alone created. As Alan Campen succinctly put it in his book *The First Information War*: "Much of what they did from August through February had not even been dreamed of in July." A spokesman for the Joint Chiefs of Staff said later, "In the first 90 days we put in more connectivity than we have had in Europe in the past 40 years."

Similar miracles were being performed for the ground forces. By the time the 430,000 American troops were in position in Saudi Arabia at the end of 1990, the coalition forces required one of the largest logistics tails in the history of warfare. The American troops needed 80,000 tons of materiel each month to sustain life. They had 700 tons of mail delivered each day along with 715,249 cans of foot powder, 2.3 million pounds of coffee and 551,654 bottles of sunscreen. The 40,000 strong British force reckoned it would need 60,000 tons of ammunition that would last just thirty days.

The extent of the logistics nightmare was little appreciated by those who simply saw the military buildup unfold. But just as the war itself was a watershed and a blueprint for the future, so the logistics tail in the Gulf was to be cited in the years ahead as a graphic illustration of the vulnerabilities inherent in modern warfare. Every order for every carton of cigarettes or air-to-air missile was processed by computer and the vast majority of the orders were carried out over an open phone line. A single hacker with a decent modem could have wreaked havoc with the allied plans. Fortunately, Saddam had little appreciation for the subtleties of modern warfare.

While the technical staff was working triple overtime to create the system, the air operating staff was creating the complex traffic that would flow through it.

"The air war was an operation of astonishing complexity," the British commander in the Gulf, Lt. Gen. Sir Peter de la Billiere, recalled. "Devised by [Brig. Gen.] Buster Glosson and his targeting teams, under the overall guidance of [USAF Gen.] Chuck Horner, it was a masterpiece of human planning and computer-controlled aggression, directed with a degree of precision which far surpassed that of any air attack in the past."

The backbone of the air war was the Air Tasking Order (ATO). Issued daily by the Joint Force Air Component Commander, the ATO was a document several hundred pages long that gave in precise detail the orders for each coalition aircraft scheduled to fly sorties. The process started each day at 0700 local time with the receipt of the commander's objectives. These were married with incoming intelligence, Bomb Damage Assessments (BDAs) and weather forecasts into a revised master attack plan that specified target names and priorities. That was readied by 2000 local time each day. Air operations then designated particular air force components to each target, including TOTs (Time on Target), the amount of ordnance each aircraft would have on board, the fuse settings, routes, refueling points, IFF (Identify Friend or Foe) squawks and radio frequencies, while intelligence refined the precise points that the aircraft should hit.

Before dawn the next day, the mass of data accumulated was inputted into the electronic brain of the air war, called CAFMS—the Computer Assisted Force Management System. This thirty-two-bit minicomputer (with a full redundancy backup) with 10MB of RAM and several removable disk drives for storage was housed in a van, of which the USAF had three in total, with a fourth system that lacked the custom-made shelter and protection and was held at the 1912 Computer Systems Group. Each system could service eleven remote terminals to be placed in allied airfields, connected by wire, satellite or microwave links. By the standards of modern home computers, the CAFMS first deployed in August 1990 looks fairly ordinary in terms of performance. By the standards of the day it was a formidable piece of equipment. But as was simultaneously being demonstrated with the communications links, it just was not enough. By December all four of the CAFMS systems were in theater, but had to be adapted "on the hoof" to work together in a single network and to serve more than the maximum forty-four remote sites laid out in the specifications. In fact, fifty-five sites in all were brought onstream. The systems also had to undergo homemade upgrades to satisfy the mushrooming demands of the Tactical Air Control Center. The efficiency rate of the CAFMS was impressive, in that it was working 98 percent of the time. Its performance however was slow, and got slower as the war went on. The ATOs that the CAFMS had to produce went from 200 pages in length in August to over 900 pages in length by the following February. Remote sites had to download this information, and that could take several hours, with several hours more required for printing. (Again, alarmingly slow compared with technology routinely available today.)

Interservice connectivity was even more difficult, with the Army and Navy not having compatible systems to read the ATOs. Thus a new network using PCs and secure voice lines was introduced, dubbed MADS (Mini-ATO Distribution System), in which the ATOs were converted to an ordinary text file and transmitted using commercially available software. This did not help naval aviators, who did not have communications access to the MADS, so in the end the Air Force resorted to a simple expedient—the ATOs were printed, as well as put on floppy disks to be flown to ships by helicopter.

Looking back at the degree of improvisation required to bring an inadequate system sufficiently up to speed to make a sophisticated air war possible gives pause for reflection. A more competent enemy might have been able to exploit the long gaps between the production of the ATOs and their ultimate dissemination. It also begs the question: why was the capability of the CAFMS system so out of sync with the warfighting capability of the USAF? In the end, of course, the system was adequate to the needs thanks to human ingenuity and sweat.

Victory often shrouds doubts and failures, but on this occasion the lesson was learned and a new, modular system with an even longer name —the Contingency Tactical Air Control Automated Planning System (CTAPS)—has been developed with interservice compatibility and instant upgrade ability built in.

But the luxuries of such judgment and hindsight were a long way off on January 17, 1991. After five months of diplomacy and desert deployment, the time had come for military might to achieve the expulsion of Iraqi forces from Kuwait. The ultimatum for the Iraqi withdrawal passed at midnight Eastern Standard Time on January 16. In the early hours of the following morning, local time, batteries of Tomahawk cruise missiles were fired from U.S. ships in the Persian Gulf. F-117A Stealth fighter-bombers streaked toward Baghdad, cloaked in a technological mantle of invisibility. Apache helicopters armed with Hellfire missiles readied an attack on a key radar site on Iraq's southwestern border. Of all the missions, this last was the most immediately critical. The American battle plan called for the destruction of all means of Iraqi communication and observation. From the outset Iraqi radar sites, communications stations, electrical plants and military command posts were to be destroyed to render the Iraqi military, the fourth largest in the world, deaf, dumb and blind. None of that could be achieved without first knocking a hole in the ring of radar air defenses around the country's borders. Within minutes, the Apaches had done just that, allowing flights of conventional attack aircraft to stream through the hole toward Baghdad. (The Stealth aircraft, self-evidently, did not need to make use of this.)

It was the secret batch of Kit 2 Tomahawk missiles launched from the USS *Wisconsin* that carried out the mission known as Pooh Bah's Party. Cruise missile technology called for a tiny video camera mounted in the missile to match precise maps that had been preloaded into the onboard computer. The topography of Kuwait and southeastern Iraq was peculiarly ill-suited to this technology. It was flat, barren and featureless. So to give the missiles proper reference points, they were routed over the mountainous regions of western Iran and then ordered to turn left and head for Baghdad. It was an egregious breach of national sovereignty to conduct a mission of war over the territory of another country without first seeking permission, but President Bush did it anyway. The fact that neither Turkey nor Syria would give permission meant that the Tomahawk missions scheduled to be flown from the Mediterranean were instead flown from the Persian Gulf.

The Kit 2s reached their targets, five of Iraq's twenty power generating plants, and proceeded to make a series of low passes over each one. On each pass, the missiles disgorged tiny spools of carbon filament that unraveled as they headed earthward. The filaments draped themselves over electric power lines and began to produce the same effect, intentionally this time, that the rope chaff had achieved in San Diego. In a shower of sparks, a significant portion of Iraq's generating capacity fell victim to the Kit 2s. In the hours to come, other plants would be attacked by more Kit 2s and conventional explosive ordnance. It was becoming rapidly apparent to the Iraqis, as it would soon become apparent to the world at large, that a different kind of war was being fought, a war in which a combatant armed with advanced technology could render forces structured along traditional Cold War lines largely impotent.

Other targets of note in those first tense hours of the air war included the AT&T building in Baghdad, which controlled the country's civilian telephone system. At 3:00 A.M. local time precisely, the designated H-Hour, two GBU-27 laser-guided bombs were dispatched from an F-117A, smashing into the building and turning it into a flaming wreck. The Al-Khark communications tower next door was brought down by another pilot. Others attacked military command bunkers, air force headquarters, and Saddam's private retreat at Abu Ghurayb. Most of the attacks were successful, but some targets were too heavily protected by steel-reinforced concrete for the bombs to penetrate. Conventionally armed Tomahawks went after the Ba'ath party headquarters, missile sites and the presidential palace.

The twin objectives were to decapitate (or at least incapacitate) the Iraqi high command and to gain swift ascendancy over the Iraqi skies. The first objective fell short of the optimum result desired, namely the death of Saddam and his chief lieutenants. Saddam's ability to dodge each and every bullet and bomb sent his way was a source of constant frustration to the allies. What they did manage was to disrupt the Iraqi command and control (C2) system through denial of electrical power, early warning systems and communications. The second objective was a more complete success. After the initial attacks by the Stealth fighter-bombers and the cruise missiles, it was the turn of conventional aircraft from the coalition forces to sweep into Iraq and obliterate as many Iraqi aircraft as they could find. But before they could do that, further damage had to be wrought on the Iraqi air defenses.

It was a move that betrayed just how wide was the gap between the allies' technological and tactical sophistication and Iraq's meager capabilities. Following the first wave of Stealth and Tomahawk attacks, a wave of BQM-74s was sent toward Baghdad, unmanned jet-powered aircraft that were only thirteen feet long, yet projected the radar image of a much larger plane, almost anti-stealth in conception. The Iraqi air defenses, which had been completely bypassed by the Stealth attacks, reacted with jubilation to the sight of all these plump, juicy targets heading their way. Whereas minutes before they had been firing wildly into the night, they now had something to lock on to. The radar systems that guided their missile and antiaircraft batteries were switched on to search out the incoming drones and a barrage of artillery was sent skyward, destroying them by the score. Iraqi delight was short lived. Just moments behind the drones lay a mass of seventy allied aircraft armed with radar-killing HARM and ALARM (the British variant) missiles whose purpose was to find and track the Iraqi radar beams, then follow the path of the beam back to the radar stations and destroy them. In that one wave, the allies rendered the Iraqis incapable of mounting an effective, coordinated air defense. They had won supremacy in the air. But the tempo of operations did not let up. If anything it increased.

In the coming weeks, allied air forces would continue to pound Iraqi targets in a relentless barrage that tore the heart out of the enemy's military machine, particularly seeking to destroy the much-feared (needlessly as it turned out) Republican Guard, the military elite of Iraq. B-52 bombers laid down old-fashioned unguided carpet bombing to destroy the Republican Guard's infrastructure, personnel and spirit. The Iraqi air force failed even to put up a token fight. In a move that was assumed to be designed to preserve some aircraft for the days after what was clearly going to be an inevitable defeat, several hundred were dispatched to Iran, where they sat out the war. In air terms, the enemy was not just deaf, dumb and blind, but toothless, too.

Overarching the whole campaign was the web of information gathering and transmission that was as vital a fuel as aviation kerosene itself. E-3 AWACS (Airborne Warning and Control System) aircraft, essentially Boeing 707s on top of which a large mushroomlike structure had been fixed to house a mass of electronic surveillance equipment, patrolled the skies above the Iraqi border. The AWACS were able to view the entire airspace of conflict. In the event there had been a fully fledged shooting war in the sky, allied dogfights would have been controlled from these aircraft. E-8 J-Stars aircraft, another version of the 707, provided the same function as AWACS on behalf of ground forces, their role being to detect enemy activity such as convoys, tank formations and Scud missile sites that the Iraqis had hidden in remote areas. These Scud missiles, while sluggish and inaccurate compared with allied weaponry, were an extremely important factor in the war.

The allied coalition, brilliantly forged by President Bush, was a fragile affair, depending for a large part of its cohesion on the absence of Israel from the fray. Forces from several Arab nations, including Morocco, Egypt and the Gulf states, had joined those of Saudi Arabia and Kuwait because they understood that Saddam had gone too far in attacking and occupying an Arab neighbor. If he got away with that, where else might his ambitions take him? Should the war have broadened to include Israel, however, then deeper allegiance, that of Arabs against Jews, would come into play and the alliance would have fractured. It was therefore vital for Saddam to try to provoke Israel into declaring war, and for this he sent Scuds raining down on the Israeli people. American diplomatic pressure and assurances that the Scud sites would be found and destroyed persuaded Israel not to retaliate.

If the AWACS and J-Stars were the eyes of the battle, yet another iteration of the Boeing 707, the RC-135, provided the ears. The RC-135 was not immediately recognizable as a 707 derivative; a fat nose stuffed with equipment and the antennae that bristled all over the external surfaces successfully disguised its provenance. The RC-135's job was to eavesdrop on Iraqi communications and spot the source of any electronic emissions such as jamming signals. The data would then be passed directly to the Tactical Air Control Center for targeting and attack. Initially, the Iraqis were casual about protecting their emitters from detection, but steady attrition of their facilities persuaded them to sharpen up. They were still outwitted by the RC-135s, which provided twenty-four-hour coverage of the battle space. According to one mission specialist, "The Iraqis were good, but they weren't good enough."

Carefully following the Powell Doctrine of the surgical application of overwhelming mass, the allied commanders were concerned that Saddam believe that the ground assault when it came would not arrive in the way it actually would. For several weeks prior to the start of the air campaign, Central Command had publicly placed a great deal of emphasis on the 4th Marine Expeditionary Brigade. In November, Vice Adm. Henry Mauz, the Chief of Naval Operations for CENTCOM, took personal charge of an amphibious landing exercise staged off the coast of Oman and code-named Camel Sand. A second amphibious rehearsal, code-named Imminent Thunder, took place later the same month.

In the Pentagon and in briefings to journalists in the region, public affairs officers carefully pointed out the exercises and the position of the Marine brigade, which eventually settled off the coast of Kuwait. Journalists were given details about how the Marines work, were allowed access to the ships and were generally encouraged to follow their eyes and noses and write stories about how an amphibious landing was a vital component of the ground phase of the war. The Americans knew that Saddam was relying heavily on the Western media for his intelligence and using the media to convince Saddam the attack was coming from the east rather than the west was considered entirely legitimate.

"We told no lies," insisted one Pentagon official involved in the operation. "The reporters wanted to believe what they saw and simply did not ask the right questions. More fool them."

Before the ground war could get under way, American and British special forces teams were inserted behind the Iraqi lines to carry out reconnaissance and sabotage. On February 23, eight special forces teams were ready for insertion. In the event, two of the teams were stood down, one because the landing site had been compromised and the second because the mission was deemed unnecessary. Of the six other teams, one landed in a supposedly clear area of rocks that turned out to be Bedouin tents and another found itself surrounded by Arab voices. Both those teams extracted immediately. Two other teams successfully carried out their missions while the final two were compromised by curious Iraqi civilians and after furious firefights were extracted to safety.

Much has been written about the heroism of the British Special Air Service (SAS) teams and their work both in sabotage and in pinpointing Scud missiles. There is no doubt that the SAS did carry out some outstanding work in disrupting Iraqi communications, destroying Scuds and generally acting as a force multiplier behind the lines to convince the Iraqis that large numbers of troops were in their rear area. This was classic SAS action, modeled on the work they had done in the Second World War behind enemy lines. Peter de la Billiere writes about the activities of one patrol, code-named Bravo Two Zero, which was inserted into Iraq on the night of January 22, in his autobiography:

> Three members of Bravo Two Zero were killed and four captured. The eighth man limped into Syria after an heroic, seven-night solo march with practically no food or water. His trek, which will go down in SAS annals as an epic of escape and evasion, was described in my own book *Storm Command*; while *Bravo Two Zero* by Andy McNab— pseudonym of the leader—gives a lively account of the adventures of the rest of the patrol. I felt deep pride in my former regiment when, after the war, Norman Schwarzkopf wrote me a letter praising in fulsome terms the contribution made by the SAS.

What both de la Billiere and McNab failed to address was the stunning incompetence that got Bravo Two Zero into such trouble. That story was eventually told by Chris Ryan, the soldier who made the extraordinary desert trek. Ryan revealed that the SAS, who want for nothing in terms of equipment, had deployed to the desert without the right gear. The quartermaster had no grenades for their launchers; stocks of Claymore mines were never issued because of a communications error, the patrol making their own instead from ice cream cartons and scrounging some from other units; silenced pistols were not available. The maps issued to the patrol were designed for air crews and the scale was so small that they were virtually useless. Maps derived from satellites were available from the Americans, but were never given to the men.

Escape maps studied by each soldier were so old they had been printed in 1928 and then updated for the Second World War; it was only just before the men left that more modern versions were issued. In fact, intelligence was so poor that the men, who had practiced building hides in the desert sands of Saudi Arabia, were dropped onto bedrock that was both flat and so hard that digging of any kind was impossible.

Even more serious, the patrol members found themselves unable to communicate with the base because they had been given the wrong radio frequencies (unlike their American counterparts, who called in both fighter cover and rescue helicopters without any difficulty in similar circumstances). To compound that error, their emergency beacons lacked the range to communicate with any overflying aircraft that might have been able to organize a rescue.

For an ordinary army patrol, much of this might be explained by the fog of war. But the SAS prides itself on its attention to detail and argues that it is the combination of perfect planning with highly developed military skills that makes the regiment so formidable. Yet here, in a real war, it was the detail that was ignored and as a result men died.

If the air war was a perfect example of the matching of technology to the task, Bravo Two Zero was an equally forceful reminder that badly applied technology can be deadly in war. While the special forces had relied on satellites for reconnaissance imagery and communications, the system that was in place proved barely up to the job at hand.

The satellites of the Defense Satellite Communications System (DSCS) operated in tandem with Britain's Skynet satellites to provide the vital communications links. So inadequate was the wire and micro-wave infrastructure available from the very start of Desert Shield that these satellites were used for communicating over very short distances; for instance one side of an airfield to another. Not the purpose for which they were designed, perhaps, but improvisation in time of war is everything. The other vital space component was the Global Positioning System (GPS), which provides anyone from weekend sailors to tank commanders in the middle of nowhere the capability to determine precisely where they are. GPS was an absolutely vital component of war in the flat, featureless desert. Special forces patrols sent into northeastern Iraq hunting Scud sites had GPS receivers with them, as did all ground forces in theater. It was a capability that stunned the Iraqis. One senior officer captured by the allies remarked on their ability to set off into the desert and get to where they were going. "If we tried that," he said, "we'd get lost."

In attacking the enemy's ability to communicate there were some missions that sophisticated air-delivered technology just could not achieve, which was why, six days into the air war, a contingent from the British Special Boat Service (the SBS is the Royal Navy's version of the SEALs) was helicoptered into a forward area only sixty kilometers from Baghdad. The two Chinooks landed close by the main road to Basra, their engines running in the event that a rapid departure was required. The SBS raiders dug up a stretch of ground below which lay the main fiber-optic cable linking Baghdad with military bases across the country. They removed a stretch to be taken back for analysis (and no doubt for souvenirs), then lay charges that punched large holes in the trunk. The operation was a complete success and the raiders were able to return to base unhindered.

Once the air war had crippled the enemy's ability to communicate, and severely damaged key elements of its ground forces, it was time for the second decisive phase of the war—the ground attack that would rout the Iraqis, neutralize their army and return Kuwait to its people, which was, after all, the goal of the entire exercise. To achieve this took a combined allied army of close to 600,000 personnel and almost 3,000 tanks. Several factors were to play an important role in what was a victory of overwhelming proportions, not the least of which was an Iraqi defensive strategy predicated on the dangerous assumption that no one had ever launched an attack through the western desert, so why would they now? Then there was Commander in Chief Norman Schwarzkopf's plan to do precisely that, the much heralded Hail Mary, or Left Hook. Iraqi defenders were massed on Kuwait's southern border with Saudi Arabia, on the coastline and on the Euphrates River on Iraq's eastern border, in the city of Basra, the points which they believed offered the only possible targets for a ground offensive. On February 24, 1991, U.S. Marines did indeed launch an attack on the Kuwaiti border defenses, feinting the direct assault on Kuwait City that had been predicted by Saddam. Completely out of Iraq's sight and hearing, the main force of allied armor deployed in a wide arc north and then east to effectively encircle the Iraqi occupying army in Kuwait.

History has already related the completeness of the allied victory in the ground war. In terms of information warfare, most of the hard work had already been done by the time the ground strike was launched, most particularly in leaving Iraqi forces on the ground almost completely out of contact with Saddam Hussein and the military leadership. Typical of an authoritarian regime, Iraq had established military communications on the Soviet model, which allowed units in the field to speak directly to their command and control center, but not to each other. How better for a dictator to prevent conspiracies and plots among well-armed potential rivals? Saddam went one step further in deploying fourteen different types of military communications equipment, each incompatible with the other. If Iraqi military units had been able to communicate with each other, their resistance could have been better coordinated, and consequent allied losses more than the remarkably low figure that resulted.

Contrary to appearances, not every element of information warfare deployed in the Gulf War was of the high-tech variety. In fact some very old-fashioned, low-tech tactics were used to great effect, which endorses the contention that information warfare has, in many respects, been part of military strategy for as long as war has existed. Hundreds of thousands of cards bearing a drawing of an Iraqi soldier surrendering, with a bubble emerging from his head indicating that he would rather live to see his wife and children than die in a pointless battle, were dropped over Iraqi positions. The card also bore a promise of safe conduct and fair treatment. Many thousands of Iraqi troops took advantage of the offer, having been already completely demoralized by consistent bombing and by the total absence of direction from their commanders. Desertion cannot have been an easy decision. Anyone caught deserting was either hanged or shot. Many also believed their own side's propaganda that said POWs would be tortured by the allies. The fact that so many did desert is testament to the fragility of military loyalty in a regime that uses fear as a guarantor of obedience. It is also testament to the degree to which the allied forces destroyed morale and to the effectiveness in a modern war of an old form of psychological operations (psyops), namely the inducement to defect.

The political fallout from the Gulf War, which brought recriminations against Bush for failing to topple Saddam (which he rightly claimed was not part of the U.N. mandate that formed the basis for Desert Storm) and the subsequent brutal repression by Saddam of the Kurdish uprising in northern Iraq, an uprising fomented by Bush himself, diluted the joy of military victory. The unsettling aftermath will forever leave a question mark over what the real point of the Gulf War was. After one of the most stunning displays of military supremacy in history, the *status quo ante* virtually prevailed, but for the eviction of Saddam from Kuwait.

For the military of all nations that took part, but particularly the United States, the Gulf War was something else. In purely human terms it reaffirmed that human beings are capable of extraordinary deeds under brutal pressure. That is the lesson of great deeds through the ages. It was a real-life test of weapons, machinery and technology that had never been subjected to the duress of battle, operated in a theater where the video camera's presence was not just passively accepted by commanders but used to alter perception at home and in the enemy's camp. Most fundamentally it marked a crossroads at which lay crucial questions about the future of warfare. The answers to those questions, which are still being pondered, will determine the future shape of the U.S. military. It is a debate between those whose bedazzlement with the technology that played such an important role in the war leads them to believe that future wars will be fought and won by those who control the electronic spectrum and who can deploy smaller forces packing bigger punches for fewer bucks; and those more cautious analysts who believe that when the glamour and sophistication of the technology is stripped away, the Gulf War was no different from any other conflict in history, namely a brutal fight to take and hold strategically important land, that the technology making that job easier does not alter the fundamental premise that any future war will still require large numbers of machines and men and women to fight and die to achieve that end.

For them the Gulf War was just another point on the arc of history. For the revolutionaries, the Gulf War, played out on video screens like a deadly electronic game, marked the beginning of a new era.

Which side does history support?

THREE

The Challenge of the Chip

T H E Gulf War was a revelation to anyone who saw for the first time a TV picture of a laser-guided bomb heading for its target. It was a miracle for those who believed it impossible for a coalition of international forces, no matter how well armed, to take on the fourth largest army in the world, on its own turf, and crush it completely with only minimal loss of life. For the generals and admirals around the world it was an alarm call that told them that the wars they had been planning to fight for decades were now a thing of the past. Warfare required new tactics that would take advantage of the Information Age.

Watching the war unfold was like walking through a window between the old world and the new. It punctuated the end of the Cold War and the beginning of the New World Order. Thoughtful observers within the military saw the strategic and tactical philosophy of forty-five years challenged in the most convincing way possible, namely in the heat of battle. They saw that new technologies applied to unfamiliar targets could alter the dynamics of traditional warfare before the first tank had even started its engine and that the United States stood at a crossroads in determining how to embrace the radically different future promised by this technology.

The years since then have been characterized by a free-flowing debate which as yet has produced no resolution, a debate between traditional war fighters who see war as a relentlessly brutal exercise that will always require the deployment of large numbers of people and machines to take and hold land, and the modernists who see war moving away from a fixed battlefield into a location that has come to be called cyberspace. This debate has been distorted and fueled by the inevitable change in defense posture mandated by the end of the Cold War. During that forty-five-year period, there were three elements to America's defense policy. The most evident was the race to produce an ever more lethal nuclear arsenal to compete with that of the Soviet Union, creating the potential for Mutually Assured Destruction (MAD).

Conventional force structure was predicated almost exclusively on the likelihood of a Soviet land invasion of Western Europe. War would be fought between competing forces of tanks, infantry and air forces, with battlefield nuclear weapons as the first step on the nuclear ladder that would ultimately lead to MAD.

The third element consisted of intelligence and covert operations in which spies ferreted out the enemy's secrets or set out to undermine the enemy's efforts to establish friendly regimes in Third World countries. This was the world of assassination, corruption, terrorism and localized conflict between surrogates for the two superpowers. From the attempts to subvert the Castro regime using Cuban exile forces in 1960 to the contra rebels in Nicaragua in the late 1980s, the United States fought wars largely at a distance. When the United States was directly involved, it used much the same philosophy that guided operations in World War II. First there was Korea, a war that ended in a truce but did not produce the overwhelming victory that the country which had overpowered Japan and Germany might have expected. If there were lessons to be learned from that, they quickly became subsumed by the reality of the Cold War, in which U.S. conventional forces did little more than conduct endless exercises on the plains of West Germany. That was where the war was going to be.

There was, of course, the nightmare of Vietnam, a guerrilla war fought and lost by conventional means. After that there were sporadic uses of conventional and special forces over the years: the botched Desert One operation in Iran that was to rescue the American hostages held in the U.S. embassy in Teheran; the Grenada invasion in 1983 spurred by the threat to American citizens from a Marxist takeover of the Caribbean island. In 1983, U.S. Marines went into the Lebanon to try to impose peace on the warring factions who had reduced the once glamorous and prosperous Mediterranean city of Beirut to a bombed-out ruin. In an act of terror which would have ghastly echoes down the years that followed, 241 Marines were killed when a suicide bomber drove a truck packed with explosives through the gates of the U.S. compound and detonated his deadly cargo. Withdrawal followed soon after. An air strike in 1986 designed to kill Col. Muammar Qaddafi of Libya and topple his regime was launched after proof was found of Libya's complicity in terrorist attacks on U.S. troops in Germany. The air strike was operationally well planned and executed, but ultimately a failure. Qaddafi survived, and again there would be echoes in 1991 when American forces tried to kill Saddam Hussein, with as little success. Some of the operations were limited successes, some were out and out failures. All were characterized by an absence of a clear guiding philosophy of how to use military force in peacetime to achieve political and security goals.

The military operation that did go well was Just Cause in 1989, when President Bush ordered U.S. troops to seize Manuel Noriega, the corrupt, drug-running President of Panama, and bring him to trial in a federal court. After a combined force of paratroopers and helicopter gunships subdued the Panamanian Defense Forces, Noriega was placed under arrest, taken to the United States and after a trial jailed for life. Just Cause demonstrated the will of the United States to use military force in pursuit of an essentially civil purpose, namely the arrest of a suspected drug runner. But other than these actions, for forty-five years American defense policy rested on American forces bolstering NATO to face off against the Soviet Union and the Warsaw Pact.

That foe disappeared with the collapse of the communist party and the dissolution of the Soviet Union in 1989. This necessarily set in train a change in the way the U.S. military and intelligence communities did their jobs. The Defense Intelligence Agency (DIA) had spent the Cold War amassing volumes of data on Soviet military capability. This meant using traditional intelligence weapons such as photo-reconnaissance, recruitment of enemy personnel and espionage to gather that data. Finding out what a new Soviet tank could do might involve a spy taking grainy photographs or video from a concealed position inside East Germany. Or "turning" a Soviet tank commander with financial bribes or the promise of eventual defection, then pumping him for information on armaments, armor, speed, upgrades, disposition of forces, caliber of those forces and so on. Thus did the United States and its allies create a vast database of information about Soviet military might, from the abilities of the humble foot soldier right up to the punch that a new SS-25 missile could deliver.

When the Soviet system collapsed, two things became obvious. First, the judgment that the Soviet Union was a near-equal in terms of military strength was wrong; compared with the armies of the West, the Soviet machine was pathetically ill-equipped for modern warfare. And second, this function of using covert means to find out what the enemy was capable of was largely redundant. Almost everything the Americans or their NATO partners needed to know about these weapons was available in the sales brochures of the arms dealers who hawked Soviet hardware around the world as the CIS (the loose Confederation of Independent States that comprised the republics of the former U.S.S.R.) desperately scrambled to acquire foreign currency. What was not available on the open market was for sale by impoverished former Warsaw Pact officers. The military does not live in isolation from the rest of society, and it was obvious to all that as the threat from the only other global superpower vanished, along with the superpower itself, so would the questions about the amount of money being spent on defense increase in volume. Talk of peace dividends may have heartened advocates of increased spending on education, welfare and infrastructure renewal, but it sent chills around the corridors of the Pentagon. Peace dividends equaled defense budget cuts.

The Gulf War changed that. Not only had the military demonstrated they could do the job they were paid to do, and do it magnificently, they also became aware that there was a new kind of warfare that could emerge from the shell of the old. The Chairman of the Joint Chiefs of Staff, Gen. Colin Powell, submitted an article to the magazine *Byte* in July 1992 that gave a very limited, and by later standards extremely simplistic, view of why the Gulf War was such a watershed.

> The Information Age has dawned in the armed forces of the U.S. The sight of a soldier going to war with a rifle in one hand and a laptop computer in the other would have been shocking only a few years ago. Yet that is exactly what was seen in the sands of Saudi Arabia in 1990 and 1991. Information systems have become essential ingredients to the success of combat operation on today's battlefield.

This was a peculiarly thin analysis from the man who had masterminded the allied victory in the Gulf. To focus on the hardware without talking about the software that would flow through those systems and where it would be directed was like discussing a car without mentioning its engine. Other analysts went that extra mile, and started talking about the role of "information" in the Gulf War as a commodity not a descriptor, and maintained that here was a future for the armed services, if only they could work out what it was. Absolutely critical in this process was a concept paper written just after the end of the Gulf War and published in November 1991 by Gen. Glenn Otis, a former commander of the U.S. Army's Training and Doctrine Command (TRADOC), where the Army's most fundamental philosophies are defined and introduced into the training that underpins life in the military.

> Many lessons have been and will be derived from Desert Storm. Some are not new: others are. One however is fundamental: the nature of warfare has changed dramatically. The combatant that wins the information campaign prevails. We demonstrated this lesson to the world: information is the key to modern warfare—strategically, operationally, tactically, and technically.

Information campaign, information as a strategic, operational, tactical as well as technical commodity. This was an intoxicating concept, and to everyone in the defense community, Otis's paper was the equivalent of a life preserver thrown to a drowning man. It crystallized what everyone had been thinking about, and looking for, into a straightforward blueprint for the future. The term "information warfare" had not yet been formally adopted, but that was only a matter of time (the term was coined in 1976 by a team working with Andy Marshall). For now, it was enough that a new kind of war, which would require a new kind of warrior equipped with new kinds of weapons (not just laptops), was out there waiting to be fought. As with all challenges, particularly those to do with justifying budgets, the U.S. military machine was equal to this one. In all four services (Air Force, Army, Navy and the Marines) there was a frenzy of activity in creating new units and specialists to respond to the demands of this, as yet undefined, new discipline. As the services and the analysts who observe military affairs began to explore the ramifications of information warfare, a new phrase took shape, "revolution in military affairs." This concept served two purposes; it gave urgency and point to the very real change in the notion of how modern wars would be fought, and it gave the budget defenders another string to their bow. As with any concept that takes root in the military mind, it was soon turned into an acronym, RMA, or Revolution in Military Affairs. In fact, the Soviets discussed the issue first in the mid-1980s and described it as a "Military Technical Revolution," which was converted by the Pentagon into RMA.

As the military thinkers gained perspective on what had been represented by the Gulf War and the rise of information warfare, it became clear that this RMA was as fundamental as any in history, of which there have been relatively few in the last 650 years. They can be summed up as follows (all dates rough estimates):

1340 The bow makes armies cheaper and therefore bigger.
1420 Artillery replaces old concepts of siege warfare.
1600 Ships carry artillery, marking the start of modern naval thinking.
1600 Efficient construction methods make fortresses defensible again.
1600 Musketry adds standoff lethality to hand-to-hand combat.
1800 Birth of modern army with rationalized equipment and staff system.
1850 Naval revolution includes metal hulls, steam turbine engines, long-range artillery, submarine and the torpedo.
1860 The railroad provides mobility, telegraph communications. Rifling and the machine bring new levels of accuracy and destruction.
1920 Tanks, carrier aviation, strategic bombing, amphibious assaults.
1945 Nuclear weapons.
1990 The microchip.

This awareness that another RMA was under way was hastened by events in the civilian world, where the computer revolution was placing machines several times more powerful than quite recent military computers on the desktops of suburban teenagers. As has been noted, the

CAFMS computer that produced Air Tasking Orders during the Gulf War was sophisticated in 1990. Six years later a game-playing ninth-grader would sneer at a computer with only 10MB of RAM. For the first time since the invention of the transistor, the American military and space projects were no longer the major market for sophisticated electronics. Now it commanded less than 1 percent of that market. In his *Byte* article, Colin Powell acknowledged where the future lay.

A downsized force and a shrinking defense budget result in an increased reliance on technology, which must provide the force multiplier required to ensure a viable military deterrent. Increasingly, military requirements are being met by off the shelf hardware and software. . . . New information technology applications will spin off to ensure that America's fighting forces maintain the edge so proudly demonstrated in Desert Storm.

Powell was right. The civilian sector was leading the way. Prices plummeted, sophistication grew, and soon the military began to understand that it was no longer on the cutting edge of this revolutionary technology. This had two effects, one beneficial and one extremely worrying. As Powell had noted, lower technology prices meant the military could buy sophisticated technology off the shelf at a fraction of the price that it would have cost had it been custom-built. The day of the million-dollar toilet seat was drawing to a close. More worrying was that the technology the services could buy in their local computer store was also available to anyone with a few thousand dollars. Computers had been able to talk to each other over phone lines for over twenty years, but conversations were slow and few people had the necessary technology. The computer revolution put greatly enhanced power into a much wider cross section of society. If information was being heralded as the new source of military power, it was also becoming available to individuals who did not share the same agenda as the Pentagon. This would, as will be explained in more detail later, expand a threat that had already started to cause deep concern in military and commercial circles, namely the rise of the computer hacker.

General Powell's earlier analysis of information technology's impact on war was destined for a widely read civilian magazine. His next effort came in a directive to the entire military establishment from the Office of the Chairman of the Joint Chiefs of Staff, and set out to define what information warfare actually is, namely:

actions taken to achieve information superiority by affecting adversary information, information-based processes, information systems, and computer-based networks while defending one's own information,

information-based processes, information systems, and computer-based networks.

For the first time, the military appeared to accept that in this new age wars need not necessarily be fought by traditional military means, personnel or equipment. To say this foreshadowed a war without soldiers would be nonsense, but the Joint Chiefs did in part suggest that war could be fought by other means (soon to be crystallized in the new acronym WBOM, War by Other Means), so if a military under pressure to downsize and cut budgets wanted to find ways to hang on to as many people and as much materiel as possible, shaping this definition in a more conventional military fashion was necessary. Thus Powell's office integrated the information warfare concept into the traditional form of warfare known as Command and Control Warfare. C2W was now defined as the integration into a strategy of the battlefield of at least:

- operations security (OPSEC)
- psychological operations (psyops)
- military deception
- electronic warfare (EW)
- destruction

This interpretation is far too narrow, as recent history demonstrates, but it was a start. (C2W is only part of what might be described as information warfare, even by the Joint Chiefs' own reckoning, in that it did not at that stage include attacks on enemy computers, as had already happened in real war. See previous chapter.) While the military struggled to come to grips with this elusive concept, defense analysts added their voices to the debate. It soon became obvious nobody had an easy answer. Martin Libicki, Senior Fellow at the National Defense University, summed up the difficulty everyone was having in a delightful metaphor: "Coming to grips with information warfare . . . is like the effort of the blind men to discover the nature of the elephant: the one who touched its leg called it a tree, another who touched its tail called it a rope. . . ."

The problem with the elephant as a metaphor for IW, as Libicki himself pointed out, is that there may not even be an elephant there at all, just a lot of tails and legs with no body of thought to connect them. If IW existed purely on an assumption, without a rigorous effort being made to define it as a discipline capable of standing on its own, then the results could be dangerous and costly. And even if the elephant does exist, without that analysis being carried out we run the risk of accepting the definition of whichever blind man can shout the loudest.

Into the middle of this debate was thrown a new situation to which the U.S. military would have to respond, one that would test some of these theories about information warfare and would demonstrate that the future held many more challenges and dangers than the post–Gulf War euphoria had promised. It would also demonstrate that the mightiest military nation in the world could be brought low by a Third World army armed with AK-47s, native drums and raw cunning.

It was called Somalia.

The Gang That Couldn't Shoot Straight

T H E U.S. operation in Somalia was born, lived and died by the television camera. It was nurtured by political opportunism and poisoned by political weakness.

In 1984 the world was stunned by BBC television pictures of the famine in Ethiopia. Concern over the fate of starving babies sparked rock singer Bob Geldof and some friends to create Band Aid, and then Live Aid, pro bono musical enterprises that raised millions of dollars for famine relief.

Similar pictures were shown around the world in 1992. This time the famine was in Somalia, a former British, then Italian colony wrapped around the Horn of Africa that was ravaged not just by nature but by internal power struggles between war lords vying for control of the country. The world felt impelled to act, although it should be noted that there were many places in Africa where the suffering was as bad, if not worse, such as in Liberia, but the cameras were not there and so the world did not feel responsible. In this way has a new TV age version of the old philosophical question been born: if there is a terrible tragedy happening in the world but there are no TV cameras on hand to record it, has it really happened?

As far as the United Nations was concerned, this was happening, it was very real, and it was the kind of situation the United Nations was designed to deal with. To respond, the U.N. required a mixture of humanitarian relief, peacekeeping and social reconstruction. The particular challenges facing this operation lay in getting enough food and medicine to this barren land while preventing the marauding bandit gangs who held sway there from stealing it. On April 24, the United Nations Security Council passed Resolution 751. This established the structure for the United Nations Operation in Somalia, UNOSOM, which was designed to bring relief to the starving people. The call for

U.S. involvement did not come until July, when the U.N. efforts were failing to impact the human misery in Somalia. There was an urgent need for airlift capability to move more food and supplies to the region and to protect the humanitarian efforts of the U.N. and other relief agencies. President Bush had seen the same TV pictures as everyone else, and issued the order on August 15 that sent American troops on Operation Provide Relief, a fixed-term assignment that would run until December 9 and whose goal was "to provide military assistance in support of emergency humanitarian relief to Kenya and Somalia." Twelve aircraft, eight C-130 Hercules and four C-141 Starlifters were assigned to ferry supplies directly to the ailing nation and northern Kenya from where it would be taken by land across the border.

The country where this relief was heading truly deserved the sobriquet godforsaken. Bounded to the north by the Gulf of Aden, to the southeast by the Indian Ocean, its 250 million square miles, the size of New England, resembled more than anything else the low desert of the American Southwest. The country had been ravaged by the same drought and famine that had brought so much misery to eastern Africa during the 1980s. Any hope the people had of ameliorating their own situation was erased by the continuing war between rival clans.

Given its strategic importance, lying as it does at the mouth to the Red Sea and thus the access to the Suez Canal, Somalia had been the focus of one of those proxy wars between the superpowers during the Cold War, and large amounts of sophisticated weaponry had been supplied to the clans fighting for control of the country. It mattered little to them what either the Soviet Union or the United States wanted to gain from their allegiance. The fact that swearing it brought them a shower of guns and supplies with which to fight was fine by them.

As the Cold War began to crumble, the fighting intensified. The regime of the Marxist leader Siad Barre fell in 1991, leaving behind a vacuum that the fourteen indigenous clans each sought to fill. The clans were deeply mistrustful of each other and even though alliances were formed from time to time, convenience not loyalty was what held them together. The people were highly aggressive and thought nothing of using women and children to help fight their battles. By the time the TV crews arrived to document the unfolding disaster, half a million Somalis had starved to death and a million more were threatened.

Operation Provide Relief tried for six months to make a difference to this tragedy. U.S. forces airlifted over 28,000 tons of supplies, but the security situation deteriorated steadily and was impeding the relief effort. The United Nations was moved to change gear in the relief operation in November after a ship carrying badly needed aid came under fire in Mogadishu harbor. On December 3, 1992, the U.N. Security Council passed Resolution 794, which carried all the intent of the previous 751, but added a crucial new element, namely that this mission sought to impose order on the battered country under Chapter VII of the U.N. Charter. Resolution 751 came under the auspices of Chapter VI, in that it was a peacekeeping operation. Chapter VII meant peace enforcement, an altogether more proactive affair. The day after Resolution 794 passed, President Bush announced Operation Restore Hope, in which American forces would provide military leadership and a large portion of the international force that would make up the United Nations Task Force, UNITAF, which replaced UNOSOM. The operation would be based in Florida at U.S. Central Command. The mission statement talked of securing

> the major air and sea ports, key installations and food distribution points, to provide open and free passage of relief supplies, provide security for convoys and relief organization operations, and assist UN/NGO's [nongovernmental organizations such as civilian relief agencies] in providing humanitarian relief under UN auspices. Upon establishing a secure environment for uninterrupted relief operations, USCINCCENT [United States Commander in Chief Central Command] terminates and transfers relief operations to UN peacekeeping forces.

This was the second point at which television cameras were to play a significant role in the developing tragedy. George Bush was now a lame duck, having lost the presidential election to Bill Clinton a month previously. He knew the PR value of letting the world see U.S. troops once again acting to make a difference, to restore hope to the troubled people of Somalia. This view was not communicated to the military units who would have to carry out this job, and if it had been it would most certainly not have been shared.

On December 9, six Cobra helicopters took off from a U.S. naval vessel positioned off the Somali coast. They had been told to provide cover for the Navy SEALs and the Marines who were landing, under cover of darkness, on the beaches of Somalia. The security briefing had mentioned the presence of armed gangs with rocket-propelled grenades and small arms, but the threat was judged to be minor.

Yet as the helicopters moved in to cover the landings, they saw lights and flashes up and down the stretch of land where the U.S. forces were coming ashore. At least one pilot reacted by arming his weapons system and preparing to fire, before an urgent message came through his headphones that the flashes were not from guns or rocket launchers but from press photographers and TV cameras. As the special forces emerged from the sea, their shining frogmen's suits were illuminated by a battery of klieg lights. Moments later reporters were throwing microphones in their faces. It might have made great television but it could have put the men directly in harm's way.

To the aghast special forces on the beach, this was a complete negation of military security, a top secret mission reduced to low farce. To
their political masters back home, in their hubris, here was the nation
that had two years earlier conquered the fourth largest military in the
world going into a Third World country to knock a few heads together
and rid a starving people of the gangsters who were plaguing them. It
was a photo op not to be missed, so the mission details were leaked to
the media. Thus did the world see the bizarre pictures of heavily armed
troops wading ashore onto territory that had already been conquered
and occupied by TV crews and reporters.

Despite these inauspicious beginnings, the operation was essentially
a success. The international force of 38,000 (28,000 of them American)
brought a degree of peace to the country, chiefly by confiscating the
main weapon used by the warring clans, the "technicals," which were
pickup trucks with a heavy machine gun mounted on the back. It was a
crude but effective weapon for hit-and-run assaults, and removing them
from the streets hindered the clans' ability to fight each other or harass
the UNITAF. More food and supplies could get through and the grim
possibility of mass starvation receded. Yet in the complete absence of a
civilian or military governing authority in Somalia, the mission had every
appearance of being open-ended. The U.N. was looking at the possibility
of installing an expensive peacekeeping force that could be there for
decades. It was at this point that the decision was made to escalate the
U.N. mission to Somalia, and perhaps in that decision lay the seeds for
the disaster that was to befall the United States.

U.N. Secretary General Boutros Boutros-Ghali campaigned hard for
an additional layer to be added to the mandate authorizing the United
Nations to establish political institutions. Suddenly the U.N. was in the
business of creating a nation from scratch. Security Council Resolution
814, giving effect to this expanded mandate, was passed on March 26,
1993. On May 4, UNITAF was replaced by UNOSOM II. The U.S.
involvement in this force was much smaller, numbering only 4,500
troops, most, but not all, logistics units. The balance was formed by a
Quick Reaction Force that had been requested by the U.N., made up of
1,150 soldiers from the 10th Mountain Division. The new administration of President Bill Clinton signed off on the mission, which was
commanded by Turkish general Cevik Bir. Tactical command of the
U.S. forces would rest with Bir's deputy, the Commander U.S. Forces,
Somalia, Maj. Gen. Thomas Montgomery. Twenty-five thousand American troops went home.

While the U.N.'s ambition to go beyond bringing peace to Somalia by
creating a meaningful civic structure, which would ultimately lead to
the Somalis being able to rule themselves, was no doubt a worthy one,
its consequences had not been thought through. Careful diplomacy
managed to bring most of the warlords to agree to talks. But the U.N.

action was a direct threat to the power base of Mohammed Farah Aidid, the warlord who dominated Mogadishu. The international forces had already been occupying his city for close to a year and he felt threatened, so he decided to force a confrontation. His band of ragtag irregulars started to harry the U.N. forces with increasing frequency. On June 5, 1993, Somali fighters ambushed a contingent of Pakistani troops and killed twenty-four of them, literally tearing them to pieces. It was a shocking event that caught the attention of the world. It also spurred the U.N. to add yet another layer of purpose to the mission in Somalia, the kind of mission creep that had characterized America's folly in Vietnam and had been avoided ever since. The U.N. effort that had started as a mission to halt mass starvation, evolved into a mission to end the widespread lawlessness throughout the land, graduated into a mission to build a nation had now become a manhunt, and the target was Aidid. He strongly denied responsibility for the June 5 attack, but Robert Gosende, the U.S. envoy to Somalia, and U.S. Adm. Jonathan Howe (Ret.), President Clinton's choice to head the U.N. mission, both maintained that his forces were responsible.

Back in the United States, President Clinton gave his approval for a three-day bombing raid of Aidid's compound. It was a signal that any attempts at diplomacy aimed at bringing a negotiated settlement to the country's political needs were dead and buried. Clinton wanted Aidid punished. "We didn't plan to kill him, but the president knew that if something fell on Aidid and killed him, no tears would be shed," reported one of the senior officials who witnessed Clinton's decision to escalate the conflict.

On June 12, Clinton explained the raids on Aidid's compound in his weekly national radio address. "We're striking a blow against lawlessness and killing," he said. At a later news conference he added, "We cannot have a situation where one of these warlords, while everybody else is cooperating, decides he can go out and slaughter twenty peacekeepers." With Clinton's blessing, Gen. Wayne A. Downing, overall commander of U.S. special operations, and Gen. Joseph P. Hoar, Commander in Chief of Central Command, put America's elite top secret Delta Force on standby for a mission to Somalia.

The Delta Force was America's answer to the international terrorism that characterized the 1970s, when daring seizures of airliners, embassies and public buildings by well-armed, Soviet-trained terrorists created a new demand for specialist forces. The model for virtually all the special forces that were to follow was Britain's Special Air Service, the SAS, whose long experience dating back to its formation in World War II gave it a head start on the rest of the world. The Americans had been dazzled as had everyone else by SAS antiterrorist exploits, which culminated in the spectacular and ruthless raid on the Iranian embassy in London in 1980. Even when other national forces went to work, such as Germany's

GSG9, which successfully liberated a Lufthansa airliner held by Arab terrorists at Mogadishu in 1977, the SAS was called in to advise and even had two operatives playing key roles in the raid. The Delta Force came into being in 1977. Its first taste of action was the Desert One mission in Iran, but hopes of a dramatic rescue of the American hostages held in the U.S. embassy in Teheran were dashed in a shambles of poor coordination, communications foul-ups and mechanical failures that resulted in a fiery inferno at the desert rendezvous. The disaster clouded the last year of Jimmy Carter's presidency and is blamed in part for his failure to be reelected.

Caustic Brimstone, as the Somalia mission was dubbed, at first sight appeared a more routine affair. Gen. William F. Garrison, commander of the Joint Special Operations Command at Fort Bragg, North Carolina, home of the Delta Force, decided on a fairly modest plan: fifty commandos would be deployed to Mogadishu to effect the arrest of Aidid. Meanwhile, the situation in Somalia became more intense with Aidid's increasingly ambitious and daring militia stepping up attacks on U.N. peacekeepers, so Garrison's plans became correspondingly more elaborate. Caustic Brimstone evolved into Gothic Serpent, which called for 130 commandos from Delta's C Squadron, a Ranger company and sixteen helicopters from Task Force 160, the Army's special operations aviation unit.

The delay in deployment of the Delta Force was due to nervousness on the part of the politicians and also to advice from Gen. Joe Hoar at CENTCOM. Hoar surmised, accurately, that there was risk at every level of the operation; militarily it would be extremely tough to find and extract Aidid successfully; diplomatically the United States would appear to be the bully who preferred to use force over diplomacy as a way to bend a lesser people to its will. There were also people in the Clinton administration aware of the risks involved in a special forces mission that was not 100 percent successful, National Security Adviser Anthony Lake among them. He had been in Vietnam when a similar mission, ordered by President John Kennedy to kidnap South Vietnamese President Ngo Dinh Diem, had gone horribly wrong. He wondered if the burden could be spread a little and asked the British government if they would send in an SAS team to get Aidid. The British politely refused. The SAS assessment passed to the Ministry of Defence was that any kidnap mission was unlikely to succeed. Real-time intelligence on Aidid's whereabouts would be very difficult to get; covert insertion of a team in such a hostile environment would not be easy and so the element of surprise that is essential in such operations would be at risk; and anyway the British felt Aidid was simply not worth the price in political capital and soldiers' lives if things went wrong. The British assessment proved eerily prophetic but already the mission was approaching critical mass in Washington. On August 8, with doubts about the wisdom of a special operations mission still strong among members of the administration, and with U.S. envoys on the ground in Somalia pressing for Aidid to be taken out of play, the decision was made for them.

Aidid's force had become adept at using command-detonated mines. Unlike pressure mines that detonate when a person or vehicle is on top of them, these are set off remotely using wire or radio signals. On that day in August, one such mine killed four U.S. soldiers. Feelings in the administration polarized. Some felt the deaths were the cue for a whole-sale withdrawal of all American troops. Joint Chiefs Chairman Colin Powell disagreed. His status elevated by the Gulf War victory and by Clinton's trust in him, Powell's voice carried a great deal of weight. He advised the President against cutting and running. "We had to do something or we were going to be nibbled to death," an aide to Powell recalled. President Clinton approved the sending of the covert force, now dubbed Task Force Ranger, to the Horn of Africa. On August 22, another command-detonated mine blew up under a U.S. Army jeep, wounding six soldiers. Clinton was on vacation at Martha's Vineyard when he received the news. He gave the immediate order for the Delta Force to go. "We were going to set Aidid aside," a senior Clinton adviser said, using a neutral phrase for what was essentially going to be a kidnap-ping.

The covert force arrived in Somalia on August 26. The normally relaxed dress code of the Delta Force, whose hair was kept civilian length to help them blend in with the public, was now tightened to help them blend in with their fellow soldiers. Hair shaved to a bristle, they were indistinguishable from the Rangers. They immediately set to work, and just as immediately ran into trouble. Covert operations run on intel-ligence. Without accurate information on the exact disposition of an enemy, a soldier who is expected to work in the darkest of the world's back alleys and react to a threat with immediate and savage force is crippled.

In Mogadishu, a CIA cell had set up a network of twenty local agents and a sublevel of informers. Their job was to brief the commandos on the local people and environment and in particular to find Aidid's hideout so he could be snatched. The fact that the Delta Force went to work within a matter of days of its arrival did not bode well for its education in the ways of the Somali people and the militia the force had been sent in to fight. Nor did the first piece of news Gen. William Garrison received from his CIA advisers reassure the team. The chief Somali agent on the ground had a bizarre way of passing the time: he enjoyed a good game of Russian roulette. Fate decided it was his turn to lose, and he shot himself in the head.

Four days after their arrival, the Delta Force set out on its first mission to snatch Aidid. Its helicopters roared in low over the rooftops and the maze of narrow twisty streets of Mogadishu. Ropes snaked from the aircraft and the crack commandos rappeled swiftly to the building below to carry out their mission. The building was empty. They moved swiftly on to target number two and arrested a man who looked remarkably like Aidid. It turned out he was a prominent member of the U.N. relief mission. The lack of local knowledge, so severe that the soldiers could not tell one Somali from another, was becoming dangerously exposed. At one point some soldiers even thought Aidid had infiltrated their base and was working undercover in their mess hall. In fact, as Aidid later told a news conference, he had instinctively taken advantage of such ignorance and had hidden out among his own people. "They never came close," he would say of the American search.

In Washington Defense Secretary Les Aspin was apoplectic over the bungled raid. "We look like the gang that can't shoot straight," he railed. There were three more abortive operations to follow, each of them hampered by poor intelligence. Many other missions were planned, then aborted. One of the problems was that the CIA team was relying on electronic eavesdropping to find Aidid, and had brought with it a mass of the very latest technology from America. Aidid's troops were, however, using very low-tech walkie-talkies and talking drums to signal each other, and the CIA's equipment was incapable of dealing with either. The technology gap had started to work against the Americans.

On September 7 General Garrison decided to expand the operation. If they couldn't get Aidid, they would arrest his top lieutenants and cripple his command structure that way. As Garrison wrestled with the extremely difficult mission assigned him by his political masters, Aidid's forces were becoming more aggressive. On September 9 a mob of men, women and children (everyone was a fighter to the Somalis) attacked American and Pakistani patrols in Mogadishu. U.S. Cobra helicopter gunships opened fire on the crowd, killing at least 100 people, among them women and children.

The massacre horrified Clinton. As a student he had demonstrated against U.S. tactics in Vietnam, in which women and children were killed by American guns. These were the tactics that Clinton was to write that he "loathed." Now a mission he had ordered was doing the same thing. On September 12, he hosted former President Jimmy Carter at the White House prior to the next day's high-profile signing of the Middle East peace accord between Israel and the Palestinians. In talks that went on late into the night Carter told Clinton that Aidid had been in touch with him personally to protest his innocence over atrocities attributed to him. Carter stressed to Clinton that there could be no peace without a political accommodation with Aidid. The President started to second-guess his own policy. So did his senior advisers, who were conscious of the growing impatience in Congress with the worsening situation in Somalia. A request by General Montgomery, the U.S.

Commander in Country, for four M-1 Abrams tanks, fourteen Bradley Fighting Vehicles and some heavy artillery had been approved by CENT-COM Commander Joseph Hoar, who had seen the difficulties faced by the American troops firsthand during a visit that week to Mogadishu. On September 23, Defense Secretary Aspin, one of the advisers cautioning the President about Congress's mood, rejected the request, fearing the signal a greater concentration of force would send about American intentions. It was clear to Hoar that as Aidid's militia was gaining strength in the city, and as the politicians refused to send more muscle to deal with it, a severe examination of policy was in order. There was a change in policy, but it was not one designed to bring the military any comfort. A diplomatic solution to the crisis would be sought, but at the same time the Delta Force was to continue its efforts to snatch Aidid.

On October 3, one of the CIA's local informants reported that Aidid's top aides were to meet in a house near the Olympic Hotel, which was in the part of the city where Aidid was particularly strong. "That's really Indian country. That's a bad place," General Montgomery told Garrison when informed of the mission. Garrison decided on some extra insurance. He ordered his four AH-6 Little Bird helicopter gunships to carry rockets on the mission and to shoot anyone who looked threatening and ask questions afterward. In Mogadishu, the informant who had reported the meeting stopped his car outside a house near the hotel, got out, opened the hood, poked around in the engine, closed the hood and drove off. To the reconnaissance team watching from a Hughes 530 helicopter, this was the prearranged signal that indicated the house where the meeting would take place. Delta Force rushed to their transport, only to be stood down moments later by a message from the CIA. The informant had confessed to being too scared to pull up outside the correct house. The target was actually a block away.

The assault force took off. The Little Birds went in first, commandos and Rangers clinging to the skids as the helicopters stirred up clouds of dust and sand around the target building. The assault force hit the ground running with only one casualty, a Ranger who left his helicopter prematurely from forty feet up and was critically injured in the fall. The force secured the perimeter of the building while the Delta commandos went in. For once the intelligence had been accurate. Inside they found a large group of Aidid supporters, including some very senior aides, as promised. They herded the prisoners into the courtyard and radioed for evacuation. "We're ready to get out of Dodge," the officer in command of the assault called in. It was impossible to land helicopters to evacuate the force and their prisoners because of the narrowness of the streets, so a convoy of twelve vehicles was dispatched to collect them.

It was the rocket-propelled grenades, known as RPGs, that proved the decisive weapon at this point. As Aidid's militia raced to the battle scene, ambush units headed for the main roads between the Olympic Hotel and the American base ready to harry the American retreat or intercept any reinforcements. Helicopters hovered overhead providing cover, as the assault force and prisoners piled into the vehicles sent to extract them. At that moment an RPG smashed into the Black Hawk helicopter known as Super 6-1. It was crippled and went into a spin, crashing nose first into an alleyway, killing the two pilots and wounding the five passengers, among them three Delta snipers. In accordance with contingency plans already worked out by Garrison, the Quick Reaction Force was alerted to provide reinforcements from the U.S. base near the New Port while Rangers in another Black Hawk were sent in to recover the wounded from the stricken helicopter. The assault force was also ordered to provide extra ground support.

A Little Bird roared in and landed, the pilot maintaining control with one hand, firing a machine gun at approaching Somalis with the other. His co-pilot dashed into the wreckage and extracted two of the wounded snipers, bundled them into the back of the Little Bird and took off. A Black Hawk dropped fifteen Rangers by rope before being hit by another RPG. It was severely damaged but managed to return to base. By now there were close to ninety American soldiers in the vicinity of the wreckage.

This was urban warfare at its worst. Swirling sand cut visibility to close to zero in places. Gunmen, who knew every inch of the area, appeared in every window and hid behind every bit of available cover. What should have been a fast, coordinated snatch operation had collapsed into a dirty, exceptionally violent firefight where the advantage was rapidly moving toward the local guerrillas.

Militia were coming at the Americans from all directions, hosing them with fire. Air support was incapable of pinning militia gunmen down, there were just too many of them, so the gunners aimed for the RPG crews. The initial assault force now aboard the convoy with its prisoners came under withering fire as it headed for the downed helicopter. With .50-caliber machine guns and grenade launchers on board the Humvees they were riding in, the 142 soldiers of the assault team and Rangers providing cover should have been a formidable force. But the streets were becoming choked with crowds of armed Somalis as well as women and children where hostiles were indistinguishable from innocent bystanders. At one point the convoy fired a salvo of grenades at the crowd to prevent itself from being trapped. It was a fighting retreat and the superior American firepower caused dozens of casualties, with the dead and wounded literally falling in heaps on top of one another. The assaults on the convoy continued from Somali gunmen, who raked it with AK-47 fire at every opportunity. An RPG smashed into one truck, decapitating the American driver. Three of the prisoners were killed by Somali fire. Eventually the convoy made it back to base.

At the downed helicopter, Rangers struggled with desperate frustration to free the dead pilots while the armored metal of the cockpit frame resisted their every effort. Small-arms fire poured in from all sides and it was fortunate the enemy were so poorly trained that few rounds found their mark. Even so, the attrition was terrible, with more and more of the beleaguered force taking casualties.

To the southwest of the stricken Super 6-1 and due south of the hotel, Black Hawk Super 6-4 was hit by an RPG. Its tail shattered and the craft slammed into the ground with tremendous force. It was bad enough having one helicopter down. The second made a difficult situation nearly untenable. The four-man crew of Super 6-4 was injured, but it was impossible to get any ground troops to them. The Quick Reaction Force had come under intense fire shortly after leaving base to reinforce the rescue mission, and after firing 60,000 rounds in thirty minutes retreated. Their inability to get through led the aviation commander to reverse his previous decision, and allow two Delta snipers to attend to the wounded crew of Super 6-4. Sfc. Randall D. Shugart and M. Sgt. Gary I. Gordon were on board another Black Hawk Super 6-2. Their pilot lowered them to a clearing 100 yards from the crash site and took off again to provide air cover. The confusion on the ground was such that the two men almost got lost trying to find the stricken Super 6-4, so the pilot of Super 6-2 directed them to it by hand signals.

The chances of Gordon and Shugart being able to rescue any crew member and get all out alive were slim at best. Their chances dropped to zero the moment another RPG found its target, blasting Super 6-2 through the cockpit, maiming a Delta sniper who had been covering the men below and grievously damaging the number two engine. The pilot was forced to pull out and make a crash landing in the port area. Without air cover, Shugart and Gordon were doomed, but as the wounded pilot, CWO Michael Durant, would later recall, the two men were totally cool and thought only of making him comfortable.

"What's wrong?" Gordon asked him. "Where do you hurt?"

Durant had a broken back and shattered leg. Together Shugart and Gordon lifted him gently from his twisted cockpit and laid him on the side of the helicopter away from the main Somali assault. They gave him a Heckler and Koch MP-5 submachine gun, a favorite of special forces all over the world, then went looking for a way to get him, the other crew members and themselves out of the firestorm. Shugart yelled in pain as he was hit by a Somali bullet. Gordon reappeared next to Durant. The MP-5 was almost empty. The other crew members were dead. Gordon asked, still in a calm, methodical manner, if there were any weapons in the helicopter. Durant told him of two M-16 rifles stowed behind the pilots' seats. Gordon collected them, gave one to Durant with an ammunition clip of five rounds, saying "Good luck" as he did so, then returned to the other side of the aircraft to take on the enemy. There was a volley of fire and a yell of pain. The last Durant heard of Gordon was a muttered "I'm hit. Damn it." Alone and out of ammunition, Durant lay the rifle across his chest and waited. The Somalis stopped firing and moved in to take him prisoner.

Back at the first downed helicopter the battle raged on until dawn the next day. A group of soldiers had occupied a civilian house to escape the deadly fire being directed at them. In the dark (they had not banked on the assault taking this long so they had left night vision equipment behind) they rounded up the women and children in the house and held them under informal arrest, later saying they were concerned for their safety should they have been allowed to flee into the middle of a firefight. The women claim they were handcuffed, and the militia claims the Americans were employing them as human shields. A militia leader said later he had planned to attack the holed-up Americans with mortars but held off when told of the presence of women and children. The Americans categorically denied they used the women and children as human shields. "It's not part of our mindset to take hostages," a senior officer said later, "especially women and children."

Relief came when an armored column finally made it through to the battle zone. That was not without incident in itself. At the urging of General Garrison, General Montgomery asked for help from the other nations of the multinational force. He wanted them to provide the armor his own government had denied him. The Pakistanis and the Malaysians lent four tanks and twenty-eight Malaysian APCs (armored personnel carriers). The convoy, bolstered by American Humvees carrying men of the Quick Reaction Force, came under fierce attack the moment it left the safe port area. The mission was a microcosm of the difficulty faced in multinational military operations. In confusion wrought by the constant fire and the difficulty in understanding the leader of the American squad he was carrying, a Malaysian APC driver turned south instead of north, along with a second APC. Both APCs were destroyed by rocket-propelled grenades. The stranded American team had to fight their way through the city to the main force.

The armored convoy made good the retreat by 7:00 A.M., by which time the rescue teams had finally managed to pry the dead body of pilot Clifton Wolcott from the wreckage of Super 6-1. They had stuck to their stated aim of not leaving a dead comrade behind. Even when in the aftermath of the Somalia operation it was suggested that if they had not insisted on retrieving Wolcott's body a great many of the ensuing deaths might have been avoided, the special forces stuck to their principle.

The toll of casualties read: 18 Americans and 1 Malaysian dead, 84 Americans and 7 Malaysians wounded; 312 Somalis dead and 814 wounded.

The Americans had successfully carried out their mission of capturing the senior leadership of Aidid's militia. On the face of it they had won a total victory.

But it certainly did not feel as if they had.

It felt even less like a victory when the world's press and TV carried pictures of the captured Michael Durant. He had been brutally beaten and was in agony from his wounds. More revolting still was what the Somalis did to the five dead at the site of Super 6-4, including Sergeants Gordon and Shugart. The bodies had been defiled, one of them dragged naked through the streets of Mogadishu in front of a CNN camera by jubilant Somalis. This was a deliberate insult by the politically savvy Aidid and his followers. There is no doubt that Aidid had closely followed the political debate that had prompted the tentative American deployment. A smart man, Aidid clearly understood that pressure could be applied in Mogadishu that would be felt in Washington.

President Clinton is said to have wept as he saw the pictures.

"By Saturday and Sunday we had won the war," one senior military officer recalls, "but on Monday Aidid mounted a strategic attack in the Information domain. The American people simply had not been briefed about this area of battle space, and so when bodies started appearing on TV screens, Americans said, 'Wait a minute, nobody told us this was going to happen'—and Aidid won."

The TV cameras had played their part yet again in shaping policy. Less than ten years earlier, President Reagan had reversed policy after 241 Marines had died in Lebanon. This time it took just a single graphic image to cause a similar reversal. Within days President Clinton ordered a complete withdrawal of American forces from Somalia. All missions to apprehend Aidid were called off, and in fact some weeks later, as part of the move toward a negotiated peace, Aidid rode in an American helicopter, a bitter aftermath to the suffering of those who died or were wounded trying to arrest him. As the father of one of the dead men said later, "The thing that haunts me is if it was so important to capture Mohammed Aidid when the Rangers went over there, why was it so unimportant on October 4?"

The appalling conclusion to the engagement cruelly exposed the vacillation inside the administration and the absence of clearly thought through policy on what exactly it was trying to achieve in Somalia.

It was also a salutary lesson for the military in the limitations of modern warfare. Despite the casualties, in the Pentagon's judgment, the firefights in Mogadishu that day had produced a decisive American victory. Exactly how many of Aidid's men died is still unclear but informal Pentagon estimates talk of several hundred, perhaps as many as a thousand dead and as many more wounded. A handful of Americans died, giving a kill ratio of around 75:1—a massacre in anyone's language. Yet, the television image of one soldier beaten and naked was enough to turn victory into defeat.

If there was hedging before the disaster, there was even more after. It took an angry call from a doctor at the Walter Reed military hospital in Washington, D.C., to get the President to visit the wounded of that operation who had been lying there for three weeks with no official acknowledgment of their presence or their heroism. The day after the call, Clinton did visit them, and according to some of the men was visibly moved.

"He didn't know what to say," one reported, "the men could see that."

While his reception was generally cordial, some of the wounded were hostile and refused to be photographed with him. Sgt. John Burns said, with the President standing in close earshot, "I don't want to end up in some political propaganda picture. You know—'President Visits Wounded Soldier.' "

The delay in anyone from the administration visiting the wounded was no accident. There was a clear effort to distance the President from the consequences of the disaster. Some of his close advisers even cautioned him, the Commander in Chief, against writing personal letters to the families of the American dead, saying Defense Secretary Les Aspin should do it. Clinton did in the end write a note to each one. The aides also advised him not to attend two memorial services for the dead. This advice he took. The Oval Office withheld pictures and video of Clinton meeting the wounded in the hospital and at a White House reception he gave for them. One official said later it was done in the hope that the American people would forget about Somalia and the mistakes made as quickly as possible. Others deny that, saying Clinton did not want to be accused of exploiting the heroism of the wounded men.

What the President did do was award the Medal of Honor, America's highest award for valor, posthumously to Sergeants Gordon and Shugart, and Silver Stars to several of the wounded who came back from that operation. The White House awards ceremony would be Clinton's chance to show the military that he did, despite the policy blunders, respect their courage and that he cared for their people. He was in for an unpleasant surprise. After awarding the Medals of Honor to the two widows, he invited the families to the Oval Office for a moment of quiet reflection. It was a classic Clinton gesture; thoughtful, full of compassion. The President approached Herbert Shugart, the father of Randall Shugart, and held his hand out. To his astonishment, the handshake was declined. "The blame for my son's death rests with the White House and you. You are not fit to command." Clinton reeled under the onslaught, which continued for several moments. As Commander in Chief, President Clinton could have simply accepted responsibility for the loss and sympathized with a grieving father. Instead, he took fifteen minutes out of his schedule to explain to Herbert Shugart why the death of his son was not his fault. It was a craven performance that horrified the Pentagon officials who were present. Immediately after the meeting, White House officials tried to keep the incident quiet and it was successfully suppressed until the story appeared in the London *Sunday Times*.

President Clinton had issued similar denials of responsibility days earlier to the families of some of the dead Rangers. He claimed to be surprised and angry that subordinates had ordered the raid, given that policy was shifting toward a diplomatic solution. The families later said the President's remarks showed how out of touch he was with his own military chain of command. Clinton continued to state publicly that he was the victim of events, not their master, that it was a United Nations operation and therefore their responsibility, and that his military had prosecuted an assault without his knowledge.

By the accounting of recent history, he succeeded in his damage control. The bulk of the political blame fell on Defense Secretary Les Aspin for vetoing the armor reinforcement that had been requested. The U.S. envoy to Somalia, Robert Gosende, who had been so insistent on taking out Aidid, was removed from his post by Secretary of State Warren Christopher. Christopher had endorsed Gosende's position, but after the tragedy said he had not been paying sufficient attention to Somalia. Christopher and Aspin resisted calls from Congress for their resignations. Aspin died two years later; Christopher left his job with full honors at the end of the first Clinton administration.

Congressional investigations into the affair produced strong criticism of the administration for moving beyond the neutral role of peacekeeping into taking sides. "Cardinal rules were violated," Congressman Ron Dellums of the House Defense Committee said. "We chose sides and we decided who the enemies were. It's baggage from the Cold War." No one took much notice, although the hearings in both Houses did provide those who felt betrayed by their government an opportunity to get their voices heard. Larry Joyce, whose son Casey died early on in the botched operation, told the Senate, the soldiers who died "were betrayed by an administration that gave them a no-win mission and didn't provide them with the resources and moral support they needed. Amateur hour!"

It was a charge that found sympathy in the Pentagon, where military commanders were fuming over the vacillation of the young President, a man who had never seen war, had done his best to avoid it while a student and seemed to have no real conception of what it was all about. Yet the military does not escape blame. The mission it carried out on October 3 was badly planned, probably unnecessary and based on inadequate intelligence.

The deaths of those eighteen soldiers would haunt military commanders for years afterward and inform their every decision. Even though the public seems to have regarded Somalia as being a long way away and too difficult to understand, for the Army it was painfully close. The most intense fighting since Vietnam, surpassing anything the troops in the Gulf War had experienced, created a mind-set in the military that judged any future operation on the number of likely casualties. Even though anyone who serves in the armed services is acutely aware that as part of the job one day his or her life will be in danger, the military commanders now saw future missions as viable only if they could result in minimal casualties. It was a view of the military's role distorted by media coverage of dead soldiers, a deep mistrust for a young administration with little apparent grasp of military affairs, and a degree of internal dissent over responsibility for the Somalia episode that encouraged extreme caution over future missions.

Although the Somalia story is an old one and many of its details are known, it is still taught as one of the classic tales of information warfare. It was a short and bloody conflict that, judged by any normal military standards, America had won convincingly. But in terms of perception management, Aidid had proved more sophisticated than his technologically superior enemy.

Somalia demonstrated the power of the media in molding perception. The President reflected his generation, which grew up amid the antiwar protest movements of the 1960s and 1970s. It had no direct experience of war and viewed conflict through the very narrow prism of television. To that extent, President Clinton was the new reality of political leadership and the armed forces were quick to understand that they needed a new approach if they were to have a strong role in the nation's security in the next century.

With a military that only wanted to fight wars where no one got hurt, advocates of information warfare found a ready audience for what they had to offer.

Riding the Tiger

S O M A L I A was the tactical victory that was allowed to become a strategic defeat. Victory in the Gulf had deceived the United States into thinking that in the modern age it was invincible. Somalia had demonstrated how wrong the United States was. Americans virtually ignored what happened in Africa, so the defense establishment was able to digest the consequences virtually unhampered by public scrutiny. Having been shaken by the missteps of Somalia, it was a duty the establishment carried out with extreme thoroughness. Not only was it having to contend with a world where strange threats and challenges would be the rule rather than the exception, it was also having to grasp a whole new level of menace that was unfamiliar. America was vulnerable to threats that could not be diverted by the application of massive force and technological superiority.

On June 11, 1993, the Secretary of Defense and the Director of Central Intelligence tasked a special Joint Security Commission to examine threats to the American security system and to find ways to "assure the adequacy of protection within the contours of a security system that is simplified, more uniform and more cost effective." The commission's report of February 28, 1994, examined everything from the screening of personnel to the number of layers of physical security surrounding defense establishments. There were a number of concerns spelled out in the report, not the least of which was that the multiplicity of agencies in the defense and intelligence communities, all with their own ideas and philosophies, had led to inefficiencies and waste. No real surprise there. There are many defenders of the defense establishment who say that the inherent competition between agencies with overlapping interests has produced, in the majority of cases, everything America has ever needed. Where the report started to move into new, worrying territory was in defining the external threats to America.

The threats today are more diffuse, multifaceted and dynamic. . . . The possibility of failure of democratic reform in Russia poses a constant danger. Further, Russia's ability to maintain control of its special weapons, China's supplying of equipment and technology to unstable countries, and North Korea's, Iran's and Iraq's attempts to develop nuclear weapons, have serious and far-reaching implications.

These are complex problems whose solutions depend on mixtures of diplomacy, intelligence, law enforcement and force of arms. The evidence of Somalia gives no encouragement that the United States will find easy means of dealing with them. The next part of the threat scenario offers even less reassurance.

Burgeoning ethnic and religious rivalries that cross traditional boundaries endanger both new and long-standing peace agreements, drawing the United States into an expanding role in peacekeeping and humanitarian missions.

In other words, to use a Hollywood metaphor, watch out for Somalia 2. In the years since that paragraph was written it is notable that little or no action has been taken at the highest levels to prepare for that. As will be seen in later chapters, the military is working on new tools that may assist in doing a job like peacekeeping in Somalia, but such advances are meaningless if the politicians have not learned from the last experience. With the deaths of those American soldiers a stain on the Clinton administration's record, it would take a brave official to ask the President, "Shouldn't we work out what to do next time?" The next threat foreseen by the report was one that America had already started to witness.

The bombing of the World Trade Center and the assassinations of two CIA employees in Virginia heightened our sensitivity to the fact that terrorist activities against Americans can occur domestically as well as abroad. Violent crime and narcotics trafficking in our neighborhoods also continue to threaten American lives and values.

This report was written during 1993–94, over a year before the bombing in Oklahoma City drove home the horrible reality that the soil of America was not only the target of foreigners wishing harm to its people and institutions but that some Americans were as capable of atrocity as any Middle Eastern terror group. The last sentence about violent crime and drug trafficking underlined the fact that in America's big cities, and increasingly in rural areas, too, a war was already being fought on a daily basis throughout the country, even though the news from the front was so depressingly and so routinely bad that it had receded into the backdrop of the nation's life. Yet as the Noriega experience had demonstrated, there were precedents for America taking the drug war beyond its borders. As power began to shift from the drug cartels of Colombia to the cartels of Tijuana and northern Mexico the enemy was moving closer to the country's back door.

> The Commission recognizes that the consequences of failures to protect against some of these threats are exceptionally dire. For instance, terrorists' use of weapons of mass destruction, or an adversary's foreknowledge of our battle plans could have consequences so grave as to demand the highest reasonable attainable standard of security.

It was an implicit acknowledgment that there were many ways in which America was better off in the days when it was pitted against one identifiable enemy. Now enemies are lurking everywhere. Where this gloomy scenario begins to merge with the military's post–Gulf War search for new meaning is in cyberspace, the place where war is fought with weapons of the new age—computer viruses, worms, Trojan horses —using information networks that connect us all, from the humblest home computer operator who uses the Internet to help with homework to the lairs of the ultra-secret spy agencies. The commission report said:

> Networks are already recognized as the battlefield of the future. Information weapons will attack and defend at electronic speeds using strategies and tactics yet to be perfected. This technology is capable of deciding the outcome of geopolitical crises without the firing of a single weapon. Our security policies and processes must protect our ability to conduct such infowars while denying our enemies the same advantage.

This added a whole new layer of threat. The report went on:

> If, instead of attacking our military systems and data bases, an enemy attacked our unprotected civilian infrastructure, the economic and other results could be disastrous. Over 95% of Defense and Intelligence Community voice and data traffic uses the public telephone system.

This was a new and startling threat for government and the military to address. After all, what good is the most wonderfully equipped military in the world if the whole edifice could be brought crumbling to the ground by an attack on the phone system? Although the nine-hour failure of the AT&T network in 1990 had been an accident, not an attack, it had demonstrated what can happen. Close to 70 million calls were lost in that short time, at a cost to business and individuals that cannot be calculated. It was a failure that affected one particular switch and the fault was corrected. There are switches all over the country. If someone could find a way to paralyze them, would they? The commission's reply: no doubt about it.

> Foreign intelligence services, including those of some of our "allies," are known to target U.S. information systems and technologies, using techniques that can give them access to our information system without coming into our workspaces or approaching our people.

Here then was an unprecedented challenge to America's civilian and military establishments to deal with threats of an alarming, diverse and often unidentifiable nature. Would they be up to it? The military was already in the throes of the great debate, but there were signs that it would dissolve into an argument between the conservatives and the visionaries, between those who trusted in tanks, rockets and battalions of fighting soldiers and those who shared the commission's view that a new kind of warrior would be needed to fight alongside the old, and, if necessary, take the lead.

There were echoes in this debate of the challenge and opportunities offered by aircraft at the beginning of this century. Then, as now, the new was seen as a threat to the old, with entrenched vested interests working hard to preserve the status quo. There are two particular problems with the information revolution that confront armed forces worldwide. First, the military and the intelligence agencies that serve them are institutionally deeply conservative and reluctant to address change. This matters little when all armed forces essentially advance in lockstep. The Soviet Union might develop a radar one year but this would be countered by an American missile the next year. This kind of arms race which had been in place for the past fifty years is threatened by the information revolution, which effectively destroys the possibility of steady evolution. The military must now educate current and future leaders not to hold on to anachronistic concepts and paradigms.

Second, this revolution was being created and driven by the young. In government and the military, those in authority had neither the experience nor the knowledge to understand or implement the revolution as it unfolded. On the contrary, the kind of change in management style (senior management in their thirties, reduction of hierarchy to a minimum to encourage bottom-up ideas, eighteen-month cycles of product creation and delivery to market) that was becoming common in industry was probably impossible to implement in government and the armed forces.

There was a third, largely unspoken reservation about the whole debate, particularly in Europe: The information revolution—at least as far as it referred to national security—was an idea generated and driven by

Americans. Europeans have an instinctive distrust of America's obsession with technology and the nation's apparent belief that the right "technological fix" can solve the most complex of problems. There is also a distrust of America's propensity to lurch with equal enthusiasm from one crisis to the next—a practice known sneeringly as "the crisis du jour" policy.

As one head of a European intelligence agency put it: "The way the Pentagon and others are talking, IW is going to be some kind of silver bullet. We don't see it like that. Anyway, we'll let the Americans run with the ball for a while, let them make the mistakes and then we'll move in and pick up the bits that work."

So, while the commission's analysis and warning spurred on many in the Pentagon and the government, there were others who believed that, at best, IW might be a useful new tool in the war fighter's backpack. Over the next few years most of the skeptics in America would begin to understand the perils and opportunities presented by IW. That understanding was helped by two other crises that confronted the international community—Haiti and Bosnia.

Trial by Strength

H A I T I first impinged on American public consciousness during the early 1990s, following the ousting by a military junta of democratically elected President Jean-Bertrand Aristide, a populist priest who despite his religious calling was a handy exponent of Haiti's bruising and sometimes deadly political process. The island had a decades-long reputation for brutal repression under the dictatorships of François "Papa Doc" Duvalier and his son, Jean-Claude "Baby Doc" Duvalier. For America, the election of Aristide in 1991 was a good step forward in the hemispheric march to democracy, even though there were strong private misgivings in intelligence circles about his commitment to human rights. There were rumors that he had employed as a weapon of physical and psychological repression the Necklace, once a favorite among terrorists in South Africa, in which a car tire is fitted over the victim's head, gasoline is poured into the rim, then set on fire. So Aristide was the best of a series of poor options. In September 1991, just seven months after his election, a group of military officers led by army commander Gen. Raoul Cedras seized power and Aristide fled to Washington, D.C., where he ran an extremely effective lobbying campaign for U.S. and United Nations support.

While Aristide left Haiti in the comfort of an aircraft, thousands of his fellow countrymen did not have access to such a luxury. Under their new military dictators the Haitian people suffered terribly. The country was already the poorest in the region. Under the generals, it got worse. Human rights appeared to count for nothing as the army suppressed all signs of pro-Aristide or pro-democracy sentiment. At the United Nations there was condemnation of the coup and a demand for Aristide's reinstatement. In Haiti, there was a rush for the boats as the people decided it was better to risk a sea journey to seek possible asylum in the United States in whatever craft they could lay their hands on than stay and risk death at the hands of the junta. The rush became a flood, and on May 24, 1992, President Bush declared that fleeing Haitians must be intercepted by the U.S. Coast Guard and repatriated to Haiti. It was a stern measure, but Bush knew that anything less would provoke a mass emigration from the island. His action offered Bill Clinton, the Democratic party's presidential candidate in the 1992 election, an easy target.

"I would give them political asylum here until we restored the elected government of Haiti," he promised, three days after the Bush edict. "I would turn up the heat and try to restore the elected government and meanwhile let the refugees stay here."

Eight months later, Clinton was the President-elect and beginning to realize that cheap talk on the campaign trail is meaningless in the face of realpolitik. In the run-up to the inauguration, Clinton was given an intelligence briefing based on satellite reconnaissance and information from agents in Haiti. He was told that thousands of Haitians were chopping down trees to build an armada of boats so that they could sail to America. The President was warned by his own staff that such an invasion would be a public relations disaster in the first weeks of his administration. Days before his inauguration, he reversed himself and announced he would continue the Bush policy.

"Those who leave Haiti by boat for the United States will be intercepted and returned to Haiti by the United States Coast Guard," he said in a radio broadcast to Haiti. "To avoid the human tragedy of a boat exodus, I wanted to convey this message directly to the Haitian people: leaving by boat is not the route to freedom."

In June the United Nations imposed on Haiti an oil and arms embargo and an asset freeze and set in motion a diplomatic initiative that led on July 3, 1993, to the Governors Island agreement between General Cedras and President Aristide. This called for the junta to leave Haiti by October 30 in return for amnesty. Democracy would be restored, and Aristide would return as President. Hopes that the suffering was over started to fall apart in early October when a U.S. ship, the USS *Harlan County,* carrying American and Canadian military trainers, who were to help reform the army in accordance with the Governors Island agreement, was prevented from tying up at Port-au-Prince by a small group of baseball-bat-wielding agitators. The ship was ordered by U.S. Naval Command to leave. The sight of a small mob turning away an American military vessel caused humiliation and astonishment in the United States and delight in Haiti, where the military regime had used some elementary psychological warfare to play on American fears of getting involved in skirmishes that could get out of hand. Cedras and his colleagues had studied the Somalia expedition carefully.

In May 1994, the United Nations stepped up the pressure with a total embargo on Haiti. Two days later, Clinton changed his policy on the boat people again—fleeing Haitians would still be stopped, but the boat people seeking asylum would be processed on board Coast Guard vessels or they would be taken to a third country for processing. As Bush had understood so clearly, a weakening of the position led to a surge in the numbers of Haitians fleeing the country. Hundreds drowned, and less than two months later, the hard-line policy was back in force. Asylum would have to be sought through the U.S. embassy in Port-au-Prince, clearly a risky proposition for those who chose to try it. To compensate perhaps for the mixed signals he was sending on the issue of the boat people, Clinton began to put pressure on the Haitian junta to leave the island and turn over power to Aristide or risk an invasion, and at his request on July 31 the U.N. Security Council passed Resolution 940 12–0 with two abstentions authorizing the use of force against Haiti.

In the six weeks following that resolution a steady drumbeat of allegations and accusations was directed at the junta.

"They are slowly turning Haiti into hell," State Department human rights official Nancy Ely-Raphael told a news conference.

Congressional leaders grumbled that the administration was preparing public opinion for the use of American military force to implement the U.N. decision. Be that as it may, the situation in Haiti was getting worse. Just hours before Ely-Raphael's dramatic flourish, twelve young Haitians had been found shot to death twenty miles outside Port-au-Prince. She went on to claim that the military was stepping up its abuses in an attempt to intimidate the people.

This underscored one of the main concerns of critics who said Clinton's march to war was ill-conceived. Responsibility for the atrocities in Haiti lay mainly with middle- and junior-ranking military officers, the same people who had seen Aristide's populist democratic reforms as a threat to their corrupt lifestyle. The generals had seen that a vacuum existed at the top of the revolution and filled it. They carried the responsibility for the abuses but they were riding a tiger that they could not easily control. Thus critics of Clinton's policy argued that getting rid of the generals would leave in place a thoroughly rotten substratum of the military which would have to be purged, and that in turn would mean a lengthy U.S. commitment to staying on the island.

By mid-September Clinton himself was turning up the rhetorical heat. In a televised address to the American people he said to the junta, "Your time is up." He talked of "Cedras and his armed thugs . . . a reign of terror . . . executing children, raping women, killing priests." He showed photographs of their atrocities and said that the consequences of their continued presence in power were a threat to U.S. national interests, in the form of new waves of refugees approaching America's shores, and a threat to regional security. In human terms, the death toll and the misery inflicted on the Haitian people by the junta had to end.

By now the Pentagon was ready. For weeks it had been preparing a range of different operations with the political situation of the day dictating which operation would be implemented. Delta Force, Navy SEALs and the Rangers were prepared to invade if the deadline passed and the junta resolved to stay. But with the word "Somalia" now synonymous with disaster, the planners also aimed for an operation that would as closely as possible be bloodless, from the American point of view, and, planners hoped, subdue the island without having to fire a shot. Certainly there should be no repeat of the urban warfare at which the Delta Force had been so completely outsmarted by the Somalis. To achieve these ends the planners ordered a series of information warfare strikes that would soften any resistance to the coming U.S. invasion and reassure the Haitian people that American aims were confined to the restoration of their rightful leader.

Conscious that Somali warlord Mohammed Aidid had continuous access to the radio airwaves during the peace enforcement phase of the Somali operation, planners ordered the Voice of America to step up the frequency of Creole broadcasts to Haiti and make more frequencies available for broadcasts that supported the American invasion. One step up from boosting the Voice of America signal was the deployment of Radio Democracy, an EC-130 aircraft carrying two 10kW transmitters that broadcast pro-American and pro-Aristide propaganda as the plane circled over the island. Aware that poor Haitians might not have access to radios, the Pentagon dropped thousands of radios, pretuned to the correct frequencies for the American propaganda, by parachute on the night of Thursday, September 15. The radio broadcasts were backed up by leaflet drops which would hail the impending return of President Aristide. The leaflets read: "The Sun of democracy / The Light of Justice / The Warmth of Reconciliation / With the Return of President Aristide." As an added touch, IW experts sent derisive e-mail messages to senior officers in the Haitian junta aimed at undermining their morale. Haiti's military leaders were rapidly getting the message that there was a major invasion force heading their way.

As the hours ticked away toward the deadline of midday on Sunday, into the tension stepped former President Jimmy Carter, self-appointed peacekeeper to the United States. He urged Clinton to let him go to Haiti to strike a deal with the junta that would avert the coming invasion. Put in a delicate position by his predecessor, for whom he publicly declares affection and admiration, yet anxious to employ any means of averting an invasion he knew was unpopular in Congress and with the people as a whole, Clinton agreed and sent former Joint Chiefs Chairman Colin Powell and Senator Sam Nunn, the Chairman of the Armed Services Committee, to be Carter's assistants. His instruction to them was strict: Aristide must return, the junta must go and there was to be no changing the deadline for invasion, which was to take place that Sunday. The Carter mission entered forty-eight hours of intense negotiations. Although Carter did not know it, his mission had been specifically included as a part of a strategy of information operations intended to win the battle.

On Sunday afternoon as the deadline approached, Bill Clinton phoned Jimmy Carter using the satellite phone the mission had taken with them.

"It's time to get out," he told Carter.

"We're almost there," Carter protested, "we've got this nailed."

Clinton gave him thirty more minutes.

At that point, Clinton and his team used their own information warfare tactics. The 82nd Airborne was told to pack its parachutes and prepare for takeoff. The invasion by air and sea would begin that evening. As the troops marched aboard their transport planes, CNN was tipped off that something worth filming would be happening shortly at Pope Air Force Base in North Carolina. As the planes took off, CNN started flashing the news around the world. The invasion of Haiti was under way.

There are conflicting accounts of what effect the news of the invasion force's dispatch had on the negotiations. The Clinton administration reported that the news galvanized Cedras into accepting the deal. Carter, in an interview with CNN the next day, said they were in the middle of talks when Gen. Philippe Biamby, one of the three Haitian military leaders, burst into the room to say the planes were on their way. Carter says Biamby told Cedras:

"We must immediately break off these talks . . . and go and marshal our forces to defend our country."

Carter says he was distressed by Clinton's move, as were the Haitians. When asked by the interviewer if he was saying the peace process had *not* been hastened by the report of the planes in the air, but had in fact been jeopardized, Carter rowed back. "I think earlier on they were not convinced that the Americans would actually attack . . ." Carter painted himself as the savior of the negotiations by persuading the junta not to man the barricades but agree to step down. Either way, the agreement was signed and the planes turned around.

In the end, Carter had signed off on an agreement that meant Aristide would be returned to power on October 15 but he gave away far more than Clinton had permitted, notably allowing the junta a month's grace before leaving power and not insisting that they depart into exile. This concession put the U.S. forces in a difficult position. With the junta still intact, and with orders to stay out of any local fighting, American soldiers had to stand by while pro-Aristide demonstrators were beaten to death before their eyes by members of the Haitian national police force. Clinton ordered 1,000 U.S. military police into the island to deter further atrocities. Again the administration had brought a major operation to a messy conclusion, but this time the military could congratulate itself on letter-perfect planning and execution. Although the full invasion was halted at the last minute, the IW operation had been a success and the island was occupied and held at no risk to U.S. soldiers.

If the invasion had proceeded, the Pentagon had in store what would have been the most extensive IW operation ever seen. In concert with the NSA and the CIA, the war plan called for the shutting down of all water and electricity in advance of the invasion, the destruction of gasoline stocks to immobilize the troops, the intensification of psychological operations to demoralize both the junta leadership and the police and armed forces, and the electronic "kidnapping" of local radio and television stations to insert pro-Aristide programs.

The intervention was not designed so much to win the war without killing anybody. "Nobody has ever done that," said one senior Pentagon official. "But we believed we could silence the junta's voice, render the military virtually powerless and so undermine the morale on the island that the invasion force would be welcomed as a relief and not seen as a threat. We might have had to fire a few shots but casualties would have been light on both sides."

Whether the Haiti operation had done anything to resolve the debate about the way America would fight its wars in the twenty-first century was not clear. The IW tactics used were classic psychological operations the like of which had been used many times in the past in support of conventional war fighting. If Haiti had been a more advanced society a more thorough demonstration of IW might have been possible.

That opportunity came in Bosnia, where electronic and human intelligence gatherers on the ground, in the air and in space and from a variety of nations created one of the most intensive and profitable deployments of IW capabilities ever seen.

The peacekeeping effort in Bosnia had initially been hampered by poor intelligence gathering thanks to the policies of the United Nations, which was the overriding authority in UNPROFOR (United Nations Protection Force) from 1992, when the wars between the Serbs, Croats and Bosnians brought widespread human suffering and misery to the former Yugoslavia. Hopelessly undermanned, the U.N. was expected to use Chapter VII powers to enforce peace against warring factions, particularly the Serbs, that ignored U.N. forces and where possible brushed them aside. Part of the problem was that the forces making up UNPROFOR were forbidden by the U.N. to gather intelligence.

"Intelligence is a dirty word in the United Nations," Air Marshal John Walker, Britain's Chief of Defense Intelligence, would later say. "The UN is not a thing in itself, it's an amalgam of 183 sovereign nations. If it does intelligence, it will be doing it against a sovereign UN member, so it's incompatible. But you need a military intelligence job to protect your troops. If you don't, you pay for it in body bags."

The British, who were keenly missing the American intelligence-gathering resources they had so admired during the Gulf War, soon started an ad hoc operation that infiltrated Army I (Intelligence) Corps operatives into theater under cover of U.N. blue berets. The French and Canadians were doing the same and humint (human intelligence) was swapped between them. Sigint (signal intelligence) and elint (electronic intelligence) were another matter, as Britain, like the United States, was extremely nervous about sharing secret intelligence gathered in this way with foreign nationals, particularly the Russians and Ukrainians, who played a major role in UNPROFOR. It was a particularly uncomfortable if necessary arrangement. There was a shadow war going on in Bosnia among the allied forces, to the extent that French electronic warfare units were attempting to break the American military communications codes. Whatever nuggets they could extract were sent to French defense manufacturers trying to copy American designs or create units that could crack the American communications system. Such units would have willing buyers all over the world.

The only shining achievement of the period was by the British Secret Intelligence Service (SIS), which proved yet again the special talent the British have for acquiring and exploiting human intelligence assets. Before Yugoslavia fell apart, SIS had a number of valuable sources inside the Yugoslav government, all of whom became ineffective with the breakup. Within six months of the new Serbian government taking power in Belgrade, they had seven new sources up and running.

The muddy IW picture in Bosnia changed with the signing in Paris on December 14, 1995, of the Dayton peace accords. The agreement called for a new NATO force to be deployed, putting 60,000 troops, 20,000 of them American, with the full panoply of American high-tech IW ability, to provide a full umbrella coverage of the shattered nation states of Yugoslavia. It took time to get up to speed. The last major incident of the transition from U.N. to NATO control demonstrated how IW is only as good as technology on the ground.

Two days before the accords were due to be signed, the Bosnian Serbs released two French Mirage pilots who had been shot down during punitive NATO air strikes on Bosnian Serb positions in the late summer of 1995. The NATO strikes, led by the United States, had changed the dynamic of the U.N.'s floundering operation in Bosnia and had added much needed spine to the world community's determination to stop the Bosnian Serbs' depredations in Sarajevo and other cities of the beleaguered republic. The Bosnian Serbs had denied any knowledge of the French pilots' whereabouts, and the pilots' continued absence had threatened to delay the signing ceremony.

Until the pilots' release, the efforts by the Americans and others to locate and rescue them had been dogged by bad weather and ineffective equipment. In one attempt, U.S. special forces were on standby, ready for the order to go in and extract the pilots. "Since these were French men on the ground, not Americans, the information was thin, but we couldn't wait to check. We needed imagery," a senior American intelligence officer remembers. American J-Stars reconnaissance aircraft were sending images back to the Central Intelligence Center at USAF Molesworth in England. The CIC put the imagery on its primary server, linked by secure network to the U.S. command center in Brindisi, Italy. The primary server crashed before the pictures could be put online. The CIC brought a secondary computer onstream, but this sent the pictures straight to a fax machine, which rendered them impossible to read. The operation was aborted.

"It's an illustration of how hard IW becomes the further down the pyramid you go," the American intelligence official said.

This was confirmed by a secret mission ordered by Defense Secretary William Perry to gauge the difference between what intelligence bosses in Washington were looking at and what the soldiers on the ground were dealing with, a test of whether life was really tougher as one got further down the pyramid. The results of the mission staggered Perry and CIA chief John Deutch. The mission had brought back the version of a spy satellite photograph that had started life in pristine, crisp condition in Washington but had, after repeated transmissions down the line to an air unit in Italy preparing for a mission over Bosnia, become a muddy, grainy mess. "Every time you transmit, you lose clarity," one intelligence official said. "The photo was so useless the pilot didn't take it on his mission."

The team, which went under the highly unwieldy title of the Defense Science Board Task Force on Improved Application of Intelligence to the Battlefield, also reported too great a lag time in front-line units getting vital intelligence. Perry and Deutch ordered an immediate overhaul of transmission quality and speed.

The intelligence umbrella over Bosnia matured very swiftly after uncertain beginnings, and set new benchmarks for international, and even interagency, cooperation in a field previously characterized by suspicion and hostility. A brigade-sized unit of the U.S. Army's V Corps, consisting of over 1,000 specialists, from analysts and signals officers to counterintelligence experts, worked closely with units from the CIA, NSA and DIA, backed up by space and aerial reconnaissance assets and a network of computers capable of putting up-to-the-minute maps, images and data into the hands of soldiers in the field. Traditional aerial and satellite reconnaissance was buttressed by the use of Unmanned Aerial Vehicles (UAVs) like the Predator and Pioneer, which could "see a thousand meters down the road" as Gen. George Joulwan, Supreme Allied Commander Europe, reported later, and feed live video straight to commanders in the field.

The intelligence gathered was invaluable in monitoring the various factions' adherence to the accords. Even at night, it was impossible for any of them to move troops or weapons because heat emissions from engines would be picked up by infrared sensors overhead. The NSA and Britain's GCHQ meanwhile sucked out of the airwaves virtually any electronic communication at virtually any level of command in the field, even down to walkie-talkies and field telephones. "This is gold," one admiral said.

The information dominance that the American forces sought, and won, led to some intriguing examples of how information can be a potent weapon in winning or deterring actions. Maj. Gen. Michael V. Hayden of the Air Intelligence Agency, who had been the J-2, or intelligence officer in EUCOM (European Command) until 1996, described one operation which illustrated why, in his agency's opinion, information is the new "queen" of battle and why he believes warriors of the future are the commanders who understand this. Hayden related how organizers of the NATO bombing campaign, code named Deliberate Force, over Bosnia and Croatia in 1995 had had little trouble from the Serbs' antiaircraft missile batteries. The Serbs had limited SAM (Surface-to-Air Missile) resources so they conserved them by not using them. They knew that switching on their integrated radar would be noticed by NATO electronic surveillance and a HARM (High-Speed Anti-Radiation Missile) would be sent to destroy them. Hayden says he was therefore very surprised when electronic intelligence spotted an SA-6 battery lighting up, and even more surprised when it was located a few hundred meters north of the Una River—in Croatia.

"We called the embassy in Zagreb [the Croatian capital]," he said, "and asked the attache to visit the Croatian Ministry of Defense, lay out the facts as we knew them and warn the Croats of the dangers of operating the radar during NATO air operations. The attache was armed with a great deal of data and in the end the Zagreb government admitted ownership, turned it off and moved it. This was the first example that Lt. General Ryan [CINCSOUTH] and I have seen of an 'information first strike' and the use of information as a weapon."

Hayden had displayed a masterly grasp of the concept of information as a weapon. Bosnia proved to be a proving ground for other, more aggressive, IW systems. The Americans learned that the French were using an extensive electronic eavesdropping operation to spy on U.S. businessmen seeking contracts in the region. The NATO ally was using that intelligence to alert French companies about possible opportunities. At the same time, the Americans were spying on the Russians, the Russians on the Americans and the British on everybody.

In countries where authoritarian rule has crushed free expression people cherish any opportunity to make alternative voices heard, often at risk to their lives. Throughout the Soviet era the Voice of America, Radio Free Europe and the BBC became information lifelines for beleaguered citizens of communist countries. But after communism crumbled, it was still difficult for notionally democratic regimes to accept the complete freedom of communication that citizens, leaders and observers of Western democracies take for granted. In Serbia, Slobodan Milosevic brooked no criticism of his government, and when opposition parties started to win electoral victories in Belgrade he overturned the results. Milosevic knew that controlling the media gave him a continued edge. State-run TV and radio parroted the Milosevic line in all broadcasts, but a few brave, determined people demonstrated that freedom is a force that cannot easily be held in check. They operated the radio station B-92 and started broadcasting opposition to Milosevic. He tried to shut them down, but the world was watching. Radio Free Europe and the Voice of America offered to rebroadcast the weak, wavering signal, and after a brief interruption in service, Milosevic had to let the station back on the air. The signal was piped onto the Internet and soon the official media's monopoly was over.

Contrast the deep understanding of the power of information demonstrated by the young information guerrillas of Belgrade with the attitude of American psychological operations officers in the peacekeeping force. In Bosnia and Croatia, radio stations operated by rival factions spewed a steady stream of hate that created further divisions in the war-torn communities. Seeing this, a reporter asked the American psyops team running the radio station set up in the region by NATO what they were doing about it.

> Would the station run news bulletins to contradict the garbage pumped out by Serb, Croat and Muslim stations? A clear interpretation of the Dayton plan? Information about war crimes? No, the officers replied. The station will put out an anti-mine awareness campaign and "good old American rock and roll." Said the colonel: "We don't get involved in politics."

It is a bizarre example of how a failure to understand the culture and politics of the in-theater populace can lead to missed opportunities. But bitter experience eventually drove a reluctant system to take action. After months of impotent fury at the extent of the Bosnian Serb propaganda machine, SFOR (Stabilization Force) troops moved at the beginning of October 1997 to close down four television transmitters that allowed the Serbs free access to the airwaves. It proved to be a major blow to the Bosnian Serb leadership, who, at a stroke, were left without a system to create a propaganda image. The SFOR deployment did manage to set a new standard in international cooperation. Amazingly, given their track record of suspicion, the Americans started to share some of the intelligence hoard with other nations in the peacekeeping

force structure, including the Russians, suspending their previous reluctance during the days of UNPROFOR. The new collaborative spirit was
designed to improve efficiencies at every level and wipe away some of
the bitter memories of intelligence failures in recent years, particularly
in Somalia. Even in Bosnia, the shooting down of Air Force Pilot Capt.
Scott O'Grady in 1995 had been an intelligence snafu, in that detection
of antiaircraft missile sites in the area where he was flying had not been
passed on to him. His being shot down and subsequent escape made
him something of a national hero, but it could all have been avoided.
The vastly improved intelligence effort that followed would make such
incidents increasingly unlikely.

It would be tempting to see SFOR as a model for future international
peace enforcement but SFOR was a one-time situation operated under
the umbrella powers of NATO, and not the U.N., an organization which
had suffered from its own intemperance and from a shift in global
ideology. The International Institute for Strategic Studies (IISS) produced a report in 1997 that underscored how serious the prospects are
for international humanitarian and peacekeeping missions. The report
said that the people in the prosperous Western democracies "are in no
mood to sacrifice their well-being for supposed international advantage,
nor to rally to the service of a purely humanitarian goal." In releasing
the report, IISS director John Chipman added, "The Cold War tendency
to see core interests indirectly at stake in distant parts of the world has
now totally eroded. Even the brief post Cold War sense of humanitarian
obligation has begun to give way to colder realpolitik calculations of
what can be done."

Given that the U.N. was the repository for this sense of "humanitarian
obligation" Chipman referred to, the combination of this new attitude
with accumulated failure in Somalia and the early Bosnia operation
made prospects for the U.N. grimmer still. The U.S. government's General Accounting Office rammed the point home in a report to Congress
called *United Nations: Limitations in Leading Missions Requiring Force
to Restore Peace.* The report went over the history of the U.N.'s Chapter
VII operations in Somalia, Rwanda and Bosnia and concluded that unless one country was given the power to lead, such as the U.S. in Haiti
or Belgium in eastern Slovenia, the weaknesses in the U.N.'s own structure make success unlikely. The report even suggested that, with the
concurrence of the Pentagon and the State Department, "the United
Nations may not be the appropriate organization to undertake peace
operations requiring the use of force." This looked like the formation of
U.S. government policy on future operations that has not yet been put
to the test but which asks a major question. If the United Nations does
not organize such operations, who will? The United States? Europe? As
the IISS report made clear, parochialism was spreading throughout the

world. With no global cop ready to pound the international beat, there would appear to be little hope of relief for the poorer people of the world from regional thugs and local wars.

And yet, it was already clear to all the developed nations, including the United States, that peacemaking and peace enforcement was going to be a vital part of every government's foreign policy well into the next century. Somalia, Haiti and Bosnia had also underlined the reluctance of politicians to commit military forces to peace operations, however worthy, unless casualties could be guaranteed to be low and a firm exit timetable from the commitment in place at the start of any operation. Of course, the military has always been concerned to minimize casualties but the demands of a new generation of politicians with little or no experience of warfare proved a different challenge. Although the military clearly needed to train for the least likely eventuality—a full-scale conventional war—it also had to remain responsive to day-to-day problems, and provide realistic solutions that would be politically acceptable. Information warfare appeared to provide exactly the right potential that could ensure the continued relevance of the armed forces.

SEVEN

The March of the Revolutionaries

MILITARY thinkers turn for basic guidance to two of history's great-est strategists: Sun Tzu, the Chinese philosopher who lived around 500 B.C., and the nineteenth-century Prussian military theorist Carl von Clausewitz. In trying to get to grips with information warfare, it was clear that Sun Tzu was equal to the challenge.

> Know the enemy and know yourself; in a hundred battles you will never be in peril. When you are ignorant of the enemy but know yourself, your chances of winning and losing are equal. If ignorant both of your enemy and yourself, you are certain in every battle to be in peril.

It would be hard to find a more succinct way of positing the challenge facing any leader in wartime. And Sun Tzu's use of words like "know" and "ignorant" correlate neatly with concepts such as information knowledge and understanding. Thoughtful analysts in the military started from this premise and used it as an overlay over the new capabili-ties and technologies at hand. They could see how data could be con-verted into information, and that information could form the basis of a strategy and thus knowledge. Ultimately, this knowledge could be the building blocks of a predictive understanding of the battle.

When analysts reached to Clausewitz for guidance, they were less successful. Clausewitz was the man who came up with the famous phrase "war is the continuation of politics by other means." That phrase still stands the test of time, but the rest of Clausewitz's philosophy starts to look outdated. Martin Van Creveld, the noted military scholar at the Hebrew University in Jerusalem, declared that from the end of World War II onward, Clausewitz has become ever more irrelevant because he believed that war was based on what he called the "wonderful trinity" of government, military and people.

This is a very traditional view of war that held good up to 1945: nation fighting nation on a series of battlefields, each supported by a civilian population which bends its shoulder to the wheel of national will in factories and fields. This trinity becomes irrelevant in the face of a preemptive nuclear strike and Mutually Assured Destruction.

Guerrilla warfare also falls outside the Clausewitzian paradigm. While it is indeed a continuation of politics by other means, it is obviously not fought by a conventional military force in harmony with a government and its people. Often a partisan uprising succeeds at the particular expense of the people. While the Gulf War actually does fit the Clausewitz mold rather well, future conflicts that do the same will be the exception rather than the rule.

To take that argument to the extreme, consider the expanding influence of multinational companies, which nowadays can be said to have more power than nation states. After all, it is big companies who provide the jobs that employ the people, whose decisions regarding issues such as where to build new factories can make or break regions if not countries, and whose money fuels the political system. Future war could be fought between such companies. If you doubt that, think about the depths of the verbal, legal and PR conflict between General Motors and Volkswagen over the activities of Jose Ignacio Lopez, the brilliant cost-cutter who was hired away from GM by Volkswagen. He was accused by GM of taking with him vital and proprietary company information to use for the benefit of Volkswagen, an alleged act of economic espionage. The subsequent row was settled, but only after a very bruising battle that included elements of covert intelligence operations and deception. It did not come to blows, but was it any less of a conflict? With the proliferation of computer internetting, would a similar conflict one day lead to attacks on corporate databases?

Then there is the potential for conflict in environmental catastrophe, again something Clausewitz could never have foreseen. Thomas Homer-Dixon of the University of Toronto believes it eminently possible for radical climate change, disaster or population shift to lead to war in the Third World. This theory that war had moved beyond Clausewitz may appear self-evident when analyzed through the perspective of terrorism, intercorporate conflict, nuclear Mutually Assured Destruction and environmental collapse. But in the U.S. military Clausewitz remains something of an icon, and to challenge his theories is tantamount to challenging the raison d'etre of significant portions of the armed services. Gen. Frederick M. Franks, Commander of TRADOC in 1994, ruefully admitted that calling Clausewitz into question had caused him some difficulty. In a speech entitled "Winning the Information War," delivered to the Association of the U.S. Army Symposium in Orlando, Florida, General Franks said:

> I am not quoting Clausewitz . . . because maybe some of this will go beyond Clausewitz. I got in a little trouble at the War College a couple of months ago when I suggested that. Yet I believe we must not be so narrow as to confine our theories of land warfare to one theorist when our nation's interests are global.

It is telling that the commander of the U.S. Army unit responsible for establishing training and doctrine should have had his wrists slapped for suggesting Clausewitz was on his way out. Franks stuck to his guns and pressed for a coherent response to the challenges of tomorrow's technology. In his opinion, the Gulf War was a "Janus war," looking back to the ways of the old, looking forward to the ways of the new. Yet in discovering where those new ways would lead, he believed it was vital to keep in sight the five fundamental principles of fighting a war, expressed as questions every commander should ask him or herself:

- What is my mission?
- What is the enemy doing?
- How can I keep him from knowing what I am doing and lead him to believe I am doing something else?
- Where am I vulnerable and does the enemy know it?
- Where is the enemy vulnerable and how can I exploit it and win at least cost to my soldiers?

Five critical questions, a continuation of Sun Tzu by other means. Those questions should be the same for a commander with an army of bowmen as for a commander in today's military. Each question requires a proper flow of information to be answered correctly. In that sense, the concept of fighting a war has not changed, only the means and speed of acquiring and transmitting information. From the first use of the telegraph during the Civil War to the first battlefield use of a computer, the Univac 1005 installed by the 25th Infantry Division in Cu Chi, Vietnam, in 1966, armies have constantly struggled to be one step ahead of the enemy in knowledge. But what Vietnam demonstrated, and the Gulf War (and later Bosnia and Somalia) proved, is that it is not just information in the battle space that can have an impact on the fighting of a war. In Vietnam, television pictures of what was really going on there played an enormous part in shaping public policy. Oddly, in the Gulf War the TV pictures were no less shocking, often more so for being live, but such was the skill with which the military had learned the lesson of Vietnam, namely that more or less unfettered media access can work against military objectives, that in 1991 the media became easily diverted by the incredible shots of bombs and missiles homing in on their targets that their military handlers laid before them. The public was wowed by what it saw, namely, war as a video game. (The challenge to the media, and to their handlers, would of course have been greater if the war itself had not been so completely lopsided.)

The lesson that had to be learned from the Gulf War for General Franks was not how to gather more information—patently there were systems in use that were producing mountains of information—but what use was made of it once it had been acquired. The key to that was processing and transmission. In the Gulf, there was too much information flowing in to process in a timely fashion. The planners assigning missions to the air forces at their disposal could no more than scratch the surface of the targets available to them because they could not work their way through the data flow. A similar problem exists in Bosnia where the Predator UAV (Unmanned Aerial Vehicle) gathers vast amounts of data on points of potential conflict that have to be analyzed by large numbers of people. Military contractors are now working on systems that will crunch the information as it comes in, reducing the time lag and enabling much more efficient use of forces, particularly aircraft.

More fundamentally, Franks saw in the new information technologies an opportunity to radically alter the vertical hierarchical structures that typified conventional military organizations and which "stand in the way of rapid information transfer." Franks's belief that individual soldiers and commanders in the field now have the potential to be freed by technology to do their jobs faster with more available intelligence and less reference up the chain of command evolved into the seminal pamphlet 525-5 issued by TRADOC in August 1994. A key conclusion of that document read:

The ability to move information rapidly and process it will likely change the way we command military operations. It will greatly influence force organization, command procedures, and staff systems. Maneuver, combat support, and combat service support leaders horizontally linked by common information will, for the first time, have a means to visualize how they will execute in harmony, integrated by a shared vision of the battlespace. Individual soldiers will be empowered for independent action because of enhanced situational awareness, digital control, and a common view of what needs to be done.

This makes some giant assumptions, as in "shared vision of the battlespace" and "a common view of what needs to be done." Establishing a common view was precisely the problem. The range of opinions about the nature of future warfare went from the traditional war-fighting beliefs of the Clausewitz brigade to the cyberwarriors who talked in terms of "empty battlefields," meaning America would never again fight a war where two armies meet on a field of battle. Instead, wars would be fought down telephone lines, via satellite transmissions, through various computer bugs, worms and viruses designed to destroy an enemy without firing a shot.

Michael L. Brown, a former Army officer who held senior advisory positions in the Office of the Secretary of Defense, the Supreme Allied Commander in Europe and the Secretary of the Army, and who is now with SAIC (Scientific Application International Corporation) in McLean, Virginia, split this argument over the role of information in future warfare into two distinct areas. He believes the revolutionaries fall into what he calls the "Strategic Attack Paradigm," while the conservatives support the "Operational Attack Paradigm."

The Strategic Paradigm foreshadows war by remote control. As well as the computer-assisted attacks just mentioned, it also includes the use of precision-guided bombs and missiles to knock out strategic elements of the enemy's telecommunications, transportation, water or power supply networks. Brown described the Operational Paradigm as the conventional military concept that war is fought between armies and that victory is achieved by the complete destruction of the enemy's forces (a notably absent feature of the Gulf War, which left Saddam Hussein with a serviceable military machine). The use of information technologies in this paradigm is subservient to the greater cause, and is designed to enhance a commander's ability to shorten his decision-making cycle (the OODA loop, which stands for Orient, Observe, Decide, Act) dramatically while making the enemy's decision-making cycle correspondingly longer through attacks on his lines of communication and his sensors, such as radar.

In viewing this debate, it is tempting to see everything through the prism of revolutionary technology, by which the conservatives in the military come across as reactionary and Luddite. Just as in civilian life, it is true that the older members of the military society are going to be least likely to understand the possibilities of information technology. They are least likely to have a computer at home, for instance. That does not mean they actually resist technological advance. Indeed, most are all too willing to embrace the new capabilities offered by the revolution in information technology, but only in service of, and subservient to, the traditional aim of armies fighting and destroying armies.

There are deep thinkers in the Pentagon who by virtue of their high positions and far-reaching responsibilities have the breadth of vision to understand the scope of the challenge. People like Adm. William Owens, who as Vice Chairman of the Joint Chiefs of Staff played an important role in leading military thinking toward a broader approach. He developed the notion of "system-of-systems" in which information about the enemy is hoovered up by a vast array of sensors from satellites to Unmanned Aerial Vehicles (UAVs) and then transmitted through a network to provide a real-time overview of the entire battle space. Commanders then use that information as they wish. To understand the resistance to such thinking, even though this system-of-systems would be a military application of information technology, it is worth noting the comments before Congress of senior commanders at a hearing that took place just two weeks after Owens's retirement in February 1996. Their point was a none too subtle dig at Owens, saying too much focus had been placed on new technology at the expense of conventional hardware like tanks and ships.

"My concern is that we are creating a force that ten years from now [will have] a lot of headquarters and little combat capability," Gen. John Sheehan, CINC Atlantic Command, told the Senate Armed Services Committee. The CINC CENTCOM, Gen. Binford Peay, testified he would allocate more spending to the "grit and grist" of military hardware than on information technologies. "Technology is not simply a panacea for everything."

Despite this rearguard reaction from the warriors in the field, the commands pressed on, trying to formulate the war-fighting doctrines of the future.

The result was a classic scramble for power, money and influence among the services. All sensed that IW represented a new opportunity to win scarce resources at a time of declining budgets. The cyberknights, who were true believers in the future of IW wars, helped drive the process. These cyberknights maintained that they were driven by national security concerns, while the more skeptical admirals and generals followed along because for them IW was a low-risk investment. If it worked, they would get a share of the pie. If it failed in all its ambitious goals, and information warfare did not materialize, then there would still be some benefits to the traditional war fighter. Information warfare could always be a force multiplier and demonstrate what the 1994 Defense Science Board report on IW called "Information in Warfare," which has always been of value to the battlefield commander.

There had been a similar struggle between the services when space was thought of as the final frontier in the 1950s and 1960s. That had created near chaos and had wasted both time and billions of dollars as each service fought for dominance of the new arena. Now, IW presented a similar challenge and each service sought to define its role and to quickly argue why it should be the primary influence in the infosphere.

In the Navy, Adm. Jeremy Tuttle prefaced a report on the future of U.S. naval warfare, called Sonata, with the words, "We have crossed the threshold of the Information Age." A subset of Sonata was Copernicus, an evolving doctrine of command and control (C2) warfare introduced by the Navy in 1990. The newest iteration of Copernicus in 1995 addressed IW in a way that indicated the Navy would be embracing IW as "the integrated use of operations security, military deception, psychological operations, electronic warfare and physical destruction to deny information to, influence, degrade or destroy an adversary's C2 capabilities, while protecting friendly C2 capabilities against such actions." Copernicus also introduced the Battlecube Information Exchange System, a concept in which the various dimensions of the battle space, including subsurface, surface, air and space, would be pulled together into a cube by the information capabilities offered by sensors, computers and advanced communications. Everyone inside the cube would share a common tactical picture and a communications network that maximized the acquired intelligence and streamlined the passage of information from sensor to shooter.

The Air Force Chief of Staff, Gen. Ronald R. Fogelman, and the Air Force Secretary, Sheila Widnall, asked the U.S. Air Force Scientific Advisory Board to examine all the technologies that would apply to the Air Force of the future and recommend how to create it. The *New World Vistas* document was a guide to a breathtaking future of high-tech war fighting in air and space that would make the USAF "the most capable and respected Air and Space Force in the world in the 21st century." The use of the word "Space" indicates a future where air dominance will be joined by space dominance as the bottom-line requirement for American military strategy. "Global presence with weapons capable of destroying or disabling anything that flies as well as most unarmored ground targets will drive a new warfare paradigm." That was the view of the panel tasked with reviewing the possibilities for directed energy weapons. It foreshadows a time when space-based or space-relayed lasers will have the capability of passing over the airspace of any nation and controlling it. The USAF has added "Information Control" as the third primary mission that it plans to achieve. This would give the United States an unassailable edge in future wars.

The Army's vision of the future was called Force XXI, and was credited in 1996 by Defense Secretary William Perry as being the most far-reaching overhaul of all three main services, in that a top-to-bottom revision of standard procedures, from acquisition through to battlefield communications, was undertaken, with an experimental force, the 1st Brigade of the 4th Infantry, acting as the Army's guinea pig. This experimental force would be the working laboratory for the Army's adaptation to what became known as "the digitized battlefield"—in other words the application of information technology to the soldiers' war-fighting capability—over a series of Advanced Warfighting Exercises (AWEs) going on well into the next century. The goal is:

The Force XXI Army will be able to locate enemy forces quickly and precisely, whether those enemies are agrarian war lords, industrial armies or Information Age peers. Force XXI armies will know the precise location of their own forces, while denying that kind of information to their foes.

From the start, however, there was a feeling that the Army was trying to put a high-tech shine on an old pair of boots. The Army was focusing on "Information in Warfare" as a force multiplier and not on information warfare in its present sense. Jim Winters is a former Army officer who now heads the Information Operations Division at TRADOC, part of the Space and Information Operations Directorate of the Deputy Chief of Staff for Combat Development. Although Winters is now a civilian, he is running a key military department with responsibility for integrating information operations across all lines of command in the Army, testament to his capacity to see the big picture. At his office at Fort Monroe on the Chesapeake Bay just north of Norfolk, Virginia, Winters is working to produce a doctrine for information warfare that fits across all Army commands. "This is an Army that shows frequent facility to adjust and adapt," he said. "We have in the Army a process whereby we are directly addressing the need for change. Force XXI is driven by a leadership that recognizes there is a need for change. But we see it as a process of evolution, not revolution."

This would make sense if there was enough money to pay for evolution, because it means developing systems for tomorrow's wars while also paying for systems to fight today's. As the Quadrennial Defense Review released in May 1997 made clear, there is not enough money for that, and as long as political opportunism and bureaucratic inertia ensure that big-ticket items such as tanks, ships and aircraft dominate the $250 billion annual defense budget, the future will appear too expensive to pay for. This was the consensus view of what the QDR stood for. Major new aircraft systems—the Joint Strike Fighter and the F-22—stayed intact, and the majority of cuts were to be found in base closings and force structure. The strategic posture remained that the United States should be ready to fight two wars, equivalent in size to Desert Storm, at the same time, even though there was serious doubt whether a military that had taken so many cuts since 1989 could actually fulfill those missions. Final decisions would rest with Congress, and then only after an independent National Defense Panel had reviewed the QDR. One of the members of that panel was Andrew Krepinevich, executive director of the Center for Strategic and Budgetary Assessments. In his testimony to the Senate Armed Services Committee hearings on the future of tactical aviation, this expert on air warfare made clear his view that the Pentagon cannot have its cake today and still be able to eat it tomorrow.

The focus of our current modernization program—the F-22, the F/A-18E/F, and Joint Strike Fighter aircraft—seems to presume that the long-term future will resemble the recent past; i.e. that an adequate forward-basing structure will exist in future contingencies, that our tactical air forces will be accorded access (indeed, early access) by prospective host nations in the theater of operations, and that these bases will remain sanctuaries from enemy attack. (NOTE: These aircraft all have short-range capability, so operating from bases or carriers close to the battle space is vital. If those bases/carriers are not available, the aircraft are useless.) It also seems safe to assume that our carriers, operating close to the shore, will remain very difficult to detect and engage.

Before we procure over 4,000 tactical aircraft at a cost that will likely exceed $300 billion—aircraft that are intended to operate well into the new century—we need to know how our tactical air forces, thus modernized, will meet what are likely to be the very different challenges they will confront in 2015 and 2020, as well as the more familiar challenges of today.

Krepinevich was asking that the politicians and Pentagon chiefs forget their normal, Cold War, paradigms for ordering new hardware, which included not only the need to fight a war that may never require the use of a tactical air force but also the need to preserve jobs and industries in important political constituencies, and instead focus on interpreting the challenge of an unknown future. The consequence of getting it wrong, he told them, would be devastating.

We confront this era of transformational change, both geo-political and military-technical, within an environment of declining resources for defense. Consequently, there is the risk that if the wrong transformation path is chosen (or if no attempt is made at transformation), we will find it difficult, if not impossible, to buy our way out of our mistakes.

Yet while the conventional military was arguing about who would dominate the infosphere and what new traditional weapons platform could be squeezed from an overstuffed budget, another, quieter, revolution was going on. Into laboratories, computer companies, think tanks across America came people who had begun to understand that the information revolution was not just about computers and battle space management. These men and women believed that microtechnology combined with computers could produce a real revolution not just in how wars might be fought but in the weapons that could be employed on the new battlefields.

Part 2

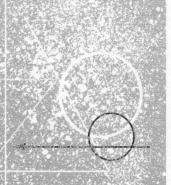

EIGHT

From Double Tap to Double Click

THICK cloud created a mantle of blackness beneath which the soldier lay still and quiet in the trench. It had taken two painstaking hours of crawling with agonizing slowness to get there from the cover of the woods a mile to his south. The enemy was now 300 meters away bunkered down around their mobile missile launcher. The standard crew, including drivers and infantry support, would be fifteen strong. Occasionally the soldier heard voices drift toward him above the insect noise of the night and grinned at the thought of how his own officer would have dealt with such poor security.

He eased his Head-Up Display over his eye and pressed some buttons on a miniature keyboard strapped to his forearm. An image jumped to life in front of him. It was of the missile site, lit in the ghostly greens and whites of night vision optics. The view was from above. He could see the glowing forms of twelve sleeping men and three guards, who casually patrolled the perimeter of their encampment. The object of his search eluded him. He punched a few more buttons. The view shifted as the four-inch-long micro-aircraft carrying the camera and the transmitter that relayed the pictures to him altered position at his command. The craft was powered by tiny whisper jets, inaudible to anyone on the ground. It was able to hover because of the minute sensors and actuators spread across its wings that maximized lift from every square millimeter of surface. The surface of the wing would appear to be rippling like a live muscle.

There. The case carrying the biological warheads was slotted into the back of the launcher, within easy grasp of the fire team. He ordered the remote camera to zoom in. Antiwobble technology in the tiny camera kept the image still. Confirmed. A few keystrokes and he sent an e-mail to his commander five miles away. A few more computer commands and he had sent a still frame of the picture as well, together with the GPS

(Global Positioning System) coordinates. He moved slowly and carefully. The computer strapped to his back was so slim he hardly knew it was there. He inched his weapon above the level of the trench in which he lay. The laser sight/range finder/video camera attached to the top was connected to his computer and to the VDU (Visual Display Unit) in his helmet. He steadied the weapon. Now he had a ground-level view of the target. He studied the terrain between him and the launcher. Dead flat. Only a few bushes to negotiate. He brought the weapon back, laid it down next to him and extracted a thin case from a deep pocket in his pants. He opened the case carefully to reveal a line of what looked like flying beetles. They were actually tiny flying machines. He could not call in any kind of artillery on the site for fear of blowing the biological warheads up with everything else and creating appalling fallout. Instead these tiny devices would be his weapon. He pulled a thin lead from the box and plugged it into his forearm keyboard. On his helmet VDU he brought up the live image of the sleeping men and the sentries. The GPS information, derived from triangulating the data from his weapon sight and from the aircraft overhead, was fed into the box and thence into the beetles. A green light indicated all was ready.

One by one he lifted the tiny devices from their lair and laid them on the flat earth next to him. Six of them. He hit the red button on the box and all six hopped into the air in unison, hummed very quietly to a height of twenty feet, then sped toward the target. Reverse data fed from the box sent the tracking signal back to his VDU so he could watch their progress. As they got closer to the target area they dispersed, programmed to come in from different angles and at different intervals. He held his breath. He had banked on the sentries being too dopy to notice something like a big bug landing; after all there were bugs aplenty in this part of the world. But it was always impossible to be sure. Landing on a sentry's head would be a disaster. One by one they found earth, ringing the encampment. He slowly let his breath out.

He pulled a thin mask over his face, checked his oxygen supply, then, with one more glance at the live picture from the UAV (Unmanned Aerial Vehicle), he flipped the cover from an orange switch and depressed it. Back in unison once again, the bugs exploded, more with a puff than a bang, releasing clouds of gas that instantly incapacitated all the men around the missile launcher. The gas from those downwind of the site blew away unused, but that was okay. Even one would have done the trick. The important thing was to be sure. He waited ten minutes. The gas had a very short life and would swiftly disperse, but he was patient, watching the image from the UAV. No movement. He e-mailed a quick report to his commander, who e-mailed him back the go command. He sent a tiny electronic beep that would resonate in the earpieces of the other members of his squad fanned out behind him at twenty-yard intervals. The beep would alert them to his incoming e-mail.

Each man sent back a confirming signal that registered as numbers in his eyepiece. All present, correct and online. "Site secured. Follow me. Take care. Go on green." He punched a button and sent the signal that would illuminate a green go light in his team's VDUs.

With his weapon at the ready he advanced out of his trench, treating the deathly quiet site he was approaching as if it was full of alert hostiles. He switched to live video and squirted it back to base. As he approached, he switched the output back to the UAV so both he and the base could get a full view of the sweep up. He and his men converged on the site. All the enemy soldiers were unconscious and would stay that way for hours. When they awoke they would feel so wretched they would be incapable of fighting for days, by which time their services would not be needed. The site was secured. He detailed two soldiers to collect the warheads, then e-mailed for support. A powered hang glider that had been circling a mile away headed toward the missile launcher and landed, its whisper-quiet motor idling as the soldiers took the warheads from the box, checked they were not armed, then slid them into the slings attached to the glider. It was done in less than a minute. The soldiers turned the machine so it had the necessary six feet of clear ground to take off, then he keyed in an e-mail and the hang glider departed, controlled by an operator back at the base. Another soldier was still crouched at the missile launcher, his forearm keypad connected to the launcher's console by a lead. His fingers flew over the keys for a few moments, then he disconnected. The virus he had planted in the launcher's control circuits would destroy it the moment someone tried to activate a launch. His terse report, "Launcher Bugged," flashed across the team leader's eyepiece. He rattled a quick e-mail back to his team. "Go. Now." The team melted back into the darkness, their mission accomplished. Not a word had been spoken, not a shot had been fired. Weapons that could kill hundreds, perhaps thousands of people had been removed from the enemy's control and a missile launcher had been deactivated.

Fiction? Yes it is, in that the above event never happened. But most of the equipment described is either in production or in various stages of development. The soldier's computer, eyepiece, weapon and laser sight/range finder/video camera were field-tested in early 1997 (see the discussion of Land Warrior below), and a UAV that small is on the drawing board. The technology that enabled it to fly has been tested and works, and while the flying beetles are further out toward the more fanciful end of the spectrum, serious consideration is being given to create something along those lines.

So while the military appears mired in a debate over its future mission, it should not be assumed that weapons development is standing still. The process of developing new military technology, or, as the case is nowadays, adapting new technology to military use, has to continue, even if nobody has a clear idea yet of how such technology meshes with the armed services of the future. It is not inconceivable that the argument over the shape of that future force might be swayed by the advances being made in technology. For forty-five years the U.S. military had developed new weapons and machinery at a leisurely pace. From the moment a planner first saw the need for a new kind of aircraft, a new missile or a new vehicle to the moment that device rolled off the assembly line a decade, sometimes two decades might have passed. And even then, the new product would more than likely contain some revolutionary technology that pushed the envelope of knowledge outward. That new technology would eventually filter into the civilian market and become incorporated into consumer products.

Nowadays that idea-to-product cycle has been stripped down to eighteen months to two years, and it is the civilian marketplace where the revolution is happening.

In the comfortable Cold War environment, defense ministries could go to their industrial base, state the requirements and provide the cash for industry to come up with a technical solution. This happy arrangement meant that the military was frequently driving progress, with many advances trickling down to the civilian sector. The information revolution has turned that dynamic on its head. It is the civilian sector that is driving change and the military that benefits from the trickle-down. As the pace of the information revolution moves ever faster, the challenge for the military is to adapt and catch up. Failure not only means being left behind but it also means that real power might transfer to the country with the more flexible military.

There are two broad principles at work in the way the military applies new technology to the business of fighting a war. The first is to use advances in information technology to get inside and disrupt an enemy's OODA (Orient, Observe, Decide, Act) loop; that is, to enable yourself to make decisions faster and more efficiently while at the same time destroying the enemy's ability to make decisions. The second is to move the fighting soldier further and further away from the point of contact with the enemy while making the firepower that soldier can deliver much more intense and accurate. To put it succinctly, to increase survivability and lethality.

The arguments in favor of this are economic and humane. A modern soldier, equipped with increasingly sophisticated weapons and ancillary equipment, represents a significant financial investment on the part of the Army in terms of training and hardware, an investment worth protecting. With the Army's post–Cold War shrinkage, every asset, human and technological, has to deliver many more bangs for the bucks. And the days of treating soldiers as cannon fodder are gone. Commanders want to keep their troops alive. Therefore the key phrase becomes "standoff." Rather than use the soldier as a hand-to-hand combatant, give him the technology that allows him to stay back from the enemy and call in strikes by air or artillery.

In the trench warfare of 1914–18 millions of men on both sides were thrown onto enemy wire and machine guns in order to take and hold an objective. It was immensely wasteful in human lives, a human catastrophe on an epic scale. Now imagine that one side could use standoff technology to call in precision weapons to destroy a section of the enemy trenches without having to send a single soldier against it. The swing in philosophy from seeing the soldier as little more than a target for the enemy guns to a life that must be protected is profound. The Gulf War saw early demonstrations of the standoff principle when special forces from both the United States and Britain used laser beams to call in air strikes on Iraqi missile sites. They did not have to go in and sabotage these sites in person; mission accomplished, they could simply melt into the night without the enemy even knowing they had been there.

The technology that embodies this principle is called Land Warrior, a system being developed by the Hughes Aircraft Company and partners that include Motorola and Honeywell and which is due to be in full service by 2000. Masterminding the development is a thirty-year veteran of the Army, Maj. Gen. (Ret.) Don Infante, a former commander of the U.S. Army Air Defense Artillery Center at Fort Bliss and the project manager for the Patriot missile system.

"This is the first weapons system to be mounted on a human," he says. "We've put systems on planes, battleships, tanks, but never taken a system and put it on a man. We have neglected the soldier for so long."

The heart of Land Warrior is a slim-line computer and radio shaped to fit the soldier's lightweight pack frame. The radio is wired to the computer, which is connected to a laser sight/range finder/video camera on the modified M4 carbine and to an audiovisual display in the helmet. The weapon effectively becomes the soldier's secondary visual system, sending signals to the helmet display and, through the radio, onward to the command post. Thus the soldier can view enemy operations without putting himself into the line of fire—he simply has to hold the gun above his trench or around the corner of a building and, if necessary, fire the gun as well. All cables are integrated into the frame of the carrying pack, which is designed to flex with the soldier's movements. Also in the frame is a GPS node that gives the soldier his precise coordinates on the ground. This data, combined with input from the gun-mounted laser range finder, which is effective up to 2.5 kilometers (1.5 miles), is fed into the computer, which calculates precisely the coordinates of an enemy position. The radio then sends those coordinates back to the base along with a still video image. The current version of Land Warrior cannot as yet send live video but that will happen when wider bandwidth technology comes onstream, as it will. Lightweight, flexible body armor protects the body, while in the helmet four sensors detect incoming laser threats and set off alarms in the audiovisual displays. The helmet's eyepiece/display comes with day and night versions, the latter also offering night vision capability. Power comes from either rechargeable or disposable batteries.

"In future land battles, we are aiming to increase the soldier's battle space and make him more survivable, with the 'shoot round corners' capability and his ability to make direct contact with fire support systems and air support systems," Infante says. "Increased survivability goes hand in hand with increased lethality. The more lethal you are, the more survivable you are. You've got to keep him alive."

The Land Warrior still presupposes a conflict being fought along conventional lines, soldiers going against hard targets that have to be destroyed on the field of battle—a technology that plays into the conservative view of future war. Yet it takes little or no adaptation of the concept to create the perfect system for fighting battles in urban environments wherever in the world U.S. troops are called to keep or enforce the peace. "If you had these systems in Somalia," Don Infante says, "ten U.S. soldiers could have taken on a ragtag battalion of Somalis all night in time for reinforcements to reach them." Police systems in America are looking at a version of the system called Law Warrior as an escalation of the capabilities of SWAT teams who have to deal with increasingly well-armed criminals. It is a technological edge law enforcement sorely needs.

Where the Land Warrior really qualifies as a modern piece of military equipment is in its ability to be upgraded. The move in civilian and military technology is away from fixed systems that have to be discarded when they become outdated. Take the humble computer modem as an example. Until 1996, every time a faster modem was released, you had to dump the old one and buy the new. Then U.S. Robotics introduced a modem where the internal card or external box held the basic technology but the software that governed the modem's operation and speed resided on the computer hard drive. Thus when faster technology emerged all you had to do was change the software. In a sense the military has been doing that for years, upgrading old Boeing 707s to perform sophisticated functions that could never have been dreamed of in the early 1960s when the 707s ruled the skies. Now systems like Land Warrior are deliberately designed to be upgradeable so that they will always operate at maximum efficiency.

As technology improves, there will inevitably be hardware revisions as well. One design being worked on in Britain and California is an "intelligent" uniform that will analyze any wounds its wearer receives, then transmit the data back to medics. The fabric will be made from polyester woven with fiber-optic wires that will act as sensors. A spectrometer will detect the presence of blood, while blood loss will be measured by the degree to which the figure-hugging polyester contracts as the body shrinks. It will even be capable of rudimentary first aid, with automatic tourniquets at the top of arms and legs that would inflate to stem bleeding from a serious wound. Future designs being considered will include automatic painkillers and antiseptics that the uniform will dispense into the wound. The next stage will involve tiny microphones that will measure the passage of a bullet or shrapnel into the body and thus relay to the medical teams the seriousness of the wound. Less clear is whether a soldier's uniform will one day resemble the powered exoskeletal armored devices that science fiction writers have dreamed of for years. Essentially they magnify the wearer's movements, so that a soldier could run faster and jump higher than any human has ever done. It is believed, but not confirmed, that experiments on exoskeletal armor were conducted over thirty years ago in a secret black program funded by the Pentagon, but were dropped because at that stage it was impossible to control the sensitivity of the armor's magnified movements. The implication is that in trying to remove a hair from someone's shoulder their head would be ripped off.

Land Warrior is more than a weapons system, it is an information system, too, a component in what Adm. William Owens described as the "system-of-systems." The next step was to leverage the information flow coming from forward units deploying Land Warrior into intelligence available to all units in the theater. That called for the next layer in the overarching system, what became known as the electronic battlefield. The job of developing this battlefield was given to TRW's Data Technologies Division in Carson, California, whose Applique system creates an unprecedented flow of information across the battlefield.

Take the electronics out of information and warfare reverts to the days of runners, signal flags and scraps of paper. Someone had to collate all that information, create a picture of what was going on, make a decision based on that information and then send out the orders—by runner, flag or scrap of paper—to take the required action. Even adding layers of electronics into that process, from flashing lights, telegraph, radio-telephone, right up to satellite and microwave transmissions, does not alter the fact that someone has to collate the information and prepare a picture of where the enemy is by drawing on a map and then communicate orders by radio, by which time the picture has probably changed. Indeed, as the lines of communication become more sophisticated there is more information to process and the job of building that picture takes more time. In the tactical area of the battlefield anything that extends the OODA loop spells danger. TRW's job was to build a system that allowed the input of vast amounts of information that could be made available in almost real time to the war fighter.

Applique is a wireless intranet for the battlefield in which information is transmitted to the war fighter via radio to receivers mounted on the bumpers of Humvees, tanks, trucks or Bradley Fighting Vehicles and then into computer terminals mounted inside. Taking inputs from satellite intelligence, human intelligence, including data from Land Warriors in the front line, and signals from AWACS, Applique crunches the data and produces a picture of the battlefield that appears on the monitors of the computers in the field. With pinpoint precision Applique can relay the position of enemy units and vehicles so that a tank commander can, for instance, see where his enemy is long before his enemy sees him. GPS transceivers on each vehicle also identify the location of friendly forces relative to the enemy so the picture is complete, which will help reduce incidents of fratricide, when friendly fire kills soldiers on the same side. (Thirty-five of the 146 Americans killed in the Gulf War died from friendly fire.) Additionally, because base commanders and war fighters are seeing the same picture at the same time, the process of seeking and receiving permission for a strike is greatly accelerated and success rates enhanced. Conventional tank fighting exercises produce kill rates of one kill per thirty shots. Experiments with Applique return rates of thirty-one kills for thirty-two shots, a quantum leap in lethality.

"If I know where you are, but you don't know where I am, I'll beat you every time," is how Neil Siegel, TRW's Director of Army Systems and the project manager for Applique, describes the bottom-line brilliance of the system.

Extra refinements can be made to the information stream. For instance, the red shape representing enemy vehicles can be made to change color as the information on which it was based gets older. Text messages can be added to individual targets, accessible to the war fighter by the click of a mouse. The impact of Applique on resupply operations in the field is just as radical. Because a quartermaster can see from the Applique screen the precise disposition of supply trucks and tankers he can organize refueling and resupply with as much accuracy as the tank commander firing a round at the enemy. Better yet, a unit can alert the quartermaster to its needs via e-mail.

Applique is a truly revolutionary development and would be even more so if the United States ever had to fight a land battle again. There is no guarantee that will ever be the case, in which case Applique becomes a smart new technology for an old war, yet even if that proves to be true the cost in relative terms will have been very cheap, particularly for a system one general called "the greatest revolution in war fighting since smokeless powder." The system cost less than $300 million over seven years, a pittance compared with traditional big-ticket bits of hardware, and it does not require the addition of extra personnel—"force structure" as the military puts it. Indeed it might very well enable fewer people to achieve better results.

Applique was fully tested in the field for the first time in March 1997 during an Advanced Warfighting Exercise (AWE) at the Army's National Training Center, Fort Irwin, California. The experimental forces (known as EXFOR) were pitted against the resident opposition forces (known as OPFOR or red forces), made up of a conventional infantry battalion that takes on all challengers and provides a level of opposition that few forces anywhere in the world could muster. Could a force that relied on computer wizardry take on the best in the business and win? It was the job of the 1st Brigade of the 4th Infantry Division based at Fort Hood, Texas, to find out.

To test the future of warfare, the 1st Brigade was stripped bare of normal operating procedure and told to rebuild itself from scratch, taking the most modern equipment on offer and making it work as efficiently, or more so, than conventional structured divisions. Normally new technology goes through a lengthy development and testing process before it is put into the field. Getting it to troops while it is still at the experimental phase is a bold and innovative decision by the Army that brings military procurement more in line with the pace of technological change. The process of building an army of the future with equipment that was barely off the laboratory workbench was fraught with complexities, which obliged companies like TRW, Hughes and others to send hundreds of staff into the field to act as trainers, repairmen, troubleshooters, programmers and hand-holders to the military personnel who had to use it. The extraordinary scenes of civilians rushing around helping the war fighters get to grips with unfamiliar machinery is a harbinger of the day when the people who build the equipment will routinely be on hand to provide logistic support in the field. The whiz kid programmer may be the surrogate warrior of the future.

The 1st Brigade was fitted with the Applique system, which would handle all the functions of command, control, communications, computers and intelligence (C4I). It also had the benefit of Unmanned Aerial Vehicles (UAVs), which are essentially pilotless planes that beam back pictures of enemy activity. In all, in its role as EXFOR the 1st Brigade had seventy-two new pieces of equipment or applications to throw into the battle, including some troops using the Land Warrior system.

OPFOR was instructed not to use electronic warfare or cyber attacks on the EXFOR tactical intranet, as the equipment was so new it had not been hardened against what would be expected from a real enemy. Training exercises with Applique, Land Warrior and UAVs at Fort Hood two months earlier had already exposed some of the weaknesses of the prototypical systems, among them the kinds of computer crashes that anyone who runs a home PC can ruefully attest to. Similarly, experienced infantry soldiers who had always been taught to rely on their rifle, their instincts and their comrades found it hard to surrender some of their control to blinking lights on computer screens. At that stage, it was apparent that in the soldier's mind a bullet beats a byte every time; when things got hot for the EXFOR, they resorted to radio communications because it was quicker to yell than type e-mails to HQ. Equally HQ preferred hearing the voice of commanders in the field because they could gather much more information from tone, stress and nuance than mere words alone can convey. The verdict on the new systems was that they might be useful if they could be relied on 100 percent. Soldiers reported their Land Warrior systems broke down too easily, and they were dumped.

There was no doubting that when the elements of the new technology did work as intended during the Fort Hood exercise, OPFOR was being beaten to the punch by several minutes by EXFOR and losing tanks as a result. "They know where all our assets are," an officer from the local OPFOR reported, crediting the UAVs for that advantage. But if the UAVs delivered the intelligence, it was Applique that communicated that intelligence in real time to the EXFOR and enabled them to have what soldiers call the "God's-eye view" of the battlefield. The system was also crucially important in preventing blue-on-blue kills, otherwise known as fratricide or deaths by friendly fire. If through the smoke and dust of battle a tank commander is told by his computer that the dark shape lumbering toward him is definitely one of his own forces, then the seed of doubt that in a moment of panic can launch an attack is removed. Nor is it as easy as it used to be for units to get lost. "The first sergeant from Bravo Company is located north of Phase Line Axle, beyond the zone of separation. Get him back where he belongs!" came one fevered radio call during training at Fort Hood. Put simply, the first sergeant from Bravo Company was in enemy territory and could well have been destroyed if the Applique system had not spotted him straying from the line.

The exercise at Fort Hood was but prelude to the main event at Fort Irwin in March. OPFOR's objective in this exercise was to launch a ground assault on EXFOR and take their ground. Irwin is located high in the Mojave Desert. The terrain is bleak, landmarks few and the ground is rugged. OPFOR mechanized infantry and tanks lunged at EXFOR positions in a series of raids and thrusts that EXFOR was able to repel thanks to the blizzard of intelligence coming in from UAVs and distributed by Applique.

Tank commander Lt. Robert Grimmer of the EXFOR praised the new system, particularly for its ability to sort friend from foe in the fog of war. "Applique is by far a definite keep," he said, "it's saved my ass a lot out here." But the battle also proved much of the new technology to have severe limitations, particularly in its ability to endure the mauling it gets during military operations. One EXFOR battalion commander ran his operation from an aircraft patrolling the battlefield. His Applique set broke down in the early stages of the exercise, depriving him of vital information. Also, the arrays of sensors in the sky and on the ground produced a flood of information that choked systems and slowed down decision making. "It was a digital traffic jam out there. . . . Info discipline has gone to hell," one Army observer reported. Headquarters commanders were faced with this blizzard of data, and under the pressure of the moment, reverted to tried and tested methods of tracking the battle, such as drawing maps on acetate overlays. "We're still in the crawl and walk phase in terms of understanding the capabilities we have," Col. Tom Goedkoop, commander of 1st Brigade, said. Yet OPFOR was fought to a standstill without achieving its goal, which was chalked up as a good effort by EXFOR. EXFOR in turn was judged to have missed valuable opportunities. Leaks of initial findings by the Army's Office of Operational Test and Evaluation indicated the new digitized army had not been impressive. The leaks talked of "immature and unreliable" high-tech equipment and "no discernible evidence of an increase in lethality." The Land Warrior system was too heavy and was "still rejected by the light infantry." The leaks to *Army Times* were picked up by the national press and the promise of a high-tech future for the military began to look like a low-tech fiasco. The Army was quick to respond. The director of the OT&E Office, Philip E. Coyle III, said the exercises at Hood and Irwin were just part of a much longer process and that the Army had done "an outstanding job. . . . What the Army has done here is a model for all the military services. . . . This process is going to save years, perhaps decades, in the development of the systems involved."

Despite the brickbats aimed at much of the experimental equipment, a clear winner of the exercises appears to be the UAVs, which not only provided valuable battlefield intelligence but also proved a major distraction for OPFOR, which knew itself to be under surveillance at all times. "I've got 2,400 soldiers on the battlefield looking up into the sky this week, worried about those drones," Col. Guy Swan III, OPFOR commander, said. The senior officer at the exercise, TRADOC commander Gen. William Harzog, called the Hunter UAV a "2,500 soldier crick in the neck" while the EXFOR commander Maj. Gen. Paul Kern said the OPFOR had used so much energy attacking the UAVs that it would be worth considering the deployment of dummies just to keep the OPFOR, or in the real world the enemy, agitated.

Virtually simultaneous with the AWE at Fort Irwin, the Navy and the Marines were fighting AWEs of their own, also in California. Like the Army, the Navy and the Marines wanted to apply the information revolution to their war-fighting methods and see what emerged. The Navy exercise, Fleet Battle Alpha, tested a radical new communications structure to assess the impact of networking on naval operations, in this instance supporting the Marines in an exercise code-named Hunter Warrior. This was, to all intents and purposes, a rerun of the Gulf War in that a small Arab nation bounded on one side by the sea was invaded over land by a neighboring Arab country. The difference was that this time the United States was able to ride to the rescue while the invasion was under way. Instead of fielding a force in strength to fight the invaders on the ground, the Marines sent in a series of eight-man units equipped with Apple Newton palmtop computers and Ericsson cell phones.

The Marine units were for the main part commanded by noncommissioned officers, with their traditional platoon commanders having to sweat it out back at base.

"I've gone from double tap to double click," 2nd Lt. Brett Clark of the India 3/3 Marines said ruefully, referring to the standard practice of firing two shots from an M-16 rifle. Using laser range-finding sights and with GPS sensors on their computers, the Marines could enter precise target data into preprepared templates on the Newton and then squirt them via the Ericssons to a mock sea base occupied by the Special Marine Air-Ground Task Force (SPMAGTF) 150 miles away at Camp Pendleton north of San Diego. Commanders there coordinated the simulated fire that would rain on the advancing invaders, played on this occasion by the 7th Marines. The naval task force led by the USS Coronado provided the majority of the firepower; the incoming data from SPMAGTF would go straight into the Navy's fire control system, which would then respond. Given that there were only about fifty-six Marines on the ground against a force of some 4,000 invaders equipped with heavy armor, it would seem a tiny response to such a large force. Yet David fought Goliath to a standstill. It was the ultimate power of information—knowing where the enemy is before he knows where you are and being able to do something about it—over brute force.

While the Marines were demonstrating this power, out at sea there was evidence that the Navy had started to learn that the information revolution is about more than adding smart technologies to old ideas.

Navy Lt. Ross Mitchell, thirty-three, is in the first rank of America's new techno warriors. In December 1996, Mitchell was part of a group of young officers working out plans for the upcoming 3rd Fleet exercise, what was to become Fleet Battle Alpha. He suggested his own plan for connecting the fire control computers in all the fleet's ships in a Local Area Network (LAN) to maximize the efficiency of the fleet's firepower and minimize delay in fighting the enemy. He was told to draft a proposal, which he did over the course of the following weekend, dubbing his idea "Ring of Fire." A handful of years ago, such an idea would have filtered slowly up the chain of command over a period of months; it would have been studied endlessly, probably rejected by a senior officer who felt threatened by such innovative thinking. Even making it to fruition could have taken years. Not today. A few weeks after Mitchell's paper was submitted, the Navy designated a budget of $600,000 and assigned a team of researchers from Johns Hopkins University to help Mitchell make it happen. More extraordinary still, the Ring of Fire was in place in time for Fleet Battle Alpha in early March 1997. Ninety days from idea to execution of a radical departure in naval command and control. Just as the Army had recognized that acquisition has to be accelerated to warp speed in order to stay abreast of technology innovations, the Navy saw that the old ways had to change if it was to stay relevant. It is the kind of change that normally happens only under the intense pressure of wartime, when survival drives acceptance of innovation and risk.

The performance of Mitchell's system is still being assessed. In theory it would seem to make perfect sense. "It was essentially a concept of preplanned missions using a new system called the Naval Fire Support Weapons Control System so that all the ships in the fleet can communicate," he said. "As a ship comes on station it simply checks into the Ring and its weapons become part of the force."

That means cruisers, destroyers, frigates, submarines and the arsenal ships of the future will all be treated by the system as one big weapons array. The computers will work out which ship is nearest and whether it has the most appropriate weapons for the task at hand and then issue the fire order.

Mitchell's boss, Vice Adm. Herbert Browne, was enthusiastic about the innovation. "This is exciting stuff," he said. A little too exciting for some perhaps; during the course of the exercise it was realized that the Ring of Fire had no fail-safe systems in place to prevent missiles dropping on friendly forces. As the Army discovered at Fort Irwin, having civilian programmers on hand can short-circuit what would otherwise have been weeks of tweaking the system. After a quick discussion, the team from Johns Hopkins set to work and days later the refinement was in place, before the exercise was even over.

Throughout the exercise, the Ring of Fire was receiving data from a USAF J-Stars surveillance plane flying over the battlefield. The J-Stars, a Boeing 707 converted by Northrop Grumman to hold ground-scanning radar (in much the same way as an AWACS is a 707 converted to scan airspace), can map an entire battlefield and project the scene back to the base. On board the USS *Coronado*, fleet commanders could watch on a giant screen real-time disposition of friendly and enemy forces. The same data, combined with intelligence from a U-2 spy plane, UAVs above the battlefield, video and data from Marine units and microsensors on the ground, was streamed into the SPMAGTF's Enhanced Combat Operations at Camp Pendleton, brought together by the Joint Maritime Command Information System and then fed into a 3-D "virtual reality responsive workbench" devised by the Naval Research Laboratory to give an electronic real-time view of the battlefield.

This enabled the commanders to see where the enemy was and launch instant attacks without having to commit any ground forces. "This is going to allow us to focus our forces where we can do the most damage with the fewest American losses," Admiral Browne said. "We have an American public that will not tolerate losses, and that needs to be part of our equation."

Again, the principle of standoff was asserting itself. An enemy that can be hit with a variety of armaments from cruise missiles to five-inch guns without even making eye contact with the opposing forces and without the ability to fire back is in a considerably weakened position. Hunter Warrior is still being assessed, but the initial judgment is that it proved a quantum leap in the management of C4I, providing a degree of flexibility and a completeness of vision previously unknown in the field of battle. The second in command of the SPMAGTF, Lt. Col. Bob Schmidle, cautioned that such was the flow of information that the commanders risked losing sight of what they were doing, an irony considering how nobody had ever had a better view of a battlefield.

"I think the temptation to look inward is there. . . . It requires the same kind of discipline to realize what you need to stay focussed on, what is important, and to take other things and to shed those kinds of tasks [that are unnecessary] or to keep that information from obscuring what you need to make a decision."

Lt. Gen. Paul van Riper, commander of the Marine Corps Combat Development Command, cautioned further, saying it was one thing to get a great deal of information coming in, it was another knowing what to do with it: "They would get facts. That's all they had were facts. The interpretation of those facts depended on what the team was saying, what the analysts attributed to those facts and then what the commander attributed to the analysts."

To demonstrate that no matter how terrific the technology it still takes a knowledgeable human to draw the correct conclusion, he cited the instance where a Marine squad in the field reported they were about to be caught by the OPFOR because they could see through their thermal sights that the enemy had stopped their vehicles, dismounted and pulled out flashlights. The squad was prevented from taking precipitous action only when Col. Thomas O'Leary, the commander of SPMAGTF, learned of their interpretation of events. "Stop," he said. "They are leading their vehicles off the road so they don't run into ditches and obstacles."

Gen. van Riper also saw a danger in the brilliance of the display that it could lead to a false sense of realism. "Some commanders begin to believe the map was reality. Reality is reality. And so the question is how much better is the computer map than the grease pencil and acetate map? It's better. But you always have that danger, like any computer system, the sense that it's infallible."

None of these criticisms was intended to undermine the technology that made this widespread and effective deployment of such a small force so effective, because it did a good enough job of undermining itself. The handheld computers and mobile phones proved in many cases inadequate to the job of fighting a battle. Col. Anthony Wood, director of the Commandant's Warfighting Lab at Quantico, dismissed the idea that because a few units failed at the critical moment the concept was therefore invalidated. "They were surrogates to test the concept," he said, adding that the Marines would never go into battle with such a vulnerable system. Defenders of the new digital battlefield have a constant fight on their hands against critics who use the failure rate of the new equipment to question the viability of the entire enterprise. Yet the critics have a point. Much of the materiel deployed in these exercises has not been designed for use on the battlefield, and therefore breakdowns under harsh conditions are to be expected. The defenders must demonstrate that sensitive electronic equipment can be made sufficiently rugged to withstand harsh treatment. The soldier in the line of fire cannot be expected to welcome the order to reboot his computer at the critical moment.

In the naval exercise, it was not just the new technology that represented a revolutionary method of fighting a battle at sea. The presence of the J-Stars was in itself extraordinary. Armed services are extremely jealous of turf and under normal circumstances the data flow from J-Stars would have been managed by a land-based USAF station which would then feed the data out as needed. The USS *Coronado* was receiving J-Stars data firsthand via satellite, creating a synthesis of capabilities that augurs well for the future. The *Coronado*'s commander, Vice Adm. Herbert Browne, hopes that the experience will help create trust between the services. "Teamwork you can build. Trust you have to earn. Trust in the services is hard," he said.

Browne's wish is more than just a sentimental hope for the ending of maddening and disruptive interservice rivalry. Shrinking budgets and changing missions demand more flexibility between the armed services if the United States is to get the best defense for its tax dollars. As Browne was making that point, the Army Advanced Warfighting Exercise at Fort Irwin was still going on. In that exercise there were significant elements such as Land Warrior being tested that performed exactly the same function as the Apple Newtons, the GPS sensors, the Ericsson mobile phones and the laser range finder. Will there be cross-service cooperation in creating a system that works for everyone and is compatible with all similar systems? It is going to take a great deal more of the kind of imaginative thinking that Browne and the Navy brass had demonstrated if the future that beckons over the horizon is to be fully grasped.

But Hunter Warrior and other exercises like it were only a small part of the Pentagon's efforts to get a grip on the challenge posed by the information revolution. Within the Department of Defense, as in other defense departments around the world, a struggle had begun between those who saw IW as a way of waging future war and those who saw it simply as an adjunct to conventional warfare as it had evolved down the centuries. While Land Warrior, Applique and Ring of Fire all showed real promise in producing huge force multipliers in fighting future wars and in saving casualties, they also threatened the existing order.

Every branch of the armed forces had begun to develop its own, independent, methods of embracing IW. This was partly a laudable, service-driven effort to make the best of the new technology and partly an attempt to steal a march on their rivals for scarce cash and resources. The result was an expensive duplication of effort that did little to pro- duce a common approach to the problems posed by IW. It also did little to ensure that IW be embraced into the services in a fast and efficient fashion. Instead, interservice rivalry dominated the battle space and slowed down the integration of new systems.

There was a second reason for this rivalry. The information warriors saw IW as the potential silver bullet for the future battlefield. Classes at the National Defense University and the services' war colleges were filled with talk of a battlefield without soldiers, of an airspace filled with pilotless fighters and oceans populated not with carrier battle groups but "virtual" fleets with ships that had either tiny crews or no crews at all. To the more conventional warriors brought up on the teachings of Clausewitz or the British strategist Basil Liddell-Hart, this was heresy.

Every officer knew that no technology could replace the skills of man, that it required men to take and hold ground and that wars down the centuries had always, in the end, come down to honor and glory won by men on the battlefield. Yet here were the heretics arguing that the battle- field itself might no longer exist in the traditional sense; that wars might be fought by men and women far removed from the direct sight and sound of shot and shell that had decided every previous conflict. This fundamental undermining of the traditional military was understandably very difficult for many in the armed forces to accept.

Some were motivated by a very real concern for national security. After all, if the information warriors had their way, many of the defenses that had worked so well for so long would be swept away. There lay enormous risks. If the new warriors were wrong, America, in its head- long rush to embrace the new technology, might end up defenseless rather than better defended.

There was a second, more chauvinist, line of defense against the information warriors. Virtually every admiral and general had a career invested in the status quo. One of the great strengths of the armed forces in a democracy is that they are conservative, bureaucratic and reluctant to use the weapons at their disposal to go to war on behalf of their political masters. But those same admirals and generals depend on the status quo for the maintenance of their power, and the officers and men under them need the existing structure to advance their careers. In other words, too much was invested in the status quo for the information warriors who were arguing for a fundamental revolution to either find an accepting audience or a warm welcome from the officers whose very existence was threatened by the revolution.

But like many revolutions, what had begun as a theory had evolved rapidly into a series of developments that were slipping beyond the control of the traditionalists. While people like Lieutenant Mitchell were thinking creatively about new solutions to old problems, the world outside the Pentagon was driving events at a very different and much faster pace. As the twentieth century drew to a close, the information revolution was harnessing the technology to produce and develop new weapons and systems that would transform the way wars would be fought in the future. For the revolutionaries, these new products presented real promise of radical new ways of waging war. The traditionalists were confronted with the reality of either embracing the new world or ceding the battlefield to other nations that felt less constrained by the lessons of ancient history.

Fly on the Wall: Weapons and Wasps

T U R N left on Westwood Boulevard off Santa Monica Boulevard in Los Angeles and you soon arrive at UCLA. Amid the university's jumble of rectangular brick and glass buildings, only made pleasing to the eye by the profusion of L.A.'s famous graceful palm trees, you will find the building known as Engineering IV, a decidedly prosaic name for a truly extraordinary place. In his spacious office in Engineering IV, Professor Chih-Ming Ho is calmly changing the way the world flies, which makes the screen saver on his computer portraying the aerial antics of a float plane, the slowest aircraft in the business, deliciously ironic. To help him Ho has surrounded himself with some of the brightest young brains in the engineering world, and is receiving financial support from DARPA, the Defense Advanced Research Projects Agency. They have at their disposal several laboratories, all of which have a homemade feel to them, as if Rube Goldberg were up for a Nobel Prize. This homemade aspect to his work is inevitable; he and his team are doing experiments which cannot be measured on existing equipment, so they have had to invent it.

Professor Ho has devoted his life to the arcane subject of fluid mechanics, how air and water flow over surfaces and interact with them. Flight is entirely about the flow of air over a surface, the wing, to create lift. It seems remarkable in this day and age of rapid innovation that the basic technology of flight has not been improved upon since the late 1940s, when aluminum replaced wood as the material of choice. Engines have vastly increased in power, computers now do much of the job that hydraulics used to perform and the whole process is much safer and more economical, yet the basic concept has not changed. Planes are still cylinders with wings, tails and fins. The military has advanced the principle with the B-2 bomber, which is a single wing that flies, known as the flying wing, but one that has moving surfaces such as flaps and ailerons. The aim is to produce a plane that has no moving surfaces and then to microsize it to fit into the palm of the hand. Obstacles to achieving that are the basic inefficiencies of lift and, if there are no moving surfaces, what does one use for control mechanisms?

Ho's research started from the belief that it should be possible to minimize surface stress on the skin of wings by reducing the turbulence caused by air passing over them. It would appear to most people that air passes smoothly over wings, but in fact it doesn't. There are tiny little vortices created by the erratic behavior of air coming into contact with the skin. Preventing this would reduce drag, improve efficiency and lower costs. Ho designed a new kind of sensor that measures microturbulence; using photo-imaging, he was able actually to see these vortices. Like little worms crawling over a wing, 200 microns (twice the width of a hair) long and a few microns wide, these turbulent flows arrive and disappear in milliseconds, a constant, dynamic mass of turbulence that directly impacts the efficiency of the wing by pressing down on it. Ho reckoned that if it was possible to push the turbulent flow off the wing's surface as it appeared, then the effect would be neutralized. So he built a device, a MEMS, a Micro Electro-Mechanical System, 1 centimeter $\times$ 1 centimeter containing the sensor, a computer and a flap that pops up each time a "worm" is detected on its surface. It worked. The MEMS stopped the vortex forming. If several million of these could be deployed along the wings of a 747, it should have a profound effect on the plane's efficiency, but the cost of several million of these little devices on such a giant wing would make it prohibitive.

Yet, Ho had found a revolutionary way to reduce drag. To put it into context, the most recent practical advance in drag reduction has been the introduction of riblets—tiny grooves twice the width of a hair—onto the wings of an Airbus, an advance that has reduced drag by 10 percent and taken twenty-five years of research to achieve. Ho's invention is capable of reducing drag by up to 80 percent, a phenomenal advance. The next step is to find an economic way of applying the technology to conventional aircraft to improve efficiency.

Ho now considered what other applications he could find for his breakthrough. He decided to test the actuators on a model aircraft, a radio-controlled Mirage 3. He created a thin strip of actuators along the leading edges of the wings to see what effect they would have. It was an astonishing success. The actuators did not just amplify the effects of the model's own controls, but actually replaced them, as he discovered when he caused the plane to perform a 180 degree roll purely by using his actuators. This meant he could control the aircraft with this invention, a quantum leap from simply improving efficiency.

"This is the first time anyone has demonstrated that a macro-machine can be controlled by a micro-machine," he said. Creating a military plane which had no moving surfaces would make a stealth aircraft stealthier yet by reducing its radar shadow.

Creating a new form of military aircraft using this technology lies some years in the future. Where today's military use of Ho's new systems lies is in microplanes, tiny palm-sized aircraft that can be used for a variety of surveillance, sensing and even attack purposes. Kaigham (Ken) Gabriel, MEMS program manager at DARPA, is pushing for troops to be given microaircraft by 2005.

"We're sketching out the scenario where the combatant might have a pocketful of sensors, might have a pocketful of UAVs," he says. "So if I want to see around that hill, I just loft one of these up [to have a look]."

Scientists in universities around America are working on a variety of ideas for micro-UAVs that will turn this science fiction concept into a reality.

"When you approach technical people with this idea, their first response is that you cannot build an aircraft this small and make it useful," said Dr. Sam Blankenship, coordinator of the microflyer program at the Georgia Institute of Technology. "But many people, including us, think you can do it."

The range of practical uses for these machines is limited only by the array of sensors that may be mounted on them. In civilian life they could be used for surveys, disaster reconnaissance and relief, civil engineering projects, broadcasting or environmental research. The challenge of how to control the aircraft has been answered, up to a point, by Professor Ho; other systems are under scrutiny elsewhere.

Once the issue of controlling the wing has been resolved, the next control issue is how the machine is actually going to find its way to its destination, perform its function and return, all the while negotiating obstacles such as hills, trees, buildings and power lines. They will be out of the line of sight for long periods, so conventional remote control is not possible.

"It seems essential that micro air vehicles be designed for autonomous flight, in which they can guide themselves to and from a target without human intervention," believes Robert Michelson, also of the Georgia Institute of Technology. This will inevitably involve the addition of artificial intelligence systems to the craft's navigation systems, all the while keeping within the confines of a tiny machine.

The issue of how to power them most effectively must also be resolved. Tiny jet turbines, pulsejets and ducted fans are among the propulsion devices being examined. Fuel can either be conventional gasoline, battery or solar. One company, AstroPower Inc. of Newark, Delaware, is developing tiny solar panels that can be glued to the skin of the planes. It may be ultimately that a new form of propulsion may emerge from this research. To demonstrate that microsize is achievable,

a team from the Institut für Mikrotechnik in Mainz, Germany, has produced an operational twin-rotor helicopter smaller than a peanut, operated by micromotors.

The technology is in its infancy—this is not about building very small
planes by downsizing the technology demonstrated by the Wright brothers; tiny wings do not work in the same way as big ones, so these
scientists are effectively learning how to fly all over again.

"We may have to learn from insects and birds," Georgia Tech's Blankenship adds.

MEMS opens a window on to a new generation of technology that
will literally transform the battlefield. Tomorrow's soldier will go to war
with tiny aircraft in his backpack that he will be able to fly ahead of him
to smell, see and hear what lies over the hill or inside the next building.
Additional intelligence will be supplied by sensors disguised as blades of
grass, pockets of sand or even clouds of dust. The flood of data produced
by these tiny intelligence sources will be processed by computers both
in the rear area and with the front-line forces. This enhanced ability to
see, hear and smell in the kind of detail never before envisaged for the
battlefield should enhance the soldier's capability to kill the enemy. But
modern warfare and the information revolution are producing a different
kind of requirement that will satisfy the public and political demand for
a "clean" war. The new generation of leaders demands that wars—if
they have to be fought at all—should be prosecuted with the minimum
of force to produce the least amount of casualties on both sides. At first
glance, fighting wars without casualties might seem a contradiction in
terms but there are systems in service or being developed that allow
exactly that.

The challenge of the information revolution would always fit more
comfortably with the Air Force and the Navy than the Army. To further
complicate matters, many in the Navy accuse the Air Force of trying to
steal a march on the way to controlling IW. But while the Navy and Air
Force are competing for the Information Age roles and missions, the
Army continues to focus on traditional war. An army is designed to fight
on land, with large numbers of people and machines. You can change
some of the weaponry, and introduce radical new technologies, but going
beyond that calls into question the basic point of having an army. If the
Marines demonstrated that a few good men with cellular phones, palmtop computers and GPS could defeat an armored invasion, why would
anyone bother to fight a land war again? Standoff means not engaging
hand-to-hand, eye-to-eye—accept that, and you accept a decreasing
need for the people who do their fighting up close and personal. Standoff
means putting money into machinery that can deliver maximum punch
for minimum risk to human life.

The Navy's Arsenal Ship is just such a weapon, a floating robot operated by remote control that can carry 500 cruise missiles, making it the

most powerful ship ever conceived. And it could operate with a crew of as few as fifty, or perhaps even with no crew at all. Match that against a crew of 5,500 for a conventional aircraft carrier and the radical nature of the idea becomes plain.

While talk of building ships that need little or no human involvement deeply worries some parts of the Navy, realists can see that war has changed and they must change with it. Six of these ships will be built in the first phase of the procurement at a cost of $3.5 billion. "This is the first totally new warship concept by the Navy since the 1950s when it developed the fleet ballistic missile system," according to naval consultant and historian Norman Polmar. "It's an opportunity for Navy admirals to show they're not fighting the Battle of Midway but taking advantage of new technologies."

New technologies that mean a missile from the Arsenal Ship can be fired by a commander in a completely different location using targeting information compiled from a variety of different sources. It is war virtually by remote control. Can it be done? The Navy has already gone some way toward proving it can. During the summer of 1996 the guided missile cruiser *Yorktown* was transformed into a virtual robot ship by the application of new computers that replaced sailors at the helm, in the engine room and damage-control stations. "We basically unplugged the nervous system of the *Yorktown* and plugged in a new one," said Rear Adm. Daniel J. Murphy, the Navy's director of surface warfare programs. The computers replaced fifty sailors and proved capable of carrying out maneuvers to a degree of accuracy no different from that expected of humans. The *Yorktown* is a sophisticated vessel that requires much more human involvement than an Arsenal Ship will ever require. The fact that it could be maneuvered successfully without people proves the integrity of the Arsenal concept.

Can an unmanned ship successfully be used to send salvos by remote control? The Navy's Fleet Battle Alpha and Hunter Warrior again proved it could. For the purposes of the exercise, the destroyer USS *Benfold* carried a simulated load of Tomahawk cruise missiles, the Fast Hawk cruise missile, the Navy Tactical Missile System and Extended Range Guided Munitions, as well as missiles for Theater Missile Defense (TMD).

To test the concept, the *Benfold* was put through two scenarios, each with the same objective. In one, it acted as a conventional destroyer, in the second as an Arsenal Ship. The difference was amazing, according to Cmdr. Dave Summer of the Third Fleet. "The whole speed of battle with an Arsenal Ship is amazing," he said. "We found we could do [the same] campaign in 36 hours versus 6 days."

It was no accident that the *Benfold* was fitted to play a role in defending the fleet from missile attack as well as acting as a dumb missile platform for offshore support of a land operation. The Pentagon is thinking seriously of moving current land-based TMD capabilities for high-altitude missile interception, including the Army's THAAD missile and the Navy's Upper and Lower Tier system, onto the Arsenal Ship to make them less vulnerable to enemy or terrorist attack when deployed in theater.

One concern still dominates Arsenal Ship discussions, however, namely, its own vulnerability to attack. Although designed to have stealth capability, its presence will be no secret once it starts firing, and with a predicted top speed of twenty knots it will be a sitting duck for enemy countermeasures. This will require other Navy ships to protect it until such time as the Navy can afford to develop what is called SC-21 (Surface Combatant for the 21st Century), an Arsenal Ship that is fully equipped with 155mm guns and could also have antisurface and anti-submarine capabilities—a deadly robot ship that can defend itself.

Not everyone shares the enthusiasm to create a massively powerful ship that requires little or no human operation. In a trenchant attack on the concept, former naval officer William L. Stearman, a member of the Clinton White House National Security Council staff, said the Arsenal Ship would be far too vulnerable to low-cost attack and would be carrying far too expensive a payload (half a billion dollars' worth of missiles alone) to make it worthwhile. He accused the Navy brass of a "deep-seated infatuation with high-tech weapons systems" and said a fraction of the money earmarked for Arsenal could be used to recommission two currently mothballed battleships, which he said provided the only true platform for high-volume heavy-fire support of the kind the Arsenal Ship is designed to supply. In the same magazine article in which this attack appeared, naval analyst Norman Polmar, who had been an adviser to three Secretaries of the Navy, accused Stearman and other opponents of the Arsenal Ship of being partisans of the Air Force, who felt the role of strategic bombing was under threat, and "battleship whiners . . . who feel the grandeur of the battleship somehow equates to military effectiveness for the 21st Century."

Neither Polmar nor Stearman seemed to understand the true nature of the debate among the American admirals. The argument is not between battleships and the Arsenal Ship or even between high tech and low tech. Instead, the debate is fundamentally about the future of the Navy's capital ships in the next century. Since the Second World War, power projection for the Navy has all been about aircraft carriers. A single carrier requires up to twenty ships and submarines to provide it with adequate protection, and with eleven or so carriers in the fleet there is an automatic necessity for a continual program to upgrade submarines, fighters and destroyers. Without the carriers, the requirement for a full-force Navy that has been virtually unchanged in the past fifty years would be finished. And if the carriers were no longer needed, then the traditional role of the Navy itself would be brought into question.

The standoff concept underscores air operations, too. In the late twentieth century, standoff had become the mantra of all developed armed forces. The argument ran that the farther friendly forces could be kept from the actual point of confrontation, the fewer the casualties. This had led to a new generation of long-range precision-guided weapons. If this capability was important for the Army and Navy, it was vital for the Air Force, which wanted to keep its expensive aircraft and costly pilots out of harm's way whenever possible. The new fighter-bomber for the start of the next century is the F-22, the $5 billion-a-year budget-buster that the Air Force believes can assume the roles currently performed by the F-117A (the Stealth fighter), such as delivering precision-guided munitions, the F-4G Wild Weasel radar-killer and the RC-135 Rivet Joint electronic warfare plane. At the time of writing, a cloud of doubt hangs over the project, which calls for the first batch of thirty-two aircraft to be delivered in 2004 at a cost of up to $100 million each. The Air Force has been scrambling to justify the program on the basis that the F-22 has such a broad range of capabilities. With shrinking defense budgets, the prospect of keeping the F-22 program intact looks dubious, particularly as other parts of the Air Force are pushing their own projects at the F-22's expense. Most notable is the B-2 Stealth bomber, a weapon that has been in constant search of a mission. The development of technology which allows GPS receivers and guidance mechanisms to be attached to dumb bombs at a fraction of the cost of laser-guided bombs gives the B-2 that mission. The system can be used in any weather, a significant advantage over laser bombs, which can be badly affected by clouds, in addition to which the B-2 can operate at a great height over very long distances without refueling, making it a much cheaper per-mission system than either the F-117 or F-22. The Air Force's other controversial program is the Joint Strike Fighter, which will also provide aircraft for the Marine Corps, the Navy and Britain's Royal Navy. The JSF will be stealthy and STOVL (Short Take Off Vertical Land), enabling it to operate in a variety of terrain, from conventional runways and aircraft carriers to country roads and steel platforms set on board ships. Its versatility, speed and maneuverability will make it the finest in the world, and probably unassailable by any other nation's forces.

Yet the Joint Strike Fighter may be the last airplane of its kind ever built, for the day is not far off when America's most dangerous air strikes could be carried out by UAVs. Although in their relative infancy, UAVs have some immediate advantages over piloted aircraft. Not only does one not lose a pilot when the plane is shot down, being pilotless means the UAV can operate at speeds and G forces which human beings could not tolerate. Such considerations are some way down the road; the services have only recently woken up to the potential of UAVs in support roles. In Desert Storm and in Bosnia UAVs like Predator fitted with video cameras and transmitters sent real-time video images back to the base giving commanders vital and timely intelligence. Now they are a fixture on the battlefield as low-cost, high-return intelligence gatherers. Typical of the new generation of UAVs is the Global Hawk, made by Teledyne Ryan Aeronautical, a pilotless spy plane that does the high-altitude surveillance job of its well-known piloted predecessors the U-2 and the SR-71 Blackbird at no risk to an Air Force pilot, with an array of sensors that can capture close-up images of the ground 65,000 feet below no matter what the weather, and transmit them back to the base thousands of miles away. Better yet, the Global Hawk's price tag is a mere $10 million per plane.

Jeff Yake of the defense technology research firm Frost and Sullivan of Mountain View, California, believes that UAVs have a bright future. He says the market for them will grow in the coming years, in opposition to the general downward trend in defense budgets, from an annual military market of $340 million in 1996 to $370 million in 2002. The civilian market adds approximately another $25 million worth of demand.

"The rapid, accurate intelligence provided by the UAV's has given them a stamp of legitimacy that extends beyond the Department of Defense to encompass civil and potential commercial applications," he says.

The variety of uses to which UAVs can be put will increase as their sophistication and strength improve. In the civilian sector that could include anything from crop dusting to monitoring severe weather such as hurricanes. In the military, the Marines have already experimented with using cheap UAVs such as the Exdrone to deliver nonlethal weapons. One essential function being planned for the Tactical UAV Outrider is mine-detection. The Army is to place the Lightweight Airborne Multi-spectral Countermine Detection System on the Outrider to do the job currently done by humans. This device uses ground-penetrating radar and heat sensors to locate mines that can then be destroyed in the same way that the currently deployed Airborne Standoff Mine Detection System (ASTAMIDS) does from manned aircraft.

The issue of mines is increasingly important throughout the world as it becomes clear there are millions still embedded in the earth as a result of countless regional conflicts fought in recent decades. In countries like Angola, men, women and children regularly lose limbs or die as a result of stepping on land mines left over from old wars. Technology that can find these mines without deploying large numbers of people to do a dangerous and nerve-racking job is a bonus for everyone, not just America. The project is led by the Army's Communications and Electronics

Command's Research, Development and Engineering Center at Fort Belvoir, Virginia. Project manager Jim Campbell says the Defense Department is fast-tracking the system because of the high profile of the issue and because conventional systems are proving to be less than 100 percent effective. "We want to take the human out of mine detection," he said. "Brute force is no longer the way to hunt mines. High-tech solutions are now in order." (In the not too distant future any mines found could be destroyed by a variety of devices ranging from robots that attach themselves to the mines and then explode to microbes that actually eat explosives. These mine-eating bugs are currently being researched and tested.)

Can UAVs be used for more lethal purposes? Jeff Yake believes so. "The future of air combat is UAVs with stealthy attributes and defense mechanisms," he said, adding that advances in UAV design and adaptability will "take the pilot out of the cockpit" for combat missions in the next ten to fifteen years. Whether such a plane can ever be competition for a conventional fighter-bomber will also depend on the development of suitable munitions and on control mechanisms that will respond to threat with the speed of a human pilot.

This argument about planes and ships has been going on virtually unchanged for many years. First discussions about unmanned aircraft emerged over twenty years ago. During Operation Peace for Galilee, the Israeli invasion on Lebanon in 1982, the Israelis demonstrated the value of UAV by successfully fooling Syrian air defense and fighters. Since then UAVs have been integrated into every modern air force and have a valuable reconnaissance role. The decades-old debate had largely pitted technology against the traditional soldier, sailor and airman. The new revolution is different because microtechnology is producing an exponential leap in capability that allows weapons to reduce in size so much that they are literally forcing the pilot out of the cockpit and the captain off the bridge.

Research into new generations of weapons, sensors and control systems leads into the exotic world of MEMS, which promise to deliver tiny machines the size of insects that can see, think, act and deliver information. Imagine a device shaped and colored to look like an ordinary cockroach that can carry an explosive charge. Now imagine a UAV flying low over an enemy installation at night showering it with these tiny robots that have been programmed to seek out radar emissions. The insects crawl to their target and start exploding at key points, destroying the enemy's ability to see incoming threats. They could be programmed to eat silicon, seek out and destroy land mines, enter and survey areas impervious to other surveillance techniques, even kill individual soldiers with poison. These functions and more are just the start of what the Pentagon has in mind for these tiny war fighters. "Far-out ideas, which today have a terrible giggle factor, are likely to become our saviors," believes Thomas Jones, special projects director at Boeing.

At the Sandia National Laboratory in Albuquerque, scientists have already started to turn science fiction into science with a robot the size of a sugar cube, built from components found in everyday items like a pager (the motor), a model train (gears), cellular phone (computer chip) and a miniature sensor. Projected total cost each: $150. MARV (Miniature Autonomous Robotic Vehicle) is a crude device, yet it could be trained to follow a specific frequency emitted by a wire. The Sandia team says in the future, rather than use a UAV for delivery, such tiny robots could be dropped into enemy territory in a box called the mothership, disguised as a brick or a rock. The little machines, themselves disguised as insects, could then set about infiltrating the enemy's base, sending data back to the mothership, which would then retransmit it via satellite to U.S. commanders.

DARPA thinks the MEMS could be made even smaller so that micromachines the size of large dust particles could perform surveillance duties. Fired from a bullet or an artillery shell, or dropped from a plane, clouds of "smart dust" could descend over an enemy area, sensing the presence of nuclear, biological or chemical weapons, or human and vehicular movement, and reporting back, either individually or via a dust network. As always, power is the limiting factor once they land. The tiny spies would preserve battery life by being programmed to "wake up" at fixed intervals, react to whatever is going on around them, then go back to "sleep." This would mean a life of as much as four days.

Extraordinary as these concepts may seem, the civilian world is yet again where the action started. Automobile airbags are triggered by tiny MEMS that sense the rapid deceleration of an accident and activate the bag. New generations of these triggers, whose integral sensor is about the size of a standard printed period, have brought the cost of activating an airbag down from $30 to about $7. This defies yet another norm in the world of technology, that miniaturization is expensive. The scientists who work with MEMS try to make us understand the scale of what they do by comparing the measurements they work in with the width of a human hair. A human hair is about 100 microns wide. The electrical circuits in MEMS are a few microns wide. The moving parts on MEMS can be similarly tiny. This is engineering at the microscopic level.

The first micromachine in history was made in 1988 by Professor Yu Chong Tai at Caltech. Slightly larger than a human hair's width, this machine actually moved. C. J. Kim at UCLA made a gripper—tweezers —out of silicon and with it picked up a single-cell protozoa seven microns long, less than 10 percent of the width of a human hair.

In terms of their maturity, MEMS are in their infancy. Imagine the first transistor compared with a modern computer chip. MEMS are at the stage of that first transistor.

With intellectual and financial leadership from the government's Defense Advanced Research Projects Agency, America is pursuing an aggressive R&D program into MEMS that promises to open a window on an astonishing new world of possibilities. The program is relatively recent, and its inception gives another clue as to how the world is changing. A 1992 study by the Rand Corporation in California warned that Japan had in place a ten-year MEMS research program which received 75 percent of its funding, about $150 million, from the Ministry of International Trade and Industry (MITI). U.S. investments in MEMS lagged "by more than an order of magnitude" behind Japan, the report warned. The wake-up call was heeded, and the Department of Defense gave DARPA three years and $24 million to jump-start America's MEMS research. Ten years ago the military would have labeled the research top secret and only later allowed it out to the commercial sector. But the military mind-set has changed under pressure from the speed of innovation. Crucially from the Defense Department's point of view, America had to assert a lead in developing MEMS for the battlefield.

The guru of the micro-electro-mechanical world is Ken Gabriel, the MEMS program manager at DARPA. A revolutionary and visionary combined, he was chosen by the Department of Defense in 1992 and given a very specific mandate: bring the United States up to speed in a technology in which the United States was being fast outstripped by Japan and Europe. Gabriel's strategy has been to see where the best MEMS research is being done, either in universities or high-technology companies, add funding to accelerate the process, then push industry at large to recognize the extraordinary possibilities. Eventually, he believes, development of MEMS technology will have reached critical mass and MEMS will become a routine part of our world.

That will be a revolution even more radical than the changes wrought by the computer microchip. The microprocessor can only process information. The micro-electro-mechanical machine can gather information, process it and act upon it. Acting as coach and cheerleader to industry, where revolution is risky and conservatism safe, is a function that perhaps can only be performed by government. DARPA is a military agency and this is technology the Pentagon wants for itself. But with a sense of vision that would have been amazing twenty years ago, the Pentagon is allowing the civilians to drive the process.

"We clearly have much more of a drive to align DOD technologies with those in the commercial world for a number of reasons," Gabriel says. "One is cost, but also a recognition that, especially in electronics and information systems, technology in the commercial world moves very, very fast, and there's a real advantage to being coupled to that."

The United States has reached an impressive speed in facing the worldwide leadership challenge in MEMS technology by leveraging existing chip-making technologies at which America excels. The Japanese and the Europeans for the most part actually construct separate parts for their tiny machines and assemble them using microtools. By adopting the same technology that creates microchips, American researchers create their machines all on one piece of silicon, which adds to stability. By using or adapting existing machinery costs are kept low.

Defense applications of MEMS are many, varied, and eye-popping in the scale of what they can, or will, achieve in terms of scope and savings. Many of these devices are not yet ready for implementation, many are still very crude but the speed of technological advancement in MEMS is such that DARPA and the Department of Defense predict they will be deployed.

• Weapon safing, arming and fusing. Unexploded ordnance is a curse to armies in the field, and to civilian populations once the battle is over. The Department of Defense estimates 5 percent of all munitions fail to explode as planned. During Desert Storm, that would amount to 1,537,476 unexploded bombs and shells littering the field of battle. Thirty Americans were killed and 104 injured in the clean-up process. A shell or missile arms itself in the same way that an auto airbag does, namely as a result of a drastic change in speed tripping a device called an accelerometer. Conventional accelerometers are intricate pieces of machinery which sense the change in speed of the shell when it is fired, and arm the explosive warhead. A tiny MEMS sensor device, occupying 5 percent of the space, can perform the same function more efficiently for 10 percent of the price or less.

• Competent munitions. Most of the ordnance fired by tanks, ships and mortars is dumb; it falls wherever the laws of physics tell it to fall. For example, at a range of thirty miles, in order to achieve a 50 percent probability that a sixty-square-yard target is struck, 110 rounds have to be fired. To achieve a 90 percent probability, 364 rounds have to be fired. Using competent munitions, a cross between a dumb shell and a guided missile, those figures are cut by over 90 percent, which means ordnance requirements can be cut by 90 percent. The ordnance is made competent by adding MEMS inertial-guidance units and controls that use GPS to find the target. Existing stockpiles of ammunition can be retrofitted, which means that for a small outlay on MEMS units, America's future expenditure on ordnance can be dramatically reduced.

• Platform stabilization. Virtually any significant piece of military machinery contains gyroscopes: tanks, gun mounts, missiles, aircraft, tracking antennae, ships, in fact anything that requires stabilizing to optimize performance. A cruise missile is a good example. It contains roughly $1,000 worth of gyroscopes and accelerometers, all of which occupy a significant amount of space inside the missile casing that ideally should be used for high explosive. With just $20 worth of MEMS occupying a tiny fraction of the space the Defense Department receives a missile that is cheaper and delivers more of an explosive punch, which in turn means fewer missiles have to be assigned to each target. Multiply the cost and space savings across all platforms—for instance the UH-60 Black Hawk helicopter contains thirteen separate gyroscopes—and the financial impact on the defense budget is very significant.

• Small-scale navigation. Despite its impact on all aspects of war fighting, GPS has its limitations. Soldiers carrying GPS receivers know where they are, but the practicality of these devices is limited: at several hundred dollars each they are expensive, the batteries last only four hours at most and they can be blinded—to provide a proper 3-D positioning, each receiver has to have line of sight with four GPS satellites, so when operating in buildings or in dense wooded areas, the GPS receiver becomes useless unless the soldier finds open space to receive a signal, thus potentially exposing himself to the enemy. MEMS can be used to build a very small unit costing only $50 that uses gyroscopes to extrapolate the soldier's current position from the last input of GPS received. Until now, very few soldiers have carried GPS receivers because of the cost and their variable performance in built-up areas. When the MEMS version is produced, every soldier will be able to have GPS built into his belt buckle.

• Maintenance. The U.S. military currently maintains its equipment on the basis of how long an individual part has been in service. This is called Time-Based Maintenance, and takes no account whatsoever of the condition of the part. It sounds wasteful, but it is necessary. The risk of an aircraft falling from the sky because a part has become too old and stressed is just too great. Introducing tiny MEMS sensors into vital pieces of machinery means future maintenance need only be carried out on an as-needed basis, known as Condition-Based Maintenance. Technicians will be able to get instant readouts of machinery's condition just by asking the sensors. A study showed that the application of Condition-Based Maintenance to the H-46 helicopter fleet would save $60 million in maintenance, cut downtime by 50 percent and reduce fatal accidents by 30 percent. With sensors required for monitoring transmissions, engines, cooling systems, bearings, joints, shafts, structures and tires, the Defense Department estimates that it will need to order 50 million MEMS across the board. The cost would be $500 million, but as the H-46 example demonstrates the money would swiftly be recouped.

• Motion sensors. Sensors are used to detect an enemy's advance or to guard a military installation. Limitations currently include size and cost. Using MEMS changes the picture radically. Sensors disguised as blades of grass or bits of dirt can be scattered by the millions across the battlefield or across enemy positions. With tiny little radio transmitters or satellite connections these sensors can gather vast amounts of data about the enemy. The challenge will be to develop a system that can actually make sense of all the data coming in.

• Threat analysis. At the start of Desert Storm, there were no sensors available to detect the presence of biological agents in the air. Someone had to be infected first, a particularly expensive and unpleasant method of analysis. Equipment that could detect chemical and biological agents was swiftly pressed into service, but it was heavy, expensive and vulnerable to heat. MEMS make it possible to build very small $25 sensors that have the same capability as a laboratory spectrometer costing $17,000. This makes it practical to build MEMS sensors capable of detecting nuclear, biological or chemical (NBC) threats right into gas masks. Similar devices can be used to detect drugs or the presence of NBC threats in civilian areas. The sarin attack at the Tokyo subway station demonstrates that the need is real.

• Identify Friend or Foe (IFF). One quarter of U.S. fatalities during Desert Storm were from friendly fire, tragic accidents that happen in the heat of battle. MEMS can be used to create tiny IFF systems that are built into the surfaces of friendly fighting vehicles or into uniforms or pieces of equipment. These sensors scattered all over a tank, for instance, would instantly respond to electronic interrogation from other units in the battle space and thus, perhaps, prevent a tragedy.

• Biomedical sensors. The soldier of the twenty-first century will have a personal status monitor, created with MEMS, embedded in his body that will give an instant readout on his condition. Most deaths in combat occur within the first hour of the injury, so medics need to know instantly who to treat first. Such a monitor will give them immediate information on vital signs such as heart rate, blood pressure, oxygenation and respiration. MEMS are already being used in commercially available home blood pressure monitors. The use of MEMS in medical science is growing quickly as their versatility as sensors is exploited. Micromachines that can sense, decide and act for themselves make other significant medical applications possible, such as drug delivery in patients with chronic conditions such as diabetes. At present, such patients take insulin in predetermined amounts, no matter what their actual blood sugar levels are. If the sensor could gauge blood sugar, and the computer could instruct the actuator to deliver just the right remedial dose, the need for self-injection would disappear, as would reactive swings caused by inaccurate measures of drug. The same process could be used in

intensive care units where fine judgments about the amount of drug to
be administered have to be made.

• Data storage. As computer users will recognize, it takes ever more
space on one's hard drive to accommodate complex files such as graphics. As military systems like Applique and Land Warrior increase in
sophistication, they will need to store and transmit increasingly large
files containing 3-D digital maps, photographs, databases, even video.
There is nothing currently available that could handle all the potential
data that's required in portable form. Data storage devices made with
MEMS technology can hold as much as 100,000 times the capacity
of a CD-ROM.

• Displays. With the increased mass of information making its way
to the war fighter there has to be a corresponding increase in the efficiency of the displays used to show that data, whether it is in headquarters or in the helmet eyepiece of the individual soldier. MEMS are
integral to revolutionary new displays in which millions of individually
controllable micromirrors offer razor-sharp definition, whether on large
screens or mini 0.5 inch or 1.0 inch displays. This is technology that
replaces the cathode ray tube and liquid crystal displays (LCD) and
makes possible an array of flat-screen displays with brilliant video
quality.

These devices all offer enormous savings to the U.S. taxpayer and
dramatic improvements in efficiency to the U.S. military. But just as we
have become accustomed to the presence of computer chips in items of
quotidian familiarity, such as thermostats, vacuum cleaners or toasters,
so will we become accustomed to the miniature miracles MEMS perform, in domestic as well as military life.

But there is one area where it will be hard to ignore the impact
MEMS will have and that is in flight. MEMS have the capacity to
change just about everything we have come to take for granted about
flying.

There are two extraordinary things worth noting about the development of these new systems. The first is that the drive for change is
coming not from inside the Pentagon but from industry, which is bent
on exploiting the microtechnology revolution for its own purposes. His-
torically, defense projects were often lauded for their spinoff potential
for the private sector. Today, that situation is reversed and it is industry
that is driving the pace of change.

That new process brings with it a separate challenge for a procurement bureaucracy used to operating in a time frame measured in decades. Professor Ho's flying wing flew in prototype fifteen months after
he thought of the idea. A new Air Force fighter might take fifteen years

from completion to flight. This new pace will force the Pentagon and, eventually, every armed force to rethink how it does business. Incorporating these new systems into the military structure will require not only new procurement methods but new training and doctrine to ensure that the military keeps pace with the opportunities presented by the civilian world.

Set Tennis Balls to Stun

T H E echoes of Somalia reverberated around the U.S. armed services. If the analysts were correct in stating that America would be drawn into an increasing number of peacekeeping and peace-enforcing missions in the world's hot spots, how could the errors of Somalia be avoided in future confrontations? What had really struck home was the sight of women and children mixed in with mobs of armed Somali militia who could not care less about endangering the lives of noncombatants in their bid to attack the U.S. forces. When the Americans had been forced to defend themselves against these mobs with rifle fire or grenades they wound up killing women and children by the score. Never mind that the responsibility for putting them in harm's way lay with the leaders of the Somali militia, the fingers on the triggers were American, and if the Somali militia could do it with a clear conscience the Americans could not.

U.S. forces got a second chance in Somalia in 1995. President Clinton had withdrawn America from involvement in the efforts to rebuild the country in March 1994. By 1995 Somalia was sliding further into anarchy as infighting between warlords made the U.N.'s task of creating a working government impossible. The country's infrastructure was in ruins, electric power was intermittent or nonexistent; when a patient resolutely refuses to help itself, not even the most conscientious of doctors can help. On February 28, 1995, a multinational force led by U.S. Marines from Camp Pendleton, California, went into Somalia to effect the withdrawal of 2,400 Pakistani and Bangladeshi troops, the rear guard of the dwindling U.N. force. With twenty-three ships of the U.N. Task Force standing off the coast, 2,500 Marines and 500 Italian soldiers landed in Somalia to complete the final chapter of this perplexing episode in the history of the late twentieth century.

The Marine Corps leadership knew only too well how badly things had gone for the American forces in the fall of 1993. When it became clear that the Marines would be called on to lead the evacuation, Marine Lt. Gen. Tony Zinni started to look for ways in which the mission could be accomplished without leaving large numbers of innocents dead. He called upon the Marine Corps Experimental Unit at Quantico, Virginia, for help in putting together a package of nonlethal weapons (NLW) to take to Somalia. The possibility of using nonlethals had actually been brought to the Marine Corps's attention by members of the Los Angeles Police Department who were also reservists in the Marine Corps. The police had more experience using riot control techniques than the military and their argument was persuasive enough to convince General Zinni to call for nonlethal capability in the Somalia endgame, dubbed Operation United Shield. This was a pivotal moment in the development of NLW. Until then, NLW had been an adjunct to military and civilian arsenals as a tool for dealing with riots. Rubber bullets, water cannon, tear gas were all familiar weapons through the 1960s and 1970s as demonstrations became the foremost expression of popular feeling throughout the world. In developed democracies governments do not shoot their people in the course of a demonstration unless the life of another person or a soldier or police officer is directly threatened. Weapons that enabled police and military to disperse crowds without killing anyone were therefore highly desirable.

In the United States the requirement for these weapons was almost exclusively in the civilian sector, which was where the demonstrations were taking place. Military use of nonlethal weapons was confined to special forces operations where stun grenades and brilliant flash grenades had been used to disorient terrorists and hostage-takers in the course of a rescue operation. This use of nonlethal technology was more often than not followed by the application of extremely lethal measures that rather undermined any claim the stun-and-flash grenades could have to being nonlethal weapons. As the Cold War ended and the focus of the U.S. military's mission began to change from the application of mass in formalized battle to the more complex task of policing regional conflicts like Somalia, so did the need develop for a different kind of arsenal in the military sector. Some thinkers and strategists categorize all IW weapons as nonlethal while others consider most nonlethal weapons as IW because they are based on or use information technologies. Both arguments have merit and it is clear that nonlethal weapons have come into vogue because the Information Age makes many of them possible and because there is now a need to find ways of waging a different kind of war.

Maj. Gene Apicella of the U.S. Marine Corps was a company commander during Operation United Shield. He is a typical leatherneck, although the menace implied by the thin stripe of hair in the middle of

THE NEXT WORLD WAR

his otherwise shaved head is diluted by the rimless glasses he wears, creating an incongruous mix of studiousness and threat. Sitting at a table in the Marine Corps Commandant's Warfighting Lab at Quantico surrounded by bean bags, rubber bombs, stingball grenades and spikes for puncturing tires (called Caltrops), Apicella said the time was right for the military to develop a proper NLW capability. His reflection on the role of the Marines in Somalia sums up why General Zinni's instincts in pressing for NLW capability were correct.

"Even before this there was an understanding that you don't want to kill more people than you have to," he said. "You don't need any more enemies than you already have. Every person that dies leaves behind someone who loves them. Therefore by killing someone you have made more enemies. At the same time you have to accomplish your mission. How do you keep a checkpoint open when you have women and children surrounding you and blocking your way? They won't respond to verbal commands, and there are many more of them than there are of you. How do you impose your will? Lethality is neither needed nor justified. You have to allow those Marines who have people attempting to disrupt their mission to accomplish it without violating the rules of war. Some of these NL capabilities allow us to do that."

In the event, the Marines did not have to make much use of the NLW at their disposal. The Somalis had heard in advance of the new weapons being brought in and the manner in which they would be used and there was a significant reduction in the kind of mob attacks that had put such unaccustomed pressure on the U.S. forces in 1993. The relative absence of the mobs deprived the militia fighters of the cover they sought, so they were more exposed to U.S. sniper fire whenever they tried to harry the withdrawal operation. This unintended consequence of deploying NLW was a repeat of what had happened in Haiti the previous year when the commander of the U.S. forces there requested the addition of NLW to help with crowd control. Maj. Gen. Edward G. Anderson, Assistant Deputy Chief of Staff of the Army, reported afterward that the publicity about the deployment of NLW confused the Haitians. "It is generally accepted that the population in the area of operations will have a decent understanding of our rules of engagement," he said. "They know when we are permitted to use deadly force and when its use is prohibited. We believe the introduction of the third option, nonlethal force, was enough to confuse the Haitians' understanding of our rules of engagement." The result was that in the two weeks leading up to the Haitian elections on December 17, 1995, there was minimal trouble on the streets.

When the Pentagon announced that the Marines would be taking NLW to Somalia there was a burst of public interest in this unfamiliar new addition to the military arsenal. The list of items to be taken into Somalia announced at a Pentagon briefing included:

• Thirty-one-inch batons, similar to police nightsticks.

• Various projectiles that can be fired from a 12-gauge shotgun, including hard rubber pellets the size of BB rounds, little rubber bomb-shaped rounds with rear fins for accuracy that deliver a punch to the body, and a bean bag, small steel bearings sewn into canvas packets that deliver a similar punch when hitting the body.

• 37mm rounds fired from the M-203 grenade-launcher that attaches to the M-16 standard-issue rifle. Each round comprises three wooden or rubber cylinders that are fired at the legs of demonstrators producing bruising and discomfort.

• Pepper sprays containing oleoresin capsicum (ORC) derived from the hottest chili peppers. This has replaced tear gas as the irritant of choice in crowd control.

• Foams, either aqueous or sticky. Aqueous foam is water-based and very much akin to soap bubbles except it also contains silicon, which gives it an extra degree of firmness that prevents it blowing away in the wind. One operator working with a mobile transportable dispenser and generator with a 275-gallon water tank can lay down a barrier 200 feet long by 20 feet wide and 4 feet high. When laced with ORC it can be a serious deterrent. Cleanup is easy, using honey or corn syrup. Sticky foam, also known as the "high-tech lasso," literally gums up the target's body making it impossible to run or fight. Cleaning up this foam is much more difficult and requires the use of toxic solvents. It was originally developed to protect sensitive defense installations that might be attacked by terrorists or saboteurs, or trucks transporting nuclear weapons; fixed nozzles would incapacitate the illegal entrants and block their further access.

• Stingball grenades that explode, releasing hard rubber pellets that can also be doused in ORC.

• Tetrahedral metal spikes for disabling vehicles with rubber tires, also known as Caltrops. These were particularly effective when strewn beneath the surface of a foam layer. The Somali militia drove their vehicles straight at the foam expecting to blow through it and instead suffered multiple punctures.

What the senior Pentagon official giving that briefing did not mention was that the Marines would be taking with them the Laser Dazzler, a prototype of a device developed at the USAF's Phillips Laboratory at Kirtland Air Force Base in New Mexico and given to the Marines to field-test. The Dazzler, operational name Saber 203, projects a beam of low-powered laser light about 300 meters and is designed to frighten and deter a would-be aggressor. It is housed in a plastic cylinder that fits into the M-203 grenade launcher (see above) and is operated by a control box that snaps onto the underside of the M-16. The Phillips Laboratory sent a technical adviser to observe the Marines in Somalia and he reported back that the Saber 203 had been used successfully to prevent a confrontation that began when a large armed mob of Somalis approached a U.S. position. The adviser said a Marine had used the Laser Dazzler to "light up" a Somali in the center of the approaching mob. The bright red splash of light pulsed on the man's chest and within seconds the crowd started to disperse, leaving him standing alone, terrified and transfixed by the light. The U.S. force in Bosnia is reported to have prototype Saber 203s with them but no reports of their use have been made.

The success of the Laser Dazzler in that one incident lends weight to the argument that there is a rightful place for NLW in operations where using lethal force is a matter of last resort. General Zinni would later say of NLW: "With non-lethal weapons we can address more situations effectively and have a better chance of controlling the escalation of violence in the complex environments we are most likely to encounter. Our actions thus will be more consistent with the basic humanitarian values embraced by our nation and expected by our citizens."

The Laser Dazzler is a definite step up the technological ladder from the more basic antipersonnel weapons like baton rounds and foam. It represents a new wave of development in the NLW field sparked by the post–Cold War change in the role of the U.S. military. It actually revived an interest that had begun back in the 1960s at a conference convened by the Institute for Defense Analysis. The question to emerge was how to stop fleeing suspects without resorting to shooting them. "Why can't we use sticky stuff?" was the prescient, but at that time unanswered, question. Nothing much happened after that. Tear gas, rubber bullets and water cannon were established fixtures on the riot control front but of little use in day-to-day law enforcement. It was not until 1986, following a Supreme Court ruling in the case of *Tennessee v. Garner* that a policeman had used excessive force by shooting a fleeing suspect, that the government returned to the issue with any seriousness. Attorney General Edwin Meese convened a conference in 1987 that resulted in a report from the National Institute for Justice that recommended research into NLW. The principal avenue for research was some form of chemical dart that would administer a sleeping agent and knock out a suspect. The concept was fraught with difficulties, technical and moral. The drug of choice was a narcotic called fentanyl, but it was also a drug of choice among addicts. Would it be desirable to send police officers onto those same streets carrying quantities of such a favored drug? Then there was the question of dosage. People are of different builds and the dosage that might subdue a 140-pound criminal would not be enough for a 250-pound heavyweight. Increase the standard dose and overdosage might result. Given the numbers of people wandering the streets with all kinds of illicit drugs pumping through their veins anyway there might also be complications from inadvertently creating a dangerous cocktail in someone's body. A second agent was examined called DMSO (dimethyl sulfoxide), which can be absorbed through the skin, and can therefore be applied by firing pad pellets. This eases the moral question about whether a society should be firing darts at citizens (are guns any better?), but does not alter the same technical difficulties that are raised by fentanyl.

The process effectively stalled at this point only to be revived by the military interest in NLW. As the Marines began to understand the strengths and limitations of NLW in combat, policy about how they should be fit into the overall concept of fighting a war began to emerge. Three "prime directives" were issued that made quite clear the limited role of NLW. Gary Anderson, the Marine Corps Experimental Unit Director in Somalia, welcomed the codification of these rules:

> The first is never to put a U.S. serviceman in a position where his life will depend solely on the use of non-lethal weapons and to always back up the use of non-lethal weapons with lethal force.
>
> Second, non-lethal weapons should never be used in situations where lethal force is appropriate. I've fired both assault rifles and sticky glue guns; trust me, the assault rifle will win every time.
>
> A third and final rule is that we should never announce a policy of trying non-lethal force before resorting to lethal means. Our adversaries must always understand that they stand a near risk of death for threatening violence to U.S. personnel in pursuit of their mission objectives.

This was not, as it might seem, an implicit repudiation of NLW by the Marines. Rather it was an important underscoring of the primacy of the individual soldier's safety. It in no way reduced the military's enthusiasm for developing the potential of NLW in areas of operation where it was important for the United States to protect civilian lives and not be dragged into taking sides in civil wars in which it has no interest. At the Commandant's Warfighting Lab in Virginia, Major Apicella and his colleagues work on scenarios for the use of NLW that dovetail with the broader range of U.S. responsibilities around the world. These include the enemy using a hospital roof to mount an antiaircraft battery or rocket launcher, or using a school yard to park tanks. Dropping 500-pound bombs under these circumstances is not an option. Or what if the U.S. mission was to stop two sides from fighting without escalating the conflict or getting drawn into it? Would it not be desirable to have weapons that could disable a truck convoy without destroying it or cripple aircraft without shooting them down?

With this in mind, government-funded laboratories started to develop more sophisticated agents for use against people and machines. Somalia remained the model for the kind of challenge the U.S. military would have to face, but applications against other hostile armed enemies also started to creep into the thinking, and began to resemble the kinds of weapons only dreamed of by adventure and science fiction writers:

• Sleep agents. The idea of chemical darts tipped with fentanyl was revived. Introducing DMSO (dimethyl sulfoxide) into air-conditioning systems, or mixing it with a stimulant that would shorten DMSO's effect and delivering it as a binary grenade or shell, would knock out large numbers of people for a short time. These remain at the idea stage only with no move yet into production.

• Strobe lights. Brightly flashing colored lights tuned to the correct frequency can disorient and nauseate an enemy and even cause epileptic fits. The Lawrence Livermore Labs have actually built and demonstrated a prototype of a strobe weapon using white light. Users would wear specially designed goggles that are tuned to the frequency of the strobe so their vision would not be impaired.

• Dazzling lights and lasers. A variety of weapons that dazzle or blind an enemy have been considered and some deployed, such as the Laser Dazzler mentioned above. The British Royal Navy used lasers during the Falklands War to disorient Argentine pilots. The United States has built a weapon called the Stingray Electro-Optical Countermeasures System that blinds enemy sensors, and is working on a system called COBRA, which is a laser rifle for use against electronic and optical sensors. It could also be used against human eyes.

• Liquid stun gun. The electrified water cannon is a step up from the powerful hoses traditionally used in controlling riots and demonstrations. This version uses salt water to carry an electric charge up to twenty-five feet, to either knock out or kill. The use of electricity to subdue an opponent is familiar through the use by police of tasers, weapons that fire two hooks attached to lengths of wire. The hooks stick to the target's clothes; electric charges are transmitted down the wire to bring the target under control. Water has the advantage over tasers in that it is easier to soak someone than it is to score a direct hit with the hooks. The disadvantage is that there has to be a substantial amount of water available, which means this kind of weapon would probably only be useful if integrated into some kind of vehicle.

"Set phasers to stun" is a phrase that has entered popular culture thanks to *Star Trek*. Knowing how to create a weapon that has variable levels of power is something else. The U.S. government has asked for research to be done on selectively lethal/nonlethal weapons. The only current option in changing a single weapon from lethal to nonlethal is to change the projectile being fired, from say a 12-gauge round to a bean bag. As the British have discovered in Northern Ireland, current nonlethal projectiles such as rubber bullets can be lethal if fired at too short a range. So variability only works if the user, or the weapon, knows the distance to the target. Possibilities begin to open up if laser range finders are added to the equation.

The Lawrence Livermore Laboratories are working on a tennis ball cannon that varies its power based on the location of the target, not on the nature of the projectile. But what to fire? "Set tennis balls to stun" is not a serious proposition.

Sound emerges as one of the most likely media for such a weapon. Acoustic weapons have been around since Joshua's trumpets blew down the walls of Jericho. Sound has customarily been used as a psychological weapon to frighten the enemy, as was the case with Scottish bagpipers marching into battle. Heavy rock music was used to irritate Gen. Manuel Noriega when he was holed up in Panama City and it was used to disguise the sounds of excavation beneath the Japanese embassy in Peru during preparations for the liberation of hostages held by Tupac Amaru terrorists.

But sound can also be used to attack the body. Subsonic waves of between 1 to 3 Hz (well below the human hearing threshold, which is between 20 Hz and 20,000 Hz) actually cause certain organs to resonate and vibrate, causing extreme nausea, vomiting and loss of bowel control. Similar frequencies could shatter walls or windows. Acoustic guns capable of shooting a bird out of a tree have reportedly been developed. Jon Becker, one of the NLW experts who helped the Marines organize for United Shield, adds that the subsonic waves could be tuned to the frequency of individual organs.

"You can make someone's guts shake from the inside or their hearts explode. But how to control it? How to narrow it down?" he wonders.

The difficulty is in creating a sound wave that is not omnidirectional. While the concept of deploying a device that has a mob of angry, stone-throwing demonstrators falling instantly to the ground clutching their stomachs is not unattractive, the sound waves would also affect friendly forces. Ear protectors will do nothing to impede the waves, as the bowel-irrupting sounds act directly on the body, not the ear. Finding a way of directing the waves is, as Becker points out, an important step toward making an acoustic gun a reality. He talks of concepts that have yet to be put into practice, such as "the donut of sound," a pulsed wave that could deliver a serious punch to the body, and brings to mind images of science fiction ray guns firing similarly shaped rays at their targets. Los Alamos National Laboratory has proposed an acoustic weapon for protecting sensitive defense installations from intruders. It would be tripped into action by the presence of the intruder and thus would not require an operator.

The Russians are reported to have a very advanced acoustic weapon program, according to Janet and Chris Morris, the noted husband-and-wife defense consulting team, who visited the Center for Testing of Devices with Non-Lethal Effects on Humans in Moscow in 1994.

"We saw a ten hertz acoustic generator which could be used to deliver a pulse about the size of a baseball that could knock you down or more, depending on how you power it," Janet Morris said. The fact that the Morrises witnessed a Russian variable-power acoustic weapon in 1994 makes it extremely probable that the American military has something even more sophisticated but is keeping it under wraps. It is not unreasonable to suggest that in the shadows scientists are working on devices several orders of magnitude more sophisticated than what the public has been allowed to see.

For instance, one inventor alleges that the military has a black project under way that will result in a weapon that will send its victims to sleep; in effect it will stun them. Eldon Byrd headed the Marine Corps Non-Lethal Electromagnetic Weapons study between 1980 and 1983, experimenting with extremely low-frequency electromagnetic radiation. He found he could actually manipulate the brain to produce behavior-altering chemicals. Byrd said that these waves would send test animals to sleep or provoke their brains to produce histamines, which in humans would produce instant feelings of sickness and flu. The program was closed down early because, Byrd said, it was too successful. "The work was really outstanding," he said. He suspects the project went black and was fully developed. He cites the emergence in 1995 of a project that called for using acoustics, microwaves and brainwave manipulation to send the enemy to sleep as evidence that his work was removed from view and thrived undercover.

NLW are very attractive as a means of attacking the military infra-structure of either an enemy or a combatant in a civil war that the

United States is trying to stop. The ideas may seem exotic but they are technically feasible. Liquid metal embrittlement involves spraying, brushing or splashing liquids onto enemy bridges, storage tanks or vehicles. The spray contains metals that are liquid at or near room temperature, such as mercury, cesium, gallium, rubidium or the alloy InGa, made of indium and gallium. Once applied to the metal surface the liquid starts to form alloys with the metal under attack, creating a material that is much weaker than the original. Eventually the enemy's infrastructure starts to fall apart. Superacids also achieve the same results. These are combinations such as hydrochloric and nitric acids, which are much more corrosive than their constituent parts. An artillery shell that combines the two acids on the point of impact can deliver a superacid jelly to an enemy structure that would eat away his capacity to fight. Superacids can also eat rubber, asphalt, concrete, glass and fiber optics, which means not only can vehicle tires be rotted away but the very road on which they are traveling can, too. Other substances that attack an enemy's ability to operate include firing shells containing specialized paint that renders glass totally opaque, destroying windshields and sensors. Weapons firing very fine dust, the "grime from hell" that creates tiny microcraters on glass with the same effect as sandblasting, are currently being developed.

Another group of weapons uses chemicals to impede an enemy's mobility. Superlubricants known as "slickums" delivered as bombs would coat roads, runways, ramps, railroad tracks, even stairs and equipment with hard, clear coatings that allow virtually no traction to wheels, tracks or feet. The coating is extremely difficult to remove so the enemy's war-fighting ability slides to a halt.

"Stickums" have the opposite effect. These are polymer adhesives that do precisely as their name suggests: they trap vehicles and pedestrians like flies on flypaper. The stickums can be countered with sand or earth, but if the enemy is not prepared for it, the surprise effect could be considerable.

CAT, or Combustion Alteration Technology, actually gets at the source of the enemy's mobility, his fuel supply. CAT is the name given to the range of biological agents which alter the composition of gasoline, either by changing it into a thick gel or by actually degrading and digesting it. These products have filtered down from research in the civilian energy industry and are now being examined for use in military settings. The obvious targets are fuel dumps and storage tanks, which could be fouled with these agents by special forces on clandestine operations. Given the high concentration of these agents it would not take much to affect very large quantities of fuel, yet the practicality of using them would appear limited. Getting access to enemy fuel storage would under most circumstances appear too hazardous a mission for even the most skilled special forces.

At the Commandant's Warfighting Lab, Major Apicella has experimented with delivering NLW by a remote-controlled expendable UAV known as an Exdrone. This short-range craft has a bomb bay and a chaff or flare dispenser, which gives it the capability of dropping stingball grenades, shrieking noisemakers, tear gas or any other agent required. It also has video cameras front and back to provide the remote controller with targeting data. Given the highly experimental nature of some of the chemical NLW under consideration the opportunity to test the Exdrone as a meaningful tactical delivery system has not arisen, but it would seem a sensible option for delivering limited quantities of embrittlement agents or superacids.

Surprisingly, given the relatively benign nature of NLW, when compared with the deadliness of conventional weapons, there are a number of international conventions that outlaw their use, and in some cases possession, particularly those NLW which use chemical reactions to achieve their purpose, such as acids, slickums and stickums, as they could have long-lasting serious effects on the environment. Their use by military personnel in the course of peacekeeping, as opposed to conflict, is also potentially in contravention of U.S. Environmental Protection Agency (EPA) and Occupational Safety and Health Administration (OSHA) regulations, and may require the President to exempt the forces from penalty under the laws. These issues have yet to be tested, given that the more noxious chemicals referred to have not so far been used in theater.

Antipersonnel NLW may also be subject to regulation by international law. The area is murky because the relevant laws pertain to warfare whereas under most circumstances NLW are designed for peacekeeping purposes. The Chemical Weapons Convention would make any riot control agents such as sleeping gas illegal as a "method of warfare" without defining that phrase further. This has obliged the Clinton administration to reserve the right to use these weapons to quell civil disturbance during peacekeeping operations or during war in civilian areas outside the zone of combat.

Given the proliferation of conventional armaments, all of which have the capacity to cause hideous injuries and death to combatants, the international community seems to have spent a great deal of energy making it difficult to use less lethal alternatives. For instance, the International Red Cross is to seek a ban on the use of lasers to blind the enemy. It was a setback to the idea of using weapons that stopped short of lethality.

"People had a lot of trouble with the idea of maiming other people," Jon Becker says. "So it's okay to blind someone with a bullet to the head but not with a beam?" Major Apicella of the Marine Corps agrees with Becker that the public needs to be persuaded of the acceptability of some NLW. He cites the congressional debate during World War II on using an air burst of the A-bomb to flash-blind the Japanese. Congress rejected the idea. The consequences of the alternative became all too obvious at Hiroshima and Nagasaki. The PR problem also arises when the use of some of these more advanced weapons is contemplated in civilian law enforcement. With TV cameras recording every moment of a demonstration, it might not be politically desirable for a police force to use a weapon that has crowds of citizens throwing up uncontrollably.

The third class of NLW, electromagnetic weaponry, poses few of the international legal problems attached to the antipersonnel and antimaté-riel weapons. Pooh Bah's Party, the cruise missile attack that shorted out Baghdad's power grid in the early hours of the Gulf War, was a classic use of nonlethal means to achieve a military result. Electronic circuits are extraordinarily sensitive to disruption. A power surge can destroy a circuit board in an ordinary domestic computer unless that computer has been "hardened" by the introduction of a surge suppressor into the circuit. Nowadays there are circuit boards in just about every device used in the domestic and military worlds. The more sophisticated the device, the more susceptible it is to disruption, and it might not need a power surge to do the damage.

As one horrific incident demonstrated in 1967, stray radio frequency energy can send inadvertent signals to electronically controlled weapons with catastrophic results. On a July day that year the USS *Forrestal* was on patrol in the seas off North Vietnam. Aircraft were on deck, armed and ready to take off, when an onboard radar made a pass over the ship. A connector on one aircraft's missiles was not properly shielded and so a missile launched when the radar's radio frequency energy hit it. The missile struck another plane, causing one explosion after another as plane after plane caught fire. The disaster cost the lives of 134 men. Inadvertent radio frequency energy affecting aircraft communications and guidance systems is also the reason why certain kinds of electronic devices are not permitted to be used on civil aircraft.

Electromagnetic weaponry takes this principle of interference a major step further. Along with their more radical consequences, nuclear explo-sions also create a wave of energy called an electromagnetic pulse (EMP) that "fries" circuit boards and destroys any electronic equipment in its path. This is an immensely useful by-product of nuclear fission, as in the modern age an aggressor capable of unleashing an EMP could disable an enemy's entire electronic infrastructure. The challenge to scientists was to replicate the EMP effect without having to set off a nuclear device. This they achieved.

Nowadays the pulse that can be created is several times more power-ful than that of a nuclear explosion's EMP. If a nuclear EMP was gener-ally in the 1 to 100 MHz range, the nonnuclear EMP (NN-EMP) is in the 0.5 to 100 GHz range. This is in the microwave region and the effect of such a blast on an enemy installation such as a radar site would be to destroy the circuit boards and probably induce tremendous heat as well. Scientists testing the creation of a highly directional NN-EMP with conventional explosives found they had a winner on their hands when cars in the staff parking lot 300 yards away had been disabled by the pulse.

The Air Force is developing a cruise missile to deliver electronic knockouts on a scale even greater than what was directed at Baghdad. This technology can disable an enemy without the need to drop conventional explosives in areas where noncombatants are located, an excellent example of how NLW can serve a military objective in a humane fashion.

Until the summer of 1996, it was thought that America led the way in the development of small EMP-type devices. Then a Russian brochure advertising the sale of a small device that could be used to destroy communications or computers was obtained by American intelligence. To the consternation of the analysts, the brochure not only had illustrations of what appeared to be terrorists throwing a new type of grenade into a room full of computer terminals but it also advertised a weapon that had never been seen in the West before.

"As far as we knew, there was nothing like this available anywhere," said one intelligence official. "Now we learned that a new and apparently powerful device had been developed by the Russians and that they were prepared to sell it to anyone, including terrorists."

American intelligence gave the new Russian weapon, which has never been revealed before, the code names Beer Can or Six of Clubs because the device was a beer can and had markings on the side similar to that of the playing card. Responsibility for investigating further fell to the Springfield Research Facility of the Defense Nuclear Agency and on May 21, 1996, Gerry Carp, a senior official at the agency, prepared a detailed briefing for his colleagues and the intelligence community.

In a slide presentation headed "Targets of Radio Frequency Weapons," Carp said that the secret Russian program called for the Beer Can to be used against radar (air defense, field artillery, guidance, Army aviation), all types of radio stations, radio electronic, opto-electronic and television devices, radio electronic equipment, other means of communication including warheads of all types of missiles and smart ammunition and influence mines. Included in the briefing were three drawings taken from the Russian sales brochure that showed terrorists attacking a computer center, an attack on an airfield and on a satellite ground station.

According to the briefing, the Russians had developed a grenade, an artillery shell and a specially converted briefcase to deliver the device, which would allow the user to "dial up" the kind of frequency required to destroy a particular target.

"If validated, Beer Cans pose a real terrorist threat," said Carp.

A special team that included people from the Defense Research Agencies traveled to Russia in the summer of 1996 and managed to obtain two Six of Clubs grenades, which were brought back to the United States for testing. To the surprise of the American scientists, the systems worked as advertised, giving Russia a significant advantage in the miniaturization of this technology, which had an effective range of around one kilometer. What is of even more concern is that American intelligence has received reports that the sales brochure produced the desired effect and several of the weapons are believed to be in the hands of other countries.

"What we seem to have here is the Russians making something really dangerous and then simply selling it on to the highest bidder," said one intelligence official. "We are now faced with the problem of this reaching a terrorist. Imagine the havoc something like that could create on Wall Street or to air traffic control at Dulles or Kennedy."

The indications are that the skills and means to make similar devices are already out on the hacker market in the United States. Hackers claim they can make a HERF (High Energy Radio Frequency) gun for $300. The following was a conversation between some of America's top hackers and *Forbes ASAP*.

> Dark Tangent: A rucksack full of car batteries, a microcapacitor and a directional antenna. . . . You could park in a car and walk away. It's a poor man's nuke. . . . There are only three or four people who know how to build them. . . . If you experiment wrong, you've [micro]waved yourself.
>
> Dune: Yeah, this is a high-energy device. You could be a half mile away and take out Oracle [computers].
>
> Dark Tangent: One pulse [would wipe out Oracle]. It dumps 2 million watts in one-thousandth of a second.
>
> Dune: If we had a Cessna and a HERF gun, you could fly over Silicon Valley and—POW!—there goes Sun Microsystems—POW!—there goes Intel!

As the hacker Dark Tangent intimated, NN-EMP is not entirely harmless. Given its microwave frequency it would have the capacity to cook a human body directly in its path. As with other of the more exotic antipersonnel NLW this makes such a weapon "inhumane" under the Certain Conventional Weapons Convention.

As Xavier Maruyama argues, the 1977 Geneva Protocol stipulates that in warfare, attacks must be predicated on creating minimal loss of life: "a view to avoiding, and in any event minimizing, incidental loss of civilian life." The next article of the protocol goes further in saying that if an attacker has a choice of two equal targets he must choose that which is expected to cause the fewest civilian casualties. If one takes as a given that all war is ugly and that people are always going to get killed, any weapon that can make war as short as possible while minimizing casualties deserves support rather than opprobrium. This contradiction between the purpose of this technology and the intent of international law also applies in the case of laser-blinding technology. The law seeks to ban it, yet in a war zone like Sarajevo where snipers have wrought such death and destruction on civilian populations surely the sniper-blinding technology currently in development has the moral edge over the activities of the concealed killers. The Lawrence Livermore Laboratories have developed a system for detecting and responding to sniper fire called Lifeguard. Using infrared detectors the Lifeguard computer can detect and locate a sniper even before the sniper's bullet reaches its target, and can respond automatically with either its own deadly force, a blinding laser or dazzling light. This technology is called, with some grim humor, Deadeye.

The burst of activity in NLW development in the military sector has necessarily fed back into civilian law enforcement, where little of substance had been done until the military began to drive the process where it had its original roots. Research and development of civilian weaponry is a complicated affair, given the large numbers of police forces and investigative agencies spread across America. The National Institute for Justice is the government body which acts as the think tank for the Department of Justice and is based in Washington, D.C. The institute listens to what police forces are looking for and then commissions R&D to be carried out by commercial companies.

Police are very conservative when it comes to using weapons other than the standard-issue handgun. Their thinking reflects that of the Marines as stated earlier in that the life of the peace officer is paramount and relying on anything less than lethal force to protect it is a sure way of getting someone killed. Yet there is an obvious requirement for different technologies to meet a variety of circumstances. But needs have to be set against the reality that in war dead and wounded are expected and acceptable. In police operations, if there are dead and wounded there can be serious political consequences. That may be why prison authorities tend to be against the use of foams in quelling riots. If sticky foam sealed a prisoner's airways and killed him the consequences would clearly be serious. What if a civilian demonstrator inhaled some aqueous foam and got pneumonia? Acoustic weaponry is treated with similar concern, unless it can be deployed in such a way as to create a barrier and thus give rioters a choice between going home and receiving large amounts of discomfort. The hangover from the 1960s when police indiscriminately used gas and billy clubs to subdue rioters is with us today. The public revolts against such state violence, particularly when it can affect many innocents.

With the rise in the use of illegal drugs and the startling effects some drugs can have on the human body, much of the civilian emphasis has been on finding new and more effective ways to capture and restrain an offender. PCP, or angel dust, for instance, can give the user superhuman strength, which can repel the efforts of several officers. This led several forces to adopt the taser, which brings the suspect down with a series of sharp electric shocks. New technologies now present real alternative to the "short, sharp shock" approach. One of the most intriguing technologies involves using nets to entangle a fleeing or approaching suspect. There are two types of net under scrutiny. One is a net fired directly from a special gun that encloses the suspect and traps him, the other, a grenade fitted with a proximity fuse that goes off directly over the suspect's head and drops a net.

Once a prisoner has been arrested, the next problem facing the police officer is taking that prisoner back to headquarters for questioning. In many parts of America budget cuts have resulted in police officers on patrol without partners. Normally this means a suspect is handcuffed and placed in the back of a police car. The law states that a prisoner may not be handcuffed to the vehicle in case it crashes, nor may the rear doors be locked (although internal door handles are removed). On one terrible occasion in Phoenix a handcuffed suspect started kicking out the doors of the car, forcing the officer to pull over. As the officer opened the door the suspect kicked it hard, knocking the officer to the ground. The suspect grabbed the officer's gun and the policeman ran away. The suspect then hijacked a motorbike and its rider and rode off, shooting and killing the occupant of another police car as they rode by. When the gun was empty, the rider escaped, leaving the suspect brandishing the useless weapon at the police closing in on him. They all opened fire and he was killed. A new device now coming onto the market would have prevented that whole tragic scenario. When the suspect first started protesting, the lone officer would simply have had to press a button firing a giant rear-seat airbag that would trap the suspect in a firm pneumatic embrace.

The closest that civilian law enforcement is coming to the military NLW is in the use of light weapons to dazzle and disorient suspects. The National Institute for Justice is cooperating with DARPA in developing a handheld laser that can deliver a cylinder of light (as opposed to the conventional cone shape) that would incapacitate someone long enough for an arrest to be made. The civilians are also looking at small-scale EMP technology to disable fleeing cars. Several companies have created prototypes, all of them designed to destroy the internal circuitry that controls a car's operations. The most promising appears to be a device created by Jaycor in California in which a police vehicle fires a small sledlike device under the car that it is pursuing. The sled is then triggered, firing a mass of high-voltage wires into the car's engine, shorting out everything and bringing the fleeing car to a halt.

What this proliferation of new systems and processes produces is not the silver bullet that will bring peace to the world, but the opportunity for civilian and military leaders to take action that might satisfy the demands of the public for firm leadership while at the same time avoiding the possibility of loss of life. It is a seductive prospect. At a recent course for senior commanders studying information warfare at the National Defense University at Fort McNair outside Washington, D.C., a panel of senior generals and admirals described to the class the problems they confront today. All were agreed that committing forces to war is a tough task when neither the military nor the political leadership is allowed the latitude every other leader in history has been given—to make mistakes.

The consensus of the panel was that the leadership must be permitted to make mistakes and learn from them. That way, they believed, there would be enough courage to go around to commit force when it was needed. That this was seriously suggested as a realistic solution to the lack of experience in the senior political and military leadership is laugh-able. There is nothing that is going to change in the years ahead that will allow the media and thus the public to be forgiving of mistakes. On the contrary, the demand for perfection is going to be the order of the day and failure by any measure will not be forgiven.

NLW combined with IW thus becomes a seductive and realistic alter-native to conventional military and civilian approaches to violent action. What President, offered the alternative of committing ground troops with the possibility of high casualties or using NLW and IW that allows for troops to stay out of the line of fire, would use the former and not the latter?

It is this central fact that the American military above all other armed forces in the world has begun to understand. The recognition of reality is partly driven by the Pentagon's highly honed instincts for survival and partly by a genuine belief among some that IW and, to a lesser extent, NLW, are going to be central to the wars of the future. Individual ser-vices sense that in IW there will be funding, status and the assurance of political support for their continued existence. At the same time, as the word of the power of IW has spread through the defense community, so the number of true believers has grown. Today, every service is fully engaged in integrating IW into the strategy and the force structure. Quietly, billions of dollars have been allocated to black programs to ensure that America retains its current technological lead, and resources are being devoted to the problem with a focus that has not been seen since the beginning of the nuclear arms race.

But offense is only part of the equation. As America becomes more and more dependent on the technology, so the country becomes more vulnerable to that same technology being used against its economic and defense infrastructure. This is one of the dangerous paradoxes of IW. The better you are *at* attack, the more vulnerable you are *to* attack. That explains why so much secrecy surrounds IW in America today and why there is a genuine fear that America will be uniquely vulnerable in this new warfare.

ELEVEN

The Wrong Hands

A cultural chasm yawns between the buttoned-down, sober world of federal agents, and the freewheeling anarchy of the brilliant, twisted and destructive hacking community. Knowing how to hack is like owning the keys to the candy store. Commerce today exists as a series of digital ones and zeroes pulsing across intricate networks of wire and cable. Anyone with the perverse talent to manipulate those ones and zeroes to his own ends can create a lot of mischief, and have a lot of fun.

Justin Tanner Petersen, code name Agent Steal, and Kevin Poulsen, code name Dark Dante, were typical of the hacker breed. (Creating dramatic code names is part of the "fun" of being a hacker.) Like any of us who listen to radio station competitions which promise that "the eleventh caller will win the grand prize," they dreamed of being the eleventh caller. Unlike most of us, however, they knew how to make it happen. They broke into a Pacific Bell telephone system switching center and took control of phone calls to the radio stations. That way they could guarantee to be the eleventh caller. It was brilliantly conceived and executed, and totally illegal. In the late 1980s the two of them had snared two Porsches, two trips to Hawaii and a minor fortune in cash. Poulsen ended up serving a prison term for the scam after Petersen testified against him in return for immunity. Petersen was later arrested in Texas on hacking charges and for driving a stolen Porsche across state lines.

And that was when he started working for the government.

The government needed help. A government agency cannot create and nurture an operative with the skills and mind-set of a hacker. The hacker exists on wits, adrenaline, iconoclasm; the government feeds on conservatism, conformity and stability. So when a hacker is causing trouble, perhaps it is understandable that the only person the feds can think of to track the hacker is another hacker. That is why Petersen was promised a sweetheart deal by federal agents who kept him out of a Texas jail cell if he would help them "investigate certain individuals," as the official paperwork puts it. Nobody knows who the "individuals" were, but it was widely believed that Petersen was the bloodhound the feds put on the electronic trail of legendary superhacker Kevin Mitnick (Condor).

"The Feds turned Petersen against Mitnick. He essentially became a bounty hunter for the FBI," a computer security consultant told *Information Week*. Petersen actually admitted his role in hunting Mitnick in an interview with the hacker's underground magazine *Phrack*. "What a loser," he said, arrogance in his voice. "Everyone thinks he is some great hacker. I outsmarted him and busted him." Mitnick was indeed caught and jailed for his hacking exploits, although others would claim a share of the credit for that. Petersen was on the FBI's books for two years, from 1991 to 1993, living in an FBI apartment and using computers it had supplied to track down electronic outlaws like Mitnick.

It was akin to giving a pyromaniac a gallon of gasoline and a box of matches. Later Petersen would say he had been targeted by a hacker who was harassing him and his family electronically because he was working for the government, and that was why he used the government's computers to raise enough money to help him get away from the pressure. Either way, he siphoned $150,000 out of the Heller Financial Bank in Glendale, California, sending it to an anonymous account in another bank, a theft aided by a judiciously timed bomb threat to divert the bank's attention. The feds and their computers had unwittingly aided a crime. The red face quotient was very high. There were plenty of people ready to tell the FBI they had blundered. "[Using Petersen] was about as dangerous a thing as the FBI could do," said M. Lewis Temares, Dean of the College of Engineering at the University of Miami. "He who strikes first will strike second, especially if he only gets a slap on the wrist."

Petersen was caught in 1994 and given more than a slap on the wrist. He was sent to prison, concluding a cautionary episode which demonstrated that turning poachers into gamekeepers is not always the right way for law enforcement to get the job done. In the *Phrack* interview, Petersen described some of the secret information he and Poulsen had found on their famous hack into the headquarters of the Pacific Bell telephone company.

"Very dangerous in the wrong hands," the interviewer commented.

"We are the wrong hands," Petersen replied.

Agent Steal had sold out Dark Dante, just as he ended up tracking and helping bust Condor (Kevin Mitnick). This is not a community whose chief ethos is loyalty.

"Most hackers would have sold out their mother," Petersen once said.

Dark Dante, or Kevin Poulsen, lacked Petersen's cynicism, but was just as brilliant a manipulator of digital information. He had been at it since his teenage years, using a very basic Radio Shack TRS-80 computer to hack into military command centers. He was first arrested in 1983, but was not charged because of his age. He was recruited by Stanford University, which wanted to exploit his talents, and a life of stability beckoned, even leading to Pentagon security clearance for work on some military projects. But the lure of danger was in his soul, and soon he was breaking into phone company buildings, literally as well as electronically, to learn the secret keys to the system. "To be physically inside an office, finding the flaws in the system and what works, was intellectually challenging," he would later recall. "It proved irresistible. It wasn't for ego or money. It was for curiosity. A need for adventure. An intellectual challenge and an adrenaline rush. It was fun. And at the time it seemed pretty harmless."

It is an experience that sounds not dissimilar to someone describing drugs. The judge's order that released Poulsen in 1996 implicitly acknowledged that fact when it was made a condition of his freedom that he not go near a computer for three years. His parents even put their home PC into storage.

He had been exiled from cyberspace.

The roots of the culture that spawned Mitnick, Poulsen, Petersen and legions of other hackers lie in the early 1970s with phone "phreaking"— stealing phone time from the telephone companies by generating tones to bypass the billing system—and developed into full-blown hacking with the advent of the personal computer. It started as something rebellious and antiauthoritarian. Hippie leader Abbie Hoffman teamed up with phone phreaker Al Bell in 1971 to publish the *Youth International Party Line* and declare that "liberation of communications" was a necessary step toward "mass revolt." Yet such phony-sounding appeals to 1960s-style radical chic are not what primarily drove the hacker community that erupted in later years. The majority of those in the netherworld of electronic vandalism most resemble kids who used to build their own pipe bombs in dad's garage. There was never really any point to it, except to see if it could be done, and get something of a thrill (and hopefully no more) in the process.

In the early days, phone phreaking was a fun thing to do, a way to beat the system, save a few bucks and most importantly demonstrate a cool talent. In countries where tones would not be used for many years to come, like Britain, you could tap the hook of a pay phone to emulate the pulse signals created by the rotary dialer and thus make calls without incurring charges. The first hero of the phreaking generation was an American, John Draper, who in 1970 conclusively proved the Law of Unintended Consequences. He discovered that a toy whistle given away with a box of Cap'n Crunch cereal could emulate the 2600 Hz tones required to make long-distance telephone calls, bypassing AT&T's billing system. Under the pseudonym Captain Crunch he became a legend for his exploits, and a well-known figure to police forces, who arrested him several times.

From such beginnings grew an electronic underworld, peopled by young men (mostly, but with some very notable exceptions) with much more talent than social responsibility, nurturing an abstract nihilism that was never for anything but against just about anything connected with business and mainstream life. Phreaking graduated from Cap'n Crunch whistles to "boxes" which contained circuitry that could fool and defraud phone companies when hooked up to the phone system. These boxes can still be built using diagrams easily downloaded from the Internet, or they can be bought by mail order from companies like Consumertronics in Alamogordo, New Mexico. Consumertronics's catalogue offers a vast array of gadgets that have been designed specifically to give their users free rein over the nation's telephone lines. Instruction manuals show how to penetrate a company's PBX telephone system or voice mail system and make free long-distance calls; how to hack into computer systems; even how to eavesdrop on someone's TV or computer screen from outside a building, using what is known as the Williams Van Eck system. Cellular telephony has opened up a whole new area to the phreakers, who have reverse-engineered all the complex electronics involved so that these phones can also be phreaked.

Pick up a copy of the self-described hacker quarterly 2600, named after the 2600 Hz frequency, and you will find instructions for reprogramming cell phones, advertisements for cable boxes that circumvent billing systems to give you free cable TV, and calls for help from hackers in various jails. One advertisement even offered original Cap'n Crunch whistles for sale. The depth and breadth of the information infrastructure in the hacking community demonstrates this is not some idle pastime of bored youth. The telecommunications industry is taking major losses due to the phone phreakers' using their nefarious skills to make calls without paying or by charging someone else's telephone account.

While phreaking remains a serious threat, all of these same skills were also channeled into computer hacking from the moment home computers came on the market. The reason military installations came under attack so frequently was that the Internet was founded on the connection of defense computers with each other and with universities. Companies had their own computer systems but few were attached to the public telephone system and they could not therefore be hacked. So attacking the military was the only game in town. In 1982, in one of the most celebrated hacks ever, Kevin Mitnick broke into the computer at

NORAD, the North American Air Defense Command, the heart and soul of America's early warning system. In the same year, Mitnick invaded and took control of three central phone company offices in Manhattan as well as granting himself control of all the phone switching centers in California. There is an inescapable conclusion. Someone with the power to bend so many computers to his will using rudimentary equipment can do much worse.

Mitnick and others were the very visible elements of a culture that was spreading rapidly through schools, colleges and business as the tools required for hacking became more accessible. Debates began over the "ethics" of hacking, if there can be ethics involved in the electronic equivalent of breaking and entering. Jeff Moss (aka Dark Tangent) is part of an organization called Defcon, which organizes the hacker equivalent of annual conventions. He sees hackers as the electronic society's early warning system, helping the establishment learn about flaws in its systems that it is presumably too stupid to spot itself.

Knowledgeable hackers can provide a broad base of knowledge over many technical subject areas, allowing for the "big picture.". . . Hacking is a passion that requires you to learn all there is in a certain subject area and not just treat the knowledge as part of your 9 to 5 job. It is a non-traditional way of looking at all the information, coming to different conclusions and arriving at solutions that people thought were not possible.

This was the argument that hackers are a national resource who should be used, not abused, by the system. It has its attractions, not the least of which is that hackers have constantly demonstrated they are at least one step ahead of any effort to keep them out. It was no accident that the French secret service, the DGSE, established a bulletin board (anonymously of course) called QSD specifically designed to attract hackers who love to swap tales of derring-do and exchange trade secrets. It was the latter the DGSE wanted to pick up so they could use the techniques for their own no doubt nefarious purposes. The hackers had been hacked. It was not the first time the French government turned the tables on them.

Jean-Bernard Condat was a low-level hacker who played with networks while a music and technology student at Lyons. He was picked up in 1987 by the internal security service DST (Direction de la Surveillance du Territoire) for what was described as a "minor misdemeanour." The DST turned him into a fully fledged "double agent," to be their eyes and ears at hacker meetings and to scan the networks for information about hackers. The DST then made him set up the CCCF, the Computer Chaos Club de France, modeled on the famed Chaos Computer Club of Germany. The DST even went into the promotional business, printing up T-shirts, flyers and postcards to promote the CCCF. It lasted two years, after which Condat found a way to break the relationship. By 1995 he had become fully legitimate, appointed as systems operator for the Compuserve service in France.

The French had used customarily nefarious means to pick the hackers' brains. The hackers would be just as happy to share their knowledge, no doubt for a price, but co-opting their skills has its dangers, as the case of Justin Petersen so clearly shows. The CSI's *Current and Future Danger* pulled together some very cogent arguments as to why hackers should not be used as a defensive tool. The publication quotes noted computer security expert Professor Gene Spafford of Purdue University as saying a hacker, by the very nature of his experience, is not trustworthy.

"Whoever you are hiring has already demonstrated a willingness to flout the law, ignore other people's property and/or willingly intrude where they aren't supposed to be. Does this make them better suited for protecting your systems or data? Are arsonists better at installing fire alarm systems?"

It was a point reinforced by John O'Leary of the CSI. "If you have a billing dispute with a hired hacker, what happens?" he asked. "How do you know what else they have left in your system. . . . Do your customers know that you have hired hackers to try breaking your security and thereby gain access to them? You might be putting yourself in legal trouble with your customers if you have precipitated an incursion that compromises the confidentiality, integrity or availability of their information."

The issue of information integrity raised by O'Leary would become one of the hottest debating points about online activities in the years to come. Can you trust what appears on your screen? Several incidents led sensible people to trust nothing unless it was both branded and secured by a reliable organization. The incident that provoked one of the most interesting reactions involved an e-mail campaign in which people working in the information business received a very menacing, yet credible, posting. It came in late 1993, early 1994 from something called Blacknet, which purported to be a buyer and seller of trade and government secrets, an online espionage service which promised monetary rewards to those who passed on anything from cutting-edge computer chip designs and government secrets to plans for kids' toys and cruise missiles. Rumors and intelligence were also much in demand. With elaborate encryption systems and cutouts in place to protect the identity of those prepared to sell their information, and with an obvious knowledge of which secrets commanded currency in the technology marketplace, the pitch had an edge of credibility that sucked a lot of people in.

It was a hoax, according to some observers, by a hacker named Tim Vey (Cypherpunk), although many at first suspected it was a government sting to trap peddlers of confidential information. In a sense Blacknet, while quite amusing in its own way, backfired on the hacker community, as it gave ammunition to those who wanted to impose controls on the maelstrom of online information as well as to those in business with a vested interest in making life difficult for hackers.

"The concept of Blacknet is real," said Jim Settle, onetime head of the FBI's Computer Crime Squad who went on to work for a Bethesda, Maryland, based information security firm. "If you had something to sell, you'd post it to Blacknet and they'd find customers for it. The capability exists to do that. The problem for law enforcement today is how to investigate this activity. Where is it? In some countries it might not even be illegal." And over at the White House, Mike Nelson, a special assistant on information technology, said soberly, "We're concerned about the overall issue of anonymity in cyberspace. This is an example of how it can be abused."

The real message of Blacknet is *Caveat Lector*—Reader Beware. The credulous will always find ways to make fools of themselves. As people in general become more aware of the dark side of information, they should, and will, educate themselves to be more skeptical.

Blacknet was a sophisticated hoax, but perhaps it was also a reflection of the unease and even bitterness in the hacker community over the fact that their world had changed forever. The increasing availability of sophisticated equipment had led thousands of young people into the game. It was fun to go "trashing"—dumpster diving—outside phone company offices looking for clues to access codes and passwords among the papers thrown out by the companies. It was even more fun to get access to the CO (central office) and break into the main computer controlling phone calls for miles around. They can eavesdrop on phone calls, misdirect them, plant obscene messages on answering services, route international calls through PBXs so someone else picks up the bill, link calls together in a daisy chain that allows a massive transnational conference call to take place without costing any of the participants a dime. It was fun, dangerous, intellectually stimulating and iconoclastic. It was flipping the bird at the establishment yet at the same time rooted in the kind of kids' gang culture that in the old days sought how to make a better and faster hot rod.

Gangs were formed by people with code names like Erik Bloodaxe, Lex Luthor, Acid Phreak and Phiber Optik. Inevitably, turf wars were fought, the most celebrated being that between the Legion of Doom and the Masters of Deception in the early 1990s.

The hacking game was opened up to even more people by the publication of techniques that previously had been known to only a few people. In 1992, two students produced the first *Hacker Chronicles*, a CD-ROM packed with databases, information on phone phreaking, hacking and other tools essential for the hacker. Version 1 sold a respectable 120,000 copies, version 2 sold 3 million copies and version 3, which was released in May 1996, had sold 7 million copies by November 1996. In the right hands, the *Hacker Chronicles* is as much a weapon of war as an AK-47 or a T-72 tank or even an intercontinental ballistic missile. Yet, there are no restrictions on its sale, no public outcry about the proliferation of such information at home and abroad and no protests at the wealth of data falling into dangerous hands.

As the hackers and phreakers roamed through cyberspace having fun, causing havoc and mayhem, ripping off the phone companies and generally becoming a billion-dollar pain in the neck, people who would never have described themselves as computer literate began to use computers, attracted by not just their intrinsic convenience for writing letters, keeping accounts and playing games but because of the Internet. At the end of the 1980s, the Internet was essentially a research and communications tool for the military and academia. There were three distinct camps of users: the utilitarians, who saw the Internet as a great research tool; the hackers, who saw it as a romper room for their reckless pursuits; and the purists, who, with more than a trace of intellectual arrogance and elitism, felt there was a kind of First Amendment purity about cyberspace, a place where they could pursue an alternative digital lifestyle unsullied by crass commercialism.

Then Windows brought the graphical user interface (GUI) to the masses and the world changed again. It was the combination of pictures, sounds, faster computers and the development of the World Wide Web that made the Internet what it is today.

The WWW was created in 1989 by English computer scientist Timothy Berners-Lee to enable information to be shared among internationally dispersed teams of researchers at the European Organization for Nuclear Research (CERN) facility near Geneva, Switzerland. He wanted a means of communication that would allow anyone wanting to access a reference to click on a word that would open that file immediately, without having to search a directory for the document. This was achieved by implanting in that word the command that opened the file. This was HTML, Hypertext Markup Language. This reduced the complexity of the Internet to a few mouse clicks. All it took to be on the Internet was a computer, a modem and an account with a company that provided access.

What followed was a stampede, as people who ten years earlier could not have conceived of communicating on a global scale from a small box in their den began to take the Internet seriously. Commerce followed them onto the Web and business started to gain a foothold in cyberspace. To Joe Public this was another aspect of the Internet just like any other, but to the purists this was sacrilege. So it was to the hackers, who saw their playground being taken over by the very establishment they had such a good time annoying.

On Thanksgiving Weekend 1994 Joshua Quittner and Michelle Slatalla, authors of the groundbreaking book *Masters of Deception: The
Gang That Ruled Cyberspace*, which offered the first real glimpse into
the minds of hackers, were themselves hacked by the self-styled Internet
Liberation Front. Their phone line was taken over and their voice mail
corrupted so that anyone calling them would receive a stream of obscen-
ities. Simultaneously the hackers, who were thought to have been angry
with Quittner and Slatalla's representation of them in their book, attacked corporate computer systems at IBM and other companies, to
send a warning about the increasing commercialization of cyberspace.
The attacks created a media sensation. This was the tip of a revolt that
was to spread in two distinct directions. The Internet Liberation Front
went the way of anarchy and disruption, while the Electronic Frontier
Foundation went the traditional American route of lobbying. But the
intent was the same, to keep the dead hand of government as much as
possible out of cyberspace and let it grow freely.

As the government itself saw the Internet as a means of promoting
itself to the people, it came under hacker attack. The Web site of the
Department of Justice was replaced by a page that had a background
of swastikas, a picture of Adolf Hitler and the title "Department of
Injustice."

The fundamental hacker belief: that information is free, that computers are made for processing, not secrecy, that authority should be mis-
trusted and power decentralized. The tone was similar to that being
heard from the increasingly organized militia movement, which fomented a deep antigovernment animus. Also hacked were Web sites at
the CIA and the Air Force, their pages replaced by calls to arms among
government haters.

The militia, armed extremist groups dressed in paramilitary uniform,
are the most evident part of a generalized mistrust of government that is
spreading at a rapid pace through communities across the nation. This
mistrust is caused by anything from legitimate disputes with the federal
government over land rights in states like Nevada, to fundamental disagreement with any attempt to limit gun ownership, to extraordinary
paranoia among people who think the government is about to hand the
nation over to a New World Order that will impose an un-American way
of life on the American people.

The United Nations is the focus of this animus; people talk of "black
U.N. helicopters" running surveillance on law-abiding folk, preparing
for the takeover. Much could be dismissed as hot air but for one appalling act that showed there were some prepared to take extreme measures
to make their message heard, namely the Oklahoma City bombing in
April 1995, when 168 died in the attack on the Alfred P. Murrah Federal
Building. The effect of this atrocity on America's self-image was significant. Terror was not just something foreign crazies committed on the

streets of Berlin, Beirut, Paris or London. Now it happened on Main Street America, and the perpetrators were American. The following year witnessed the standoff between the authorities and the Montana Free-men that lasted eighty-one days, ending in June 1996. In this instance the siege ended peacefully, but another band of heavily armed rebels had stood up to the authorities. The hatred of government was moving from talk to terror.

The Internet was increasingly used as a clearing house for militia information and logistics. E-mail, often encrypted, was also used. The U.S. Militia Web site offers for sale weaponry and explosives, survivalist gear and PR accoutrements to help spread the word, such as bumper stickers that read, "Have You Cleaned Your Assault Rifle Today?" Racist hate groups like skinheads and neo-Nazis, whose views to a large extent coincide with the militia's, were on the Internet, too. The Maryland Police and Correctional Training Commission compiled a list of 250 organizations with Web sites, including the KKK, Skinheads USA, White Tribalist, Aryan Nations and others who exploited this growing mistrust of the government while adding an anti-black/Jew/immigration spin to it. They started to use the Internet for recruitment of others to their cause.

"The Internet has quadrupled the number of white-power skins I'm in touch with," one young skinhead reported proudly.

The character of these hate groups has not been changed by the Internet. They still peddle the same vicious propaganda, modeled on Adolf Hitler and the Nazis, but they use the Internet to disseminate it. The Cyber Nazi Group (motto: "Let Your Life Be a Lightning Bolt!") even posted online hints and suggestions about how cyber-Nazis should exploit Usenet discussion groups to get their message across most effec-tively, particularly to young people.

In its public writings, the Cyber Nazi Group said its Internet mission is propaganda and psychological operations. Government intelligence sources said this group and others with the same agenda had acquired the ability to hack into the nation's information infrastructure and wage cyberwar on the government. There is no doubt the tools were there. Entry-level hacking was much easier in the early 1990s than a decade before and freely available hacking programs, like SATAN and Rootkit, provided the sort of intrusive power it once took years to develop. SATAN was actually a program written for systems administrators to test security, but word quickly got out that it could be used aggressively, too. All the signs were that the hate groups had taken advantage of this new wave in hacking power.

"This is the sector of the right wing that believes in armed revolution," terrorism expert Chip Broulet told NBC News. "Cyberwar is very real. I don't think we should laugh these guys off."

This extremist fringe could prosper because the rage, as the growing antigovernment sentiment was described, reached all the way across society, right into police stations and sheriffs' offices. The militia may have espoused armed resistance while others may have preferred the pen to the sword, but their motivation was rooted in the same argument, and there was no shortage of recruits to join the extremist elements of the militia. Inspired by the harsh rhetoric of talk radio, which provided a drumbeat of hostility toward the government, Americans who felt the government was betraying the country became increasingly ready to defy the law. Richard Mack, the sheriff of Graham County, Arizona, quit his job and sued the federal government for instructing county sheriffs how to enforce the Brady gun control law. He claimed it contravened states' rights and wrote a book called *From My Cold Dead Fingers: Why America Needs Guns.* In other states, lawsuits against the government over attempts by Washington to implement federal law were commonly fought, and often won. In Texas, the self-styled Republic of Texas group challenged the legitimacy of the state of Texas and flooded the government with false liens against people's property to emphasize its belief that the state's affiliation with the union was illegal. Eventually the Republic of Texas's leaders were pursued by state authorities and holed up in their "embassy" in the remote Davis Mountains in Texas. The siege ended peacefully, and the threatened bloodbath was averted, although one Republic of Texas member was killed by Texas Rangers after fleeing. There was an accumulation of evidence about the nation's growing disaffection with its federal leadership that Washington will ignore at its peril. Inside the Washington, D.C., Beltway, policies increasingly became an argument over the number of angels dancing on the head of a pin, with the personal power of the politicians themselves the most important issue at stake. Outside the Beltway, as ordinary people focused on their work, their families and how to pay their taxes, government appeared ever more remote and irrelevant.

Compounding the pressure building inside America's borders and across cyberspace was the changing face of threat from more traditional, and more violent, foes. For if the government's internal enemies were starting to harness the power of the Internet to spread their message, would not their external enemies do the same?

In fact, the Internet has become the communications tool of choice for terror/liberation groups across the globe. The world was alerted to this use when Subcomandante Marcos, the leader of a revolt in the Chiapas region of Mexico, used his organization's Web site to disseminate propaganda about its cause. Ramzi Ahmed Yousef, the alleged planner of the World Trade Center bombing, is a computer specialist who sent encrypted messages across the Net, and the Japanese Aum Shinrikyo group made information warfare one of its top priorities. In 1994 an assault team from the group's "Ministry of Intelligence" broke into

Mitsubishi Heavy Industries, entered the company's mainframe computer and stole huge amounts of secret data, part of a policy to undermine Japan's high-tech information base.

Some groups, whose avowed intent is to hurt America, pose the greatest threat to the United States. Their exploitation of computing power offers them vastly increased opportunities for attack. With their equally ominous accumulation of weapons of mass destruction, they pose the greatest threat of all to the American people. The bomb attack on the World Trade Center in 1993 was extremely serious, even if the death (six dead and 1,000 injured) and destruction were not as bad as they might have been. This was an indication that foreign terrorists were prepared to launch physical attacks on American soil. Americans had watched with horror on March 20, 1995, as the Aum Shinrikyo launched a sarin gas attack on the Tokyo subway, and Japanese police found evidence that the terrorists were also trying to develop anthrax biological weapons (BW). In 1996, a Senate report noted that a terror attack with anthrax, sarin or a nuclear weapon was almost inevitable. President Clinton expressed deep-seated fear about the potential for the Internet to be a force multiplier for terrorism. "Are people learning . . . from the Internet how to make the same sort of trouble in the United States that was made in Japan with sarin gas?" he asked at a press conference during a visit to Tokyo in April 1996. "Isn't it a concern that anybody, anywhere in the world, can pull down off the Internet the information about how to build a bomb like the bomb that blew up the federal building in Oklahoma City?" The administration's anxiety was underscored a year later in April 1997 when Defense Secretary William Cohen warned at a conference on terrorism, "This scenario of a nuclear, biological or chemical weapon in the hands of a terrorist cell or a rogue nation is not only plausible, it's quite real."

What is even more frightening is that the attack, when it comes, will have been planned and probably financed with the help of American systems, citizens, and in some cases the American government. Terrorists have always tried to use the freedoms of a society like the United States to subvert or destroy it. It adds special pleasure to the act, and it has been one of the most enduring challenges to a free society not to curb those freedoms in the face of a terrorist threat. Today's Arab terrorists, often described as Islamic fundamentalists although that blackens the name of the vast majority of decent and devout followers of Islam, are using every device possible to leverage their power in the United States, even resorting to defrauding the government out of food stamps as a source of income. Welfare fraud allegedly netted close to $100 million for Middle Eastern terror groups. Other groups like the PKK (Kurdistan Workers Party) feed off America's drug habit for their funds. Sympathetic Arab businessmen in America are a well-known source of cash; the Virginia entrepreneur Mousa Abu Marzook was expelled to Jordan after Israel complained that he was a major financier of Hamas terrorist cells.

Universities make easy bases for support groups. A think tank based in Tampa, Florida, called the World and Islam Studies Enterprise (WISE), was affiliated with the University of Southern Florida, and its director, Ramadan Shallah, was an instructor at the university.

On October 26, 1995, in Malta, a man standing outside his hotel was shot and killed by a young man who escaped on a motorcycle. The murdered man was Fathi Shikaki, head of the shadowy terror group Palestine Islamic Jihad ("jihad" means "holy war"). Within days, the respected teacher Ramadan Shallah quietly left Florida to take Shikaki's place as leader of the Islamic Jihad. It was a complete shock to everyone at the university, and the federal authorities immediately moved in to investigate. They confiscated a mass of paperwork and computer disks from the organization to look for links with terrorist activities. Shallah's foundation is suspected of being part of a network of similar groups who raise money for terrorist groups whose avowed aim is to hurt the United States. In May 1997 the university declared that the authorities had found nothing illegal in their search, despite copious, yet apparently innocent, contacts between WISE and factions in the Middle East connected to Islamic Jihad. Needless to say, the university had been stung by the incident, particularly when people started to refer to it jokingly as "Jihad University." But the mystery remains—what happened to all the millions of dollars that flowed through WISE and several Islamic committees and foundations? How could Shallah have been working quietly in Florida and not leave a single shred of evidence that his organization had been working as a front for the Islamic Jihad? In 1994 WISE and similar organizations had been profiled in a PBS documentary, "Jihad in America," which demonstrated that they all conduct business very openly, holding symposia and conventions just like any American trade association or corporation, except the call was not for bigger sales or government action. It was, as the program title indicated, for "jihad in America."

The FBI fears these very visible organizations shield another network that hides in the shadows made up of terrorist cells that can strike when directed by masters in the Middle East. Obviously, making the links between them is not proving easy.

Slowly but very surely terror groups around the globe, who used to communicate via their patrons in the intelligence agencies of the former Soviet bloc, now use the Internet for communications, education and propaganda. The Irish Republican Army's political wing, Sinn Fein, runs an extremely sophisticated Web site, as do groups as diverse as the Tupac Amaru in Peru, the Euskadi ta Askatusuna (ETA) independence group in Spain's Basque country and the Somaliland Separatist Movement.

The former Bulgarian spy agency, DS, runs a training course on how to create computer viruses for bringing down enemy computers. Cyberterrorists share access codes to corporate computers so they can wreak havoc on—and leech funds from—the capitalist system. Even Saddam Hussein has a home page on the Internet, established to celebrate his sixtieth birthday in 1997. It can be found at http://196.27.0.22/iraq. There is no evidence the Iraqi leader is using the Net for nefarious purposes, chiefly because there is no Internet access in Iraq. Anyone sending him an e-mail will have to wait for an answer while the e-mails are printed out in Jordan and then sent 500 miles by road to his office.

Germany, burdened by the evils of its past, is choosing to meet the challenge posed by the marriage of terror and the Internet head-on, as radical leftist Angela Marquardt discovered when she posted on her home page a hotlink to a Dutch Internet provider on which the left-wing newspaper *Radikal* had set up shop. *Radikal* was a serious worry to the German police, as it offered readers advice on making bombs and derailing trains ("A Short Guide to Hindering Trains"). Remembering the terror inflicted on Germany by the ruthless Baader-Meinhof gang in the 1970s, Germany is intent on halting at its borders all Internet material not only of an anarchist bent but pornographic as well, and is demonstrating a willingness to go after the executives of companies like Compuserve who play host to material the Germans find scandalous. Their attitude is in marked contrast to the freedoms enjoyed by American netizens. It bodes ill for the ideal of Net freedom if Germany succeeds in its aims. It bodes ill for us all if terror groups exploit that ideal to kill and maim innocent people.

If it comes, where will the attack fall? The bombings of the World Trade Center and the Oklahoma City federal building gave no clues about the new age of cyberterrorism that beckons. They were conventional attacks using conventional materials. Specialists have speculated about infrastructure attacks, such as those laid out in the Rand Corporation's "Day After" exercise, which explored the possibility of an IW attack by Iran on America's critical infrastructure. Others suggest a more refined and in many ways more sinister scenario. Barry Collin of the Institute for Security and Intelligence believes that the cyberterrorist is as far removed from what we think of as "conventional" hacking as a crack IRA bomber is from a British soccer hooligan. By matching the capabilities of extremist groups, their known capabilities and the vulnerability of America's civil and corporate infrastructure to attack, he extrapolates scenarios of frightening import:

- A CyberTerrorist will remotely access the processing control systems of a cereal manufacturer, change the levels of iron supplement, and sicken and kill the children of a nation enjoying their food. That

CyberTerrorist will then perform similar remote alterations at a pro-
cessor of infant formula. The key: the CyberTerrorist does not have to
be at the factory to execute these acts.

  • A CyberTerrorist will place a number of computerized bombs
around a city, all simultaneously transmitting unique numeric patterns, each bomb receiving each other's pattern. If bomb one stops
transmitting, all the bombs detonate simultaneously. The keys: 1) the
CyberTerrorist does not have to be strapped to any of these bombs; 2)
no large truck is required; 3) the number of bombs and urban disper-
sion are extensive; 4) the encrypted patterns cannot be predicted and
matched through alternate transmission; and 5) the number of bombs
prevents disarming them all simultaneously. The bombs will detonate.

  • A CyberTerrorist will disrupt the banks, the international financial transactions, the stock exchanges. The key: the people of a country
will lose all confidence in the economic system. Would a CyberTerrorist attempt to gain entry to the Federal Reserve building or equivalent?
Unlikely, since arrest would be immediate. Furthermore, a large truck
pulling alongside the building would be noticed. However, in the case
of the CyberTerrorist, the perpetrator is sitting on another continent
while a nation's economic systems grind to a halt. Destabilization will
be achieved.

  • A CyberTerrorist will attack the next generation of air traffic
control systems, and collide two large civilian aircraft. This is a realistic scenario, since the CyberTerrorist will also crack the aircraft's in-
cockpit sensors. Much of the same can be done to the rail lines.

  • A CyberTerrorist will remotely alter the formulas of medication
at pharmaceutical manufacturers. The potential loss of life is unfath-
omable.

  • The CyberTerrorist may then decide to remotely change the pressure in the gas lines, causing a valve failure, and a block of a sleepy
suburb detonates and burns. Likewise, the electrical grid is becoming
steadily more vulnerable.

"In effect," Collin concludes, "the CyberTerrorist will make certain
that the population of a nation will not be able to eat, to drink, to move,
or to live. In addition, the people charged with the protection of their
nation will not have warning, and will not be able to shut down the
terrorist, since that CyberTerrorist is most likely on the other side of the
world."
With terror, espionage and crime moving into the ethereal world of

cyberspace, a new and deadly challenge has to be faced by a society that is becoming ever more dependent on the bits and bytes that can be exploited malevolently by hackers and terrorists. How well can America defend itself?

TWELVE

The Back Door's Open

DOMINANCE in the age of information is a double-edged blessing. The very fact of leading the revolution in information technology creates concomitant vulnerability. This is the price to be paid for being king of the hill.

There is no question that America is the most sophisticated information nation on earth:

• Communications, from the humblest dwelling to the boardrooms of corporate titans, are conducted electronically.

• Every act of commerce beyond hard cash purchases requires an electronic transaction, be it a credit card purchase, a check clearance, an ATM cash withdrawal or the daily interbank transfer of funds.

• Commercial inventories are controlled by computers, as are the production processes that create those inventories and the transportation systems that deliver them. All those computers are hooked together so that, increasingly, humans are taken out of the production-transportation-inventory cycle.

• The electrical power that runs the machines that make those goods, the oil that flows through pipelines, the planes that fly the congested skies, all are controlled by computers.

• The life of every individual is recorded on a handful of computers, be it governmental, medical, financial, credit records, motor vehicle licensing, educational.

• The defense of the nation is organized and administered through the use of computers.

All these computers use telephone lines to transmit information, so every computer is linked to all the others. And by the Internet they are linked to every computer in the world that enjoys similar facilities, including the one on your desk.

As if proof were needed that modern life is a series of interconnections, consider three incidents that demonstrate how fragile our lives have become thanks to the reliance on electronics.

In 1991, a farmer trying to bury a dead cow shut down four out of the Federal Aviation Administration's thirty main air traffic control centers for over five hours. In digging a deep enough hole to accommodate Bessie he had cut through a crucial fiber-optic cable linking the FAA centers together. Now consider what would happen if a farmer in Alexandria, Egypt, also tried to bury a dead cow, or if someone with more malevolent intent started to dig a hole in the right area; for Alexandria is the site of a key node on a 23,600-mile cable FLAG (Fiber-Optic Link Around the Globe) that crosses the South Pacific, Asia, the Middle East and the North Atlantic, providing four continents with broad-bandwidth fiber-optic communication. It is for the most part submarine, but its nodes are land-based in places like Alexandria. How difficult would it be to cripple telecommunications and data systems on four continents by taking out one cable?

In a startling investigation into the vulnerability of FLAG, writers Neal Stephenson and Peter Leyden of *Wired* discovered that FLAG and four other trunk cables, SEA-ME-WE 1, 2 and 3, and AFRICA 1, will eventually pass through a dilapidated old building within a stone's throw of the site of the Great Library of Alexandria. Clearly a strategically placed bomb could cause havoc with global communications. It is to be hoped that backup and redundancy will minimize the future threat.

The second incident occurred in 1990, when a tiny little glitch in some AT&T software which the company sent to all its switching centers brought the entire system to a halt for nine hours. When software is standardized across a multitude of platforms, be it at a corporate or a national level, the lack of redundancy creates a large degree of vulnerability, which is one of the reasons security experts are nervous about the predominance of the Windows 95 operating system on Intel chip–powered computers. Without redundancy, what affects one affects all.

The third incident happened on May 13, 1997, when Bank of America employees were conducting "routine maintenance on half of an electrical substation the bank operates at a data processing center at Market Street and South Van Ness in San Francisco, [and] a worker accidentally shut off power to the online unit." The online unit con-

trolled the bank's ATM system, which crashed when the power went
out. Power was immediately restored, but the ATM system took two
hours to reboot, putting 1,529 ATMs in Northern California out of
action as well as blocking interbranch account requests. The outage was
caused by one small human error. One switch disabled 40 percent of a
bank's ATMs. What if someone really wanted to hurt the bank? It would
seem not difficult to do.

All that stops anybody from roaming free in cyberspace is computer
security, which can mean anything from a rudimentary password system
to a costly array of firewalls that electronically block access to a system.
So the life of America, as it courses through copper wires and fiber-optic
lines or is beamed across the sky via satellite and microwave transmissions, can be accessed and affected only by those who have permission
—or by anyone who knows how to circumvent the security. As has
already been seen, the defense establishment is under constant attack
from people who invade or make mischief on computers used in defense
activities in this country.

Are similar attacks being launched on civilian targets? Without a
doubt. The number of companies and organizations attacked in the last
ten years is seemingly endless, and those are just the attacks that have
been reported. Banks are notoriously leery of publicity about any breach
in their security because banking is built on confidence, which suffers if
it becomes known that bands of hackers routinely break into the electronic vaults. This is precisely what happened to Citibank. A group of
hackers in St. Petersburg, Russia, broke through the bank's security
systems and started to siphon out money to accounts in San Francisco,
Amsterdam, Germany, Finland and Israel. In the end all but $400,000
was recovered from the $10 million heist thanks to the efforts of the
SAIC security firm based in McLean, Virginia, which traced the path of
the hack. An attack of this magnitude was made possible by the fact that
some of the hackers had worked in Citibank affiliates for three years and
had carefully seeded the computers with "backdoor" programs that let
them into the system when requested. The chief St. Petersburg hacker
was Vladimir Levin, a twenty-four-year-old mathematics graduate from
St. Petersburg University. He was arrested when transiting through London on his way to a computer conference in Holland. More arrests
followed, in America, Holland and Israel. Initially it was thought Levin
was the mastermind behind the attack, but in October 1995 St. Petersburg police rounded up a gang of what were described as "cybergang-
sters" and confiscated computers, sniper rifles, assorted guns and
ammunition and close to $500,000 in cash. These were the alleged
masterminds of the enterprise, using pawns like Levin to do their hacking. It would seem inevitable that this was not a singular incident, that
other banks must have been hit, but we will probably never know for

sure. According to FBI testimony to the Senate, this is the only case of computer theft from a bank ever reported.

John Reed, the chairman of Citibank's parent company, Citicorp, is a man who deserves credit for going public with news of the attack. By doing so he alerted the world to a major threat and helped law enforcement agencies learn more about a crime they had had little experience in dealing with. But the revelation put his company at risk because it shook customers' confidence. Reed said they were "scared to death" by the attack. "There's a huge disincentive to reporting these crimes," declared Mark Rasch of SAIC. And as if to demonstrate why, within weeks of the Citibank hack being made public, twenty of the bank's biggest customers were approached by other banks promising better security. Their claims would probably not stand up to scrutiny. An expert on banking at one of America's leading computer security companies says that one major American bank loses $500 million a year through hacking alone. Banks will always be targets for computer criminals for the same reason they were for Willie Sutton, the bank robber who said, "That's where the money is."

Despite their obvious vulnerability, the banks themselves seem remarkably ignorant about the threats they face. In the early 1990s, a major American security company was involved in advising a dozen of the major New York banks about the threats posed by hackers and cyberwarfare. All the banks decided to upgrade their computer systems and several discovered that viruses at the level of the basic code had been inserted into their systems at least ten years earlier. Although Russia is losing the information war, the old Soviet Union had considerable sophistication in developing virus-infected machine code.

"It appears that the Soviets had infected Wall Street with a virus that would have taken down the banking system in the event of war," said the investigator.

Yet, when those same banks upgraded their systems, none took the trouble to run background checks on the people doing their work. A survey conducted by the same security company revealed that 98 percent of the banks that were company clients had employed Russian computer experts who had left their home country after the end of the Cold War to do the work.

And it is not just the banks that are the potential targets in the nation's economy. In 1994 the New York Stock Exchange asked a German hacker to demonstrate how an attack could affect its systems. To its amazement, it was not the Exchange's "secure" systems he tried to invade, but the computers that controlled the air conditioning in the systems' rooms. Having gained control of those he then had the power to raise the temperature to the point where machines would start to crash and the trading systems fail. In the world of networks, spotting vulnerabilities requires minds that can think laterally and creatively. All the evidence so far suggests those minds belong to the hackers, and as the power of home computers grows it is becoming easier for more hackers to get into the game.

The all-important question for America's future security is whether the risk from hackers is one of annoyance and inconvenience, or a much larger threat posed by malicious individuals who can do serious damage to America's way of life. The answer to that is not conclusive, but nobody, from the President on down, is taking any chances. When the Clinton administration became conscious of the possible threat to the nation's infrastructure from an attack on the information infrastructure, it talked of the potential for an "electronic Pearl Harbor." When it started thinking about how to deal with the threat it spoke of needing a new Manhattan Project.

That it would use such potent symbols makes clear that computer security is being taken very seriously indeed.

What jolted the administration into action was an exercise carried out for the Secretary of Defense by the Rand Corporation of California entitled *Strategic Information Warfare: A New Face of War*. This report demonstrated the possibility that warfare could take place on American soil, a jarring change of perspective to a people who perceive war as something that happens on a distant foreign field or desert. The chief area of vulnerability addressed by Rand was the electronic interconnectivity of key elements of the nation's infrastructure and how determined hackers could create havoc with the daily life of the country. For the first time an administration learned that hacking might be more than just a nuisance to be dealt with by police. Instead, it learned, it could pose a potent threat to the national security of the United States.

The exercise Rand performed was comprehensive and, if accepted as a real possibility, extremely unnerving. They brought together some of the top people in security, intelligence, defense and academia over five months in 1995 and played a war game based on the "Day After" concept. This posits a series of ongoing future crises and invites the gamers to jump straight in and deal with them. They then move to the consequences of the crises (the Day After) and the actions taken to cope with them. Then the exercise moves to the Day Before and asks gamers to propose ways in which the crises could have been defused.

The scenario for the game took current situations and magnified them to their logical conclusions: a weakening Saudi Arabia, a strengthening Iran, strong Islamic nationalist governments in Algeria and Libya, Russia in the sway of increasingly well-organized criminal mafia groups; and overarching it all, a Global Information Infrastructure (GII) consisting of widely available Internet access whose potential for exploitation has been recognized by Iran and others. The catalyst for disaster is a split in OPEC between aggressive oil producers like Iran and Iraq, who want to drive up the price of oil by cutting back production, and Saudi Arabia, which desperately needs to maintain current flows to prop up the ailing economy. Iran emerges as the dominant force in the Persian Gulf and begins military moves against Saudi Arabia.

Simultaneously, infrastructure breakdowns start occurring in nations friendly to Saudi Arabia. In Egypt, 90 percent of the electrical power in Cairo goes out for several hours. In the United States, the public telephone network in Northern California and Oregon fails. At Fort Lewis, in Washington state, the base phone system crashes as a result of a mass dialing attack from the Internet; the system is swamped by computer-generated automatic calls. In Saudi Arabia, an oil refinery suffers a serious malfunction, which leads to an explosion and fire. In Maryland, a new high-speed Metro-Superliner train is misrouted and crashes, killing sixty people. In Britain, the Bank of England reports it has found sniffer programs in its central computer. (A sniffer program is planted in a target computer to search out passwords and codes, and transmit them back to the hacker.) These reports lead to fears that the integrity of the main funds-transfer system has been compromised. The British and American economies start to suffer as markets plunge, oil prices rocket. In all cases, evidence mounts that these events are the result of concerted attacks on computers controlling the national infrastructures of the countries involved.

The crisis develops, with U.S. military responses hampered by worm attacks on computerized battle plans and on military phone systems, further attacks on economic assets like banks, disruption of TV transmissions, and computer attacks on Saudi Arabia and Egypt. In the meantime, a global peace group that has been using the Internet to promote world peace and nonaggression sparks protests against U.S. military involvement in the Gulf. The group demonstrates its power by blocking a network evening news broadcast and substituting its own message, calling for massive civil disobedience. The ability of the United States to protect allies and vital national interests is being undermined by cyberattacks on key elements of the U.S. military and civilian infrastructure, by an economy spiraling into crisis after repeated cyberattack and by a swelling popular revolt against war.

At a meeting on May 15, 2000, in New York, the Iranian ambassador is heard to say that "the United States, as the technologically most advanced power on the planet" was highly vulnerable to "twenty-first-century attacks by states and others who had mastered contemporary computer and telecommunication technology."

This comment, like the entire scenario, is a bit of fantasy dreamed up by the Rand game organizers, but it has the ominous ring of truth to it. Nothing that Rand posited seems extraordinary in the light of what is known about people's ability to bend computer communications to their will.

So Rand got the government's attention. But what would the administration do about it?

The government's record on dealing with the Information Age is spotty. Early in the Clinton administration Vice President Al Gore had spoken of America leading the way down the new "Information Superhighway," and in general terms the government allowed the Internet to develop free of government regulation, except in the notable area of encryption, where government insistence on diluting the public's ability to use powerful data encryption methods to protect privacy and financial information, in the name of national security, is hampering development of the Internet. The use of the superhighway metaphor proved Gore knew how to press America's hot buttons, likening the Internet to the freedom of the open road with the computer as car. These are potent symbols. But for the rest, the government has shown itself to be deeply conservative and in places deeply ignorant about the computer revolution. Nowhere is this more evident than in the general attitude to security.

Throughout the government as a whole computer security is treated as an afterthought. Most security personnel are not full-time, but pressed into handling security as an additional duty. As a rule they have almost no computer security experience before being assigned to the task and receive very little training once they are doing it. In many cases, the person placed in charge of security for an office has very limited computer expertise, but receives the job because his experience is greater than everyone else's. Often the duty is rotated to another person after two or three years, thereby sacrificing any institutional knowledge which may have been gained.

Most telling of all, there is no computer security specialist career field in the U.S. government. This small fact has enormous consequences for information security. Without a career path, even dedicated, knowledgeable individuals face strong disincentives to specialize in computer security. They are more likely to head for the private sector where they often end up contracted back to the government at rates much higher than if they had been offered advancement.

"I have seen many people leave the government in pursuit of lucre in the private sector. There's nothing wrong with this," says Dan Gelber, chief counsel of the Senate Permanent Subcommittee on Investigations, minority staff, "but we are exacerbating it to an unacceptable level. These people face a ceiling, and it's an easy choice to make when you're talking about paying for your kids' college."

Reviews of computer security procedures at government agencies expose woeful inadequacies. The Senate Subcommittee on Investigations staff requested from various agencies the name of the individual or office in charge of computer security. Most responded that they either didn't know who the individual was or even whether such a position existed. Some could only offer the suggestion that responsibility for security was spread over several areas. The staff found this vague attitude to security even existed at the Department of Justice, where each component of the department, such as the FBI, was tasked with handling its own security, with no overarching policy to guide them.

At the Department of State, an Inspector General audit of the department's unclassified mainframe security system concluded that the department had essentially no security plan at all. The Inspector General stated that the State Department could not even reliably ascertain if information had been compromised.

The extent to which the government of the country is laid bare has also proven embarrassing for the United States Congress, which in 1995 was penetrated as part of a special audit by the Price Waterhouse accounting firm. It was almost laughably easy for the attackers, who did not even have to use any sophisticated code-cracking programs to get into computers that helped run the day-to-day business of Congress and in some cases computers belonging to members themselves. They did it a second time, routing this hack through the Library of Congress. And this time as well they got in just by guessing passwords. The password owners had committed the cardinal sin of choosing names of family members and pets.

Although all government agencies say they are paying more attention to information security, in many cases it appears to be only lip service. Many have built "information warfare" into preexisting offices, but then fail to fund it. Everyone agrees there is a threat, but few seem to be doing anything about it. One agency assembled ten individuals for an information warfare briefing, but in the end admitted that only one of them was actually working full-time on intelligence collection and threat analysis.

One intelligence analyst for science and technology conceded that a new focus on information warfare would require major new redirection.

"Don't wait for the intelligence community to provide a threat estimate," he said. "It will probably take the intelligence community years to break the traditional paradigms, and refocus resources on this important area."

This despite the fact that CIA Director John Deutch ranks information warfare third only behind the proliferation of weapons of mass destruction and terrorist use of NBC (Nuclear, Biological, Chemical) weapons in his list of long-term threats to American national security. Requests from Congress for a threat assessment on information warfare similar to the ones prepared by the CIA on chemical or nuclear weapons have not met with much success. Senator Jon Kyl of Arizona requested just such an assessment in an amendment to the Intelligence Authorization Bill for FY1997:

the President shall submit to Congress a report setting forth the results of a review of the national policy on protecting the national information infrastructure from strategic attacks. The reports shall include the following:

A description of the national policy and architecture governing the plans for establishing procedures, capabilities, systems, and processes necessary to perform indications, warning, and assessment functions regarding strategic attack for foreign nations, groups, or individuals or any other entity against the national information infrastructure.

The amendment required that the intelligence community provide a threat estimate within 120 days of the bill's effective date. Deutch complained that the timetable was too ambitious, and requested an extension, an unusual response, but understandable given the absence of any historical data to work with. A former high-ranking White House science and technology officer said of the intelligence community, "Usually they just can pull the information out of the box that holds the data —as of today, however, the box is just empty!"

Jim Christy of the Air Force Office of Special Investigations lamented, "We still have too many people studying the coal output of Bulgaria, and not enough studying the new threats."

Congress recently expressed its displeasure with the intelligence community's slow response. "To date, Congress has not received the requested report and overall it is clear that the Administration's response to this statutory requirement has been lackluster at best."

It is not altogether the fault of the intelligence agencies that they are unable to collect the necessary data to create a national threat assessment. First, outside the Air Force, there is still no mandatory reporting of intrusion incidents. Defense installations are a favored hacker target, but even the attacks that are discovered are not reported or collated in any organized fashion. This nonreporting represents a significant loss of data that could be used to evaluate the nature and origin of attacks.

Second, from a legal and organizational direction, intelligence and law enforcement agencies are poorly positioned to deal with cyberthreats. The structures were designed for the Cold War, not cyberspace. The laws limiting CIA and NSA involvement in domestic cases, and the FBI's division of labor between counterintelligence and criminal investigations, create divisions that are irrelevant to the virtual world. If a hacker routes his attack through different countries before crashing a U.S. government computer, is it a case for the FBI? CIA? NSA? All three? Until this outmoded geographical division of labor in intelligence is addressed, cases will continue to fall through the cracks.

Finally, the Senate Subcommittee on Investigations staff concluded that information warfare and information security are not yet a high priority within the intelligence community. Although major rhetorical emphasis is placed on the subject, real efforts are thin.

Deputy Attorney General Jamie Gorelick suggested there had been a reticence in government to discuss vulnerability publicly for fear of exposing it to public view and possible attack, which smacks of "If we don't talk about it, maybe it will go away." Senate Armed Services Committee Chairman Sam Nunn was crisp in his reply. "Vulnerability is all over the Internet," he said. "The only people who don't know about it are the people in government with responsibility for protecting the infrastructure."

It added up to a problem that could no longer be ignored or dealt with piecemeal. On July 15, 1996, President Clinton created the Commission on Critical Infrastructure Protection, and tasked it with reporting on the threat to America's computer systems and networks, particularly those governing telecommunications, oil and gas, electricity, bank and financial operations, transportation, water supplies, critical emergency response and government.

To confront the problem of infrastructure attacks right away, while the commission carried out its study, the FBI was ordered to set up the Computer Investigations and Infrastructure Assessment Center in Washington. Veteran spycatcher and economic espionage specialist Kenneth Geide was named its chief, and at once ran into a problem. Everyone knew the nation's systems could be attacked by terrorists and foreign agencies, but had it? Geide tried to educate the leaders of companies and organizations who are part of the nation's critical infrastructure about the danger and the need for heightened security.

"Many of them recognize there's a problem here, but they ask, 'Where's the anecdotal evidence?' " he said.

Without any to offer, Geide ran his audiences through the damaging experience of August 1996 when the power grid serving nine states collapsed, depriving 4 million people of electricity. A deliberate attack had been suspected from the start, but although the culprit turned out to have been tree limbs shorting out lines, which then sent a cascade of shorts through the system, Geide stressed that this incident demonstrated how easy it was to create chaos in America today. Occasionally, he met resistance from companies who were either skeptical of the threat or reluctant to cooperate because of the secretive nature of their business, such as finance.

"We all agree that the potential for threat is there," he says. "So do we wait? Is that how our society wants to address these things? Shall we wait for the power grid to collapse?"

While the commission worked on its investigation, the Defense Science Board came out with its own report on vulnerability of the national infrastructure, and suggested there was an inevitable evolution of warfare that made it not a matter of *whether* an enemy would attack the heart of America's information-based economy, but *when*:

> The objective of warfare against agriculturally-based societies was to gain control over their principal source of wealth: land. . . . The objective of war waged against industrially-based societies was to gain control over their principal source of all wealth: the means of production. . . . The objective of warfare to be waged against information-based societies is to gain control over the principal means for sustenance of all wealth: the capacity for coordination of socio-economic dependencies. Military campaigns will be organized to cripple the capacity of an information-based society to carry out its information-dependent enterprises.

If it is accepted that, as Martin Libicki of the National Defense University puts it, "Thus does war follow commerce into cyberspace . . ." we are staring down the barrel of a gun that is going to cause immense harm to America in the future. Yet Libicki is quick to point out that the process is not as inevitable or easily achievable as some of the carefully constructed scenarios, like Rand's, suggest. He believes systems can be protected and that dangers can be overstated. The bottom line is that government must take the job of protecting the infrastructure as seriously as companies protect their information.

Libicki's suggestion that hype has begun to creep into assessment of the threat cut no ice at the Defense Science Board, whose report ominously warned: "There is a need for extraordinary action. . . . [Current practices and assumptions] are ingredients in a recipe for a national security disaster." The report said that by 2005 attacks on U.S. information systems by terrorists, criminals and foreign intelligence agencies are likely to be "widespread."

The board's alarm was echoed by initial, and informal, findings of the President's Commission on Critical Infrastructure Protection. The commission's chairman, Robert Marsh, said in June 1997, before the official release of his report, that the United States lacks the tools to defend itself against a hacker assault on its power, telecommunications and banking systems. He said there was no such attack on the immediate horizon: "We couldn't find a 'smoking keyboard'; there isn't a country out there with a war plan, but it's sure to evolve," he said. He raised eyebrows among some computer experts by adding, "There isn't a firewall that a group of experts can't get around." In the eternal war between hackers and defenders, the evidence is on Marsh's side, despite the increasing sophistication of firewall defenses. Defenders have to be lucky all the time, the hacker just once.

Marsh and his team, which included former Deputy Attorney General Jamie Gorelick and former Chairman of the Senate Armed Services Committee Sam Nunn, operated from an anonymous modern concrete and glass building in Rosslyn, Virginia, on the other side of the Potomac River from Washington. In all, there were eighteen members of the commission, ten from government, eight from the private sector, all served by a staff of forty-five, which included contractors from private industry. Over the course of their year's investigation, the commission took evidence from all the key areas of civilian and government activity as it related to the nation's vital infrastructure.

"We are entering a new era," Marsh told me. "We are subtly becoming more dependent on information technology of all kinds, and I do not think we realize the degree of dependency. I think we are becoming almost vitally dependent. This is one of the few times in history where government has tried to take the initiative before a crisis actually happens."

Marsh had learned the depth of the problem in the course of his year's work. War games with private-sector security experts and the Pentagon revealed the vulnerability that all had been warned of. Marsh noted the irony of the Pentagon, focus of the most publicized and potentially devastating hacker attacks, remaining wide open to assault. "Hackers succeeded in breaking into many of the critical systems of the Pentagon," he said, "and in most cases they were undetected. The incidents were viewed in isolation and it took a long time for people to understand that there was a determined attack taking place. It certainly woke up a lot of people." It was not as if these same people had not been woken up on many occasions previously. The problem is that without a concerted effort to reform the sclerotic bureaucracy of government it is easier to roll over and go back to sleep. Marsh's commission prepared a three-foot-high raft of proposals aimed at creating a culture of awareness and readiness in government and in the private sector.

The Marsh report was the first of its kind, an initial look at how America would have to defend itself in the infosphere that would evolve in the early part of the twenty-first century. Since its publication in October 1997, the report has become a touchstone in all the debates that have followed. It is therefore worth quoting its proposed policy for defending America's infrastructure in some detail.

Critical infrastructures underpin the security of our national wealth, our defense capability, the economic prosperity of the people and, above all, the maintenance of the system of human rights and individual freedoms for which the United States was founded and has stood since 1776. The threat of infrastructure attacks therefore has the potential for strategic damage to the United States. Accordingly, the assurance of critical infrastructures deserves national attention and leadership by the federal government.

It shall be the policy of the US to assure the availability and continuity of the critical infrastructures on which our economic security, defense and standard of living depend. The infrastructures will be defended by whatever means necessary including the full range of business, legal, law enforcement, military and social tools available.

Further, the US recognizes that assuring infrastructure is not just a government or business responsibility, but is shared by those public and private interests that own and operate the infrastructures and the government agencies responsible for defense, law enforcement, and economic security of the nation.

The interdependent nature of the critical infrastructures and their collective dependence on the information and communications infrastructure have created new assurance challenges that can only be met by a partnership between owners and operators and government at all levels. Only the owners and operators have the knowledge, access, and technology to defend their systems from the growing array of widely available information-based tools. Only the federal government has the legal authority, law enforcement capability, and defense and intelligence resources needed to deter the most sophisticated nation-state and other serious cyber threats. And only the federal government has the intelligence and related capabilities to find the tools that do harm and promulgate information about them through the infrastructures.

As a matter of urgency, an Office of National Infrastructure Assurance should be established under the National Security Council (NSC) and given overall program responsibility for infrastructure assurance matters, including policy implementation, strategy development, federal interagency coordination, and liaison with state and local governments and the private sector. Among other responsibilities, this Office will devise and establish mechanisms for the exchange of views and information between the government and the private sector, identify information requirements for infrastructure assurance, and ensure that infrastructure assurance considerations are taken into account in making other government program decisions.

The Office of National Infrastructure Assurance should ensure that a program of public awareness is implemented throughout the country to inform the American public about infrastructure protection. This will include establishment of appropriate curricula in the national education system, from kindergarten through graduate school and including professional training. The Office of National Infrastructure Assurance should also ensure that individual agencies of the federal government implement infrastructure preparedness provisions and update their security plans to include protection against Information Warfare threats.

In addition, Marsh wanted to see a broad range of other measures that would begin to establish both an IW early warning system and a defensive structure. He recommended:

• At the operational level, each area of the critical infrastructure would establish Sector Infrastructure Assurance clearing houses which would collate sector-wide intelligence and disseminate it inward to sector members and outward to a central body.

• This central body would be a new Information Warning and Analysis Center staffed by government and private-sector employees. Its task would be to see the bigger picture, as derived from intelligence flowing from the various sector clearing houses, and present an ongoing series of recommendations to government and industry. Key to its success would be the protection of sensitive information that until now some sectors, for instance banking, would otherwise be reluctant to share.

• A doubling of government R&D on infrastructure protection to $500 million annually, focusing the extra expenditure on what until now has been an inadequate effort to identify and deal with information attacks.

• A change to the culture of education, which will teach information security alongside computer technology in schools. "We must inform children that they are trespassing when they break into a computer. We need to create a climate of responsibility around computers just like we do about our own property," Marsh said.

• Partner with private industry in developing and implementing indication and warning capabilities.

Marsh recognized he was operating in an era when anyone proposing action by the government is swimming against the tide. "We are not mandating that government has a solution to infrastructure vulnerability," he said. "Instead we are trying to promote a collaborative solution to the problem. We also recognize that the government must clean up its own act so that it provides a benchmark against which an industry can be measured."

While Marsh's statements appeared perfectly reasonable and the report was well enough argued, it disappointed the activists who believed a radical approach to the problem was needed. There was a general feeling that as a blueprint for the future, Marsh had failed to deliver.

"For example, there was no discussion of the international dimension which seems to show a fundamental misunderstanding of how the Internet and the Web works," said one administration official. "There is little point in having all the controls inside our borders if the hacker from Iran or England can do virtually what he wants. International agreements are going to be critical."

The Commission report had been scheduled to be handed over to President Clinton with much fanfare with the expectation that the President would endorse many of the seventy-six recommendations contained in the report. However, the official view was that the report was so weak and the recommendations so thin that the President could not endorse it. The report was instead handed off to an Interagency Working Group, Washington's equivalent of a black hole. Marsh Committee officials tried to stack the IWG with their own people but failed and the sixteen different agencies involved swiftly staked out their own positions.

The IWG soon polarized between the Justice Department, which argued that it could control everything with the FBI running the show, and every other agency, which thought that was a very bad idea. Within Washington, the FBI is considered a Neanderthal agency, and Justice runs a close second. Also, with their focus on detecting and prosecuting criminals, neither has visible experience in preventive, proactive policy making.

Generally, administration officials were publicly supportive of the report. In November 1997, the Subcommittee on Technology, Terrorism and Government Information of the Senate Judiciary Committee held public hearings on the report and a succession of officials appeared to sing its praises and support its call for action. John Hamre, the Deputy Secretary of Defense, appeared especially enthusiastic:

> DoD is very interested and involved in infrastructure protection issues and we are, and have been, taking steps accordingly.
> • In 1995 we created a directorate within the Office of the Secretary focused on infrastructure protection, specifically in anticipation of the growing importance of these issues.
> • We actively supported the President's Commission on Critical Infrastructure Protection and the Attorney-General's efforts that gave rise to it.
> • We have established a Critical Infrastructure Protection Working Group to continue working and coordinating DoD infrastructure issues.
> • We are actively developing a Critical Asset Assurance Program focused on our own special structures, such as logistics, Space, and the Defense Information Infrastructure, as well as their interface to the related national infrastructures.

- We have established crisis reaction centers to monitor our computer networks and react to indications of unauthorized penetration of our systems.
- We have created a classified internet-like system and utilize state-of-the-art firewall protection features to protect that system.

Hamre added that "There will be an electronic attack sometime in our future. I don't think such an event is imminent, but when it comes, our ability to withstand it will depend in large measure on steps we take now and in years ahead."

While the testimony was no doubt reassuring to his audience and the public, Hamre had perhaps forgotten—or did not know about—an exercise called Eligible Receiver that took place in the Pentagon a few months earlier and exposed the shocking vulnerabilities that exist in American national security today. The full dimensions of Eligible Receiver have not been publicized before, and the Top Secret classification on everything to do with the exercise should have ensured that word of the disaster should never have leaked out. However, some senior officials believe that the weaknesses exposed by the exercise are so serious that urgent action—beyond the Pentagon's normal "business as usual" approach—is required.

Eligible Receiver was an exercise ordered by the Joint Chiefs of Staff to test the ability of the military and political structure to withstand a concerted cyber attack. To ensure realism, the Red Team of outside hackers was allowed to use only techniques and information that could be downloaded from the Web, was given no insider information and had to work within U.S. law. Over three months during the summer of 1997, the Red Team proceeded to destroy America's ability to wage war.

The attacks focused on three main areas: the national information infrastructure, the military leadership and the political leadership. In each of these areas, the hackers found it exceptionally easy to penetrate apparently well-defended systems. Air traffic control systems were taken down, power grids made to fail, oil refineries stopped pumping—all initially apparent accidents. At the same time, in response to a hypothetical international crisis, the Defense Department was moving to deploy forces overseas and the logistics network was swinging into action. It proved remarkably easy to disrupt that network both by changing orders so that, for example, headlamps rather than missiles end up at a fighter squadron, and to interrupt the logistics flow by disrupting train traffic or by so interfering with the movement and servicing of civilian aircraft that there were no planes to carry the troops overseas.

The political leadership tried to ignore and then cover up what appeared at first to be random attacks. When evasion no longer worked, the hackers began to feed false news reports into the decision-making

process so that the politicians faced a lack of public will about prosecuting a potential conflict and lacked detailed and accurate information
about what was actually happening.

The result was a serious degradation of the Pentagon's ability to
deploy and to fight. And even if deployment had been possible, the
assessment was that it would have been unlikely that the President and
his advisers would have committed U.S. forces to conflict. In other
words, a team of hired hackers, using commercially available information and artificially constrained by the law and the rules of the game,
had successfully shown that an electronic Pearl Harbor is not only possible today but could be completely successful.

"Eligible Receiver was a real shock to us all," said one Pentagon
official. "It should have been a wake-up call, but as so few people know
the details, I'm not sure who has heard the alarm and what they're doing
about it."

Even before Eligible Receiver got under way, there were clear signs of
the problems ahead. In April 1997, Defense Secretary William Cohen
warned that the Internet makes it easier for terrorists not only to gain
access to attack America but also to learn where the most vulnerable
points would be.

Two computer security experts, Howard Whetzel of Avenue Technologies, Inc., and Kenneth Allard of Cyberstrategies, Inc., set out to test
the Secretary's premise.

"We wondered what would happen if information technology skills
were turned against critical information facilities in the U.S. In particular, what information could a terrorist or a disgruntled employee get
hold of by examining public information on various Web pages? Our
hypothetical terrorist had neither specialized information nor a physical
presence within the U.S., only the computer skills and Net-surfing savvy
of a high school student."

The results of their experiment should chill the heart of anyone involved with national security. This is what their terrorist could achieve
with readily available information from the Internet:

• Identify a nuclear power plant somewhere in the United States that
had experienced enough safety and security problems to suggest deeper
vulnerabilities

• Pinpoint where fissionable nuclear materials were stored and processed

• Locate the most vulnerable point for physical attack, including the
key safety and backup systems

• Compile a remarkably complete profile of the security force, including names, addresses and even pictures of key personnel

• Document the local community's emergency response system, including notification rosters, traffic choke points and radio broadcast frequencies

Being able to gather such detailed intelligence without even having to enter America gives the "terrorist" a huge advantage. He has the target, the vulnerabilities, advance warning on security and an escape route, all courtesy of the World Wide Web. Whetzel and Allard contend that such vital information had been allowed to seep out onto the Web in bits and pieces because across a range of different organizations and agencies nobody had considered how small, seemingly benign pieces of information would fit into a larger and much more terrifying picture.

"If our democratic freedoms are not to become a suicide pact, detailed technical and security information about our power plants must be removed from the Internet—even as new defensive measures are put in place," they wrote.

The work of Whetzel and Allard, combined with Eligible Receiver, sent a series of shock waves through the intelligence and military establishment. At last there was a recognition that America, the last remaining superpower and the most technologically advanced nation in the world, was also the most vulnerable to cyber attack. Although much of the debate has occurred hidden from public view, there have been occasional insights into the thinking of those charged with defending the country. One of those insights came when Lt. Gen. Patrick Hughes, the Director of the Defense Intelligence Agency, testified before the Senate Select Committee on Intelligence in January 1998. After the usual overview, the General gave his perspective on just where the future threat to the country may lie.

Technology, combined with the creative genius of military thinkers around the world, is leading to the development and application of new forms of warfare, and the innovative modification of traditional military practices. While the US and its allies are the source of much of this innovation, others are motivated by the dominant military position of the US, and our demonstrated commitment to maintaining our military lead. This basic reality is forcing many of our adversaries (current and potential) to seek other means to attack our interests.

• Information Warfare involves actions taken to degrade or manipulate an adversary's information systems while actively defending one's own. Over the next two decades, the threat to US information systems will increase as a number of foreign states and sub-national entities emphasize offensive and defensive information warfare strategies, doctrine and capabilities. Current information on our vulnerabilities, and foreign intelligence initiatives in general, point to the following threats:

- Trusted insiders who use their direct access to destroy or manipulate the information or communications system from within.
- Modification of equipment during transport or storage.
- Physical attack of key systems or nodes, including the insertion of modified or altered hardware.
- Network penetration to include hacking, exploitation, data manipulation, or the insertion of various forms of malicious code.
- Electronic attack of various interconnecting links, sensors that provide data to the system, or other system components.
- Empowered agents including "sponsored" or individual hackers, cyberterrorists, criminals or other individuals who degrade, destroy or otherwise corrupt the system. In the most advanced case, empowered robotic agents, embedded in the system, could be used to take autonomous (timed) actions against the host or remote systems or networks (cyber war).

In the months following the delivery of the Marsh report, the Clinton administration struggled to come to terms with this new challenge. The response was typical of a Washington stuck in the old mold: principals from all the relevant agencies gathered for regular meetings at the White House to discuss how the problem could be managed effectively. While everyone saw potential for new turf and more money, there was no consensus about the way forward.

The result of the establishment of the Interagency Working Group in October 1997 was a series of attempts to draft a Presidential Decision Directive that would provide policy guidance for the administration into the next century. In the polarized debate, two camps had formed. The first was comprised of Justice and the FBI, who wanted to establish a National Infrastructure Protection Center staffed by FBI agents and DOJ lawyers. In the other camp was everyone else, who believed that an Information Sharing and Analysis Center modeled on the Centers for Disease Control in Atlanta was a better idea. The ISAC would be civilian run and have no law enforcement involvement but would act as a clearinghouse for all information about threats to the critical infrastructure or methods of defending it.

However, because ISAC would increase no department budget and would probably lessen every department's authority, the idea gained no constituency. That left Attorney General Janet Reno victorious with a plan that nobody except her department actually wanted.

The first draft of the secret presidential directive was circulated within the National Security Council in December 1997 and was scheduled to be signed by the President the following month. However, the PDD underwent several rewrites over the next three months and was finally ready for the President's signature in April 1998. The final draft contained an opening statement that for the first time since the end of the Cold War recognized a shift in focus from threats to vulnerabilities. This statement acknowledged that, in the future, threats will be difficult to discern while vulnerabilities will be easier to assess. The directive also states that America will have a comprehensive critical infrastructure defense within five years. Among the key provisions are:

- Congress must be involved in the process of developing a critical infrastructure.

- International cooperation will be encouraged to find solutions to the problem.

- There will be frequent assessments of the critical infrastructure to identify vulnerabilities. (This is controversial to industry, which does not want government bureaucrats giving it a performance measurement.)

- Market forces should be allowed to some degree to determine how the infrastructure should be defended. For example, industry could establish minimum security requirements, and the insurance industry could begin writing cover where the size of premiums might be decided by the degree to which industry had infrastructure defenses in place.

- The government should at all times be aware of privacy concerns. Overly aggressive action by the government could provoke a backlash and create the very dissent that might in turn threaten the infrastructure.

- Through procurement, governments should try to influence the development of an effective infrastructure defense.

- The private sector will be encouraged to protect itself.

- Law enforcement will play a key role in any regulatory effort.

The net effect of the PDD was to cement the Justice Department's victory. It also left the establishment of the ISAC up to industry and made clear that no funds would be allocated to a body that everybody except the DOJ believed was the right solution. It will now be up to Congress to mandate a different course.

What Eligible Receiver showed and every subsequent study has confirmed is that America today is uniquely vulnerable to cyberspace attacks. It is difficult to imagine a time in the nation's history where the country was so exposed to foreign attack. During the Cold War, the vulnerability to a nuclear first strike was matched by the certainty of Mutually Assured Destruction. There is nothing comparable in the post–Cold War era. On the contrary, it is likely that any cyber attack will initially go undetected, that retaliation will be exceptionally difficult because it will be virtually impossible to track the attacker through cyberspace, and the effects of such an attack will be devastating. This is a new form of warfare fought with new weapons and with no rules that anyone yet understands. It is vital that America grasp the problem and use its influence as the world's superpower to encourage and, if need be, compel other nations to develop a common approach. Isolationism in the infosphere is a luxury no nation can afford. Ignorance, too, will demand a heavy price.

THIRTEEN

Venimus, Vidimus, Dolavimus (We Came, We Saw, We Hacked)

A sniffer program attaches itself to a computer system and records the first 128 keystrokes of people gaining legitimate access. Within those 128 keystrokes would reside the password and log-in information for that account. The sniffer would then transmit those keystrokes back to the hacker, who would have all he needed to roam around the target system.

On March 28, 1994, a sniffer program was discovered on the computer system at Griffiss Air Force Base in Rome, New York. This was no ordinary computer system; in fact it was a network of systems that together constituted the Rome Labs, and here the Air Force conducted some of its most secret electronic business, such as researching artificial intelligence, target detection and tracking and radar guidance systems. The sniffer was in, and sending password and log-in information back to its host. The startled computer team at Rome Labs sent an alert to CERT, the computer emergency response team set up in 1988. CERT soon saw this was more than a breach of computer security—this was national security under threat. Jim Christy, the Air Force's Chief of Computer Crime Investigations and Information Warfare, and his team were called in.

The investigators gathering at Rome Labs soon found out the scale of the attack they were dealing with. Not one but seven sniffers were in computers across the network and had been for five days before anyone noticed. The hacker was systematically reading e-mails and copying files to himself. It was mainly unclassified material such as battlefield simulation plans, but it would only be a matter of time before he found more secret material to steal.

The first dilemma in such investigations is balancing the importance of the material to be safeguarded against the need to catch whoever is breaking in and stop them from doing it again. It is not easy to achieve both ends. If the door is slammed in the hacker's face, he will know instantly and go away. The optimal course is to keep the door open but restrict where the hacker may roam and watch him very, very carefully. That way, there is a chance of tracing him back to his home computer and catching him in the act. The commander at Rome Labs agreed the same strategy should be used. All but one door was locked with the intent of catching him next time he tried to get in.

It was easy to trace the first leg to where the attacker was coming from. Although Christy knew the hacker had most likely used a number of routes to get into the Rome Labs, most of the penetrations were coming through two American Internet providers. The first was cyberspace.com, based in Seattle, Washington. The second was mindvox.phantom.com, a New York network set up by a group of quasi-hackers, formerly belonging to the Legion of Doom. The notorious Legion of Doom was a hacker group responsible for numerous intrusions in the late 1980s and early 1990s, and several of its members had been convicted of computer crimes.

The chase would not be easy. The investigators could follow the hacker's activities through what is known as keystroke monitoring, roughly the equivalent of a wiretap because it records the keys that are typed when the intruder enters the system. But this is effective only on the system being attacked, it doesn't lead to much information about where the intruder is coming from. The rules about what investigators could or could not do were vague, as this kind of crime was uncharted waters. Christy consulted the legal experts at AFOSI and the Justice Department to find exactly how far he was permitted to go in catching his prey. The news was not encouraging. He wanted to tap into the online chat sessions via which computer fans swap ideas, intelligence and gossip. He hoped to glean something from there, but was told only limited surveillance of the sessions was allowed. His other route was to trace the phone calls down which the hacker sent his modem commands, but that was even more difficult because the hacker had been phone phreaking to access international phone lines at no cost, and was using this skill to route his calls through phone networks and Internet sites all over the world. He would leap from continent to continent, leaving a confusing trail through Europe, South America, Mexico and the Hawaiian Islands. Sometimes he would land in Rome, New York, and other times he would visit NASA's Jet Propulsion Lab in California or Goddard Space Flight Center in Maryland. Nobody would be able to track him that way. This New Age crime would have to be cracked using Old World methods—sweat, patience and dumb luck.

Christy got lucky. The hacker made a false move. In an attempt to launch from Rome Labs to the Army Corps of Engineers in Vicksburg, Mississippi, the hacker for the first time exposed his code names: Datastream and Kuji.

Out in cyberspace, just like on the streets, law enforcement officials cultivate informers, would-be or onetime criminals with the connections and the knowledge of the players in cyberspace. AFOSI had developed a number of informants that could cruise the Internet, searching for information about illegal computer security breaches and illegal hacking plans. Sometimes these searchers would be hackers themselves, either with a conscience or with a threat to have their parole reconsidered. Christy asked that the code names Datastream and Kuji be tracked by these cybergumshoes. The breakthrough came on April 5, 1994, only eight days after the intrusion at Rome Labs had been noticed and less than two weeks after it started. One of the informers revealed an e-mail exchange he had had with someone called Datastream Cowboy. In the exchange Datastream claimed to be from Britain and said he was sixteen years old. He had been bragging about his abilities to break into computer systems and stated that he had a preference for military computer systems sites because they were so easy to crack. With confidence bordering on the arrogant, Datastream had set up his own electronic bulletin board where messages could be placed. Thinking nobody would connect this with the Rome Labs break-in, he left a home phone number, whose codes indicated it was in England. Christy brought Scotland Yard into the hunt, and they traced it to an address in Tottenham, an area of north London. They had found the Datastream Cowboy's lair. But a sixteen-year-old boy? It was hard to believe a teenager from north London was a sophisticated spy.

Scotland Yard continued to monitor the Tottenham phone and found it was indeed being used for phreaking, a crime in Great Britain. Scotland Yard's interest went from assisting the Americans to dealing with a suspected criminal. Still, Christy needed to connect the boy with the Rome Labs attack. Monitoring the line determined that at the same time phone phreaking was initiated at the Tottenham phone number, there would be a hacker incident at the Rome Labs. But it was circumstantial at best. They needed more evidence.

They continued to watch and wait. The Cowboy had slipped up once. Maybe he would again. His confidence and skills grew the more he exploited the Rome Labs' system. On April 10 he entered the home computer of an aerospace contractor. His sniffer program had captured the contractor's user name and log-on password at Rome Labs. He then broke into the contractor's computer using an Internet Scanning Software attack. This program collects information about the operating system and other information that might provide a weak link to invade. He could crack passwords using decryption tools available on the Internet. Two days later he used the Internet Scanning Software to break into Brookhaven National Labs, part of the Department of Energy in New York. He had also been online for two hours with the aerospace contractor's system.

Christy's team had Datastream pretty much covered. It was the other password, Kuji, they wanted to track now. On April 14 Kuji was found, jumping from a site in Latvia to the same Seattle Internet provider, cyberspace.com, that Datastream used. He entered the Goddard Space Flight Center in Maryland, but was kicked out by the Air Force to prevent the transfer of sensitive data.

Not long afterward Datastream jumped from cyberspace.com to the National Aerospace Plane Joint Program Office and transferred data from Wright-Patterson Air Force Base to the same site in Latvia that Kuji was unable to penetrate. The same day Kuji initiated an Internet Scanning Software attack against Wright-Patterson.

It was quite obvious now that Datastream and Kuji were either the same person or two individuals working together. Both were using the same Internet providers and were breaking into the same facilities. On April 15, Kuji initiated an Internet Scanning Software attack against NATO headquarters in Brussels, Belgium, and Wright-Patterson AFB from Rome Labs. He was barred from NATO, but a systems administrator for NATO in The Hague reported that Datastream had penetrated the SHAPE Technical Center (NATO headquarters) from mindvox.phantom.com in New York.

Investigators then discovered that when Datastream had difficulties in obtaining access at a point, he would log on to one of the Internet providers to exchange e-mail with Kuji. These e-mail exchanges would last between twenty and forty minutes, at which point Datastream would again attempt a break-in, but the second time he would be successful. Datastream was being tutored by Kuji.

After the NATO break-in, it seemed now that there was enough evidence to obtain a search warrant for the Tottenham address. However, Christy's investigative team wanted absolute proof that Datastream was residing in Tottenham, so they decided to wait until Datastream was online with Rome Labs. This would also identify all the routes that Datastream used to connect to the Internet providers in New York and Seattle. Christy's team watched as Datastream logged onto Rome Labs, and quickly jumped to a computer system at the Korean Atomic Research Institute, downloaded files from Korea, and dumped it on the Rome Lab system.

Panic swept through the investigative team as they tried to determine whether the information Datastream had stolen was from North or South Korea. If the information was from North Korea, the North Koreans would regard the transfer as an attack or act of war by the U.S. Air Force. Negotiations at the time between North Korea and the United States regarding their atomic weapons programs were extremely sensitive, and this could potentially damage the negotiations. Waiting too long before nailing Datastream might have inadvertently caused an international row of epic proportions.

To Christy's relief it was determined that Datastream had hacked into the South Korean Atomic Research Institute. It was a close call, a signal that the investigation had to be brought to a swift conclusion. There was no telling what Datastream would do next.

Scotland Yard took out a search warrant, ready for the word from Christy. On May 12, Datastream was online at Rome Labs and the British police were given the go signal. They burst into the Tottenham house and raced to a third-floor bedroom where they surprised sixteen-year-old Richard Pryce desperately trying to log off his computer. Pryce fell to his knees, balled himself in the fetal position and began to cry.

It was not just the Air Force coming under attack. The Navy was having troubles of its own, with an Argentine hacker named Julio Cesar Ardita, a twenty-one-year-old who lived with his parents in a Buenos Aires apartment. He routinely wandered through U.S. Navy and NASA computer systems, never causing any damage but just by his presence exposing a security gap of troubling proportions.

In late August 1995 a naval computer systems administrator at the Naval Command, Control and Ocean Surveillance Center in San Diego spotted some unwelcome guests among his files. It was the unfamiliarity of the file names that alerted him, names such as sni256 or test. He opened them up to discover they were sniffer files. Someone had the keys to their computer.

Investigators ran cross-checks across other naval systems and sure enough similar sniffers had been planted, from the look of it by the same person. Computer code writers have their own idiosyncrasies as distinctive as handwriting and experts can spot patterns of behavior common to an individual. The naval investigators started to cast their net further afield, and found the hacker had planted his sniffers in Army computers as well. His list of targets was impressive: the Army Research Lab, the Navy Research Lab, the Naval Command, Control and Ocean Surveillance Center in San Diego, NASA's Jet Propulsion Lab and Ames Research Center, Northeastern University, the University of Massachusetts, Caltech, as well as sites in Korea, Mexico, Taiwan, Chile and Brazil. The common factor, apart from the distinctive code construction and the use of file names such as zap and pinga, was that the lines of access into the exposed systems all led back to Harvard University's Faculty of Arts and Sciences computer. Any attempt to nab the intruder would have to start there.

Between October and December 1995, the U.S. Attorney's Office for the District of Massachusetts secured a groundbreaking new warrant that allowed, for the first time ever, an electronic surveillance of a computer network. This is not to say no computer had ever been monitored; they had, but the arrangements were always ad hoc affairs between consenting computer owners and investigators, or everyone using the system was warned of the surveillance. This was the first court-authorized covert surveillance of a computer system in history, a quantum step forward in law enforcement's handling of computer crime.

Running the surveillance was no easy task. The U.S. Attorney was distinctly conscious of the civil liberties implications of a blanket federal surveillance of a computer through which thousands of pieces of private information passed each day. Just as it would be unacceptable to tap everyone's telephone at Harvard in the hope of finding one wrongdoer, it would be unacceptable to monitor all computer transactions through the Harvard computer to find one hacker. The investigators created a filter that would be triggered by key words appearing on the system, so that, they hoped, only the hacker's activities would be monitored. For two months they watched their hacker come and go, each time adding to their knowledge of him. In December they followed their intruder to an Internet Relay Chat (IRC) session. IRC is a free-floating conversation in cyberspace going on all the time, all over the world. Different channels offer subject matter for participants to chat about. The intruder joined the channel called #hack.br., calling himself Griton, which is Spanish for screamer. They ran a trace on the name Griton through publicly accessible Internet files and found the name had been used in postings to a bulletin board called Yabbs. Griton had invited other Yabbs users to view his own bulletin board, called Scream!, in Buenos Aires. It all started to come together. Several times the Harvard computer had been accessed from four different computer systems in Buenos Aires. The investigators alerted the Argentine police, who identified the telephone number Griton gave out for Scream! as being registered to the apartment of the Ardita family. They tapped his phone and overheard twenty-one-year-old Julio Cesar Ardita tell his girlfriend, "I've infiltrated the U.S. Navy, I've even seen inside submarines and much more. I could very easily have wiped out files and rubbed out any information."

One of the computer systems Ardita had used to channel into the Harvard system belonged to Telecom Argentina, the country's national telephone company. He had hacked into it, which meant he could be charged under Argentine law. When the police arrested him at the end of December 1995 they took away his equipment and ultimately he was charged with unlawful interception of electronic communications, destructive activity in connection with computer files and possession of unauthorized access devices. But would Ardita be brought to trial in the United States? The answer was a ringing no. None of his alleged crimes brought him within the framework of the extradition treaty between the United States and Argentina, so apart from an Interpol alert from the FBI to say he was wanted in America, he remains in his own country.

Despite this, law enforcement in America congratulated itself on a job well done. A Department of Justice press release called the investigators "cybersleuthers." "We are using a traditional court order and new technology to defeat a criminal while protecting individual rights and Constitutional principles that are important to all Americans," Attorney General Janet Reno declaimed. In light of what Ardita had done, it was his father who made perhaps the most relevant comment on the affair.

"These Yankees don't have the slightest idea about security," Julio Rafael Ardita said. "Who is at fault? Obviously the North Americans are not very clear on the security of their systems if a kid from South America can enter them. I would be ashamed to admit it."

The Attorney General was banging the federal gong, but did not address two pressing questions: What does cybercrime mean to the concept of national sovereignty if criminals from all over the world can attack U.S. interests with impunity? And why, particularly after the incident at Rome Labs, were military computers still so vulnerable?

The first question continues to tax the finest Constitutional minds. The second question was being asked at the highest levels of government, and the answer was not reassuring.

The General Accounting Office is the watchdog of the U.S. government. It is there to assess efficiency and waste in the way tax dollars are spent and its reports are often trenchant reminders to government that it could do its job better.

The report that Jack L. Brock, Jr., the GAO's Director of Defense Information and Financial Management Systems, sent to Senator Sam Nunn, the Chairman of the Senate Armed Services Committee, on May 22, 1996, was more than trenchant. It was shattering in the bleak picture it painted of computer security in the U.S. defense establishment. The bottom line, Mr. Brock reported to the senator, was that "attacks on Defense computer systems pose a serious threat to national security." How was that threat manifested? Not just by Datastream Cowboy and Griton, but by an estimated 250,000 hacker attacks on the 2.1 million computers the defense establishment owned and operated over its 10,000 local networks and 100 national networks. The word "estimated" was used because, as Brock admitted, only one in 150 attacks is actually detected. How does Brock know the system is vulnerable? Because the Defense Information Systems Agency (DISA), a part of the Pentagon that actually runs test attacks on these systems on a regular basis, successfully breaks in 65 percent of the time. To that extent, the much-used statistic of 250,000 is useless as anything but a broad guide, as it is based on the DISA's own attacks. That should not in any way be taken to mean the threat is overstated. Attacks on defense systems are very real. The report said:

attackers have obtained and corrupted sensitive information—they have stolen, modified and destroyed both data and software. They have installed unwanted files and "back doors" which circumvent normal system protection and allow attackers unauthorized access in the future. They have shut down and crashed entire systems and networks, denying service to users who depend on automated systems to help meet critical missions. Numerous Defense functions have been adversely affected, including weapons and supercomputer research, logistics, finance, procurement, personnel management, military health, and payroll.

While none of the defense computers connected to the Internet carries top secret material, what they do carry is vital to the day-to-day operations of the services and that, as the report stated, includes sensitive weapons research. Corrupting these computers corrupts America's ability to defend itself.

One security company tells a cautionary tale of how it tested Pentagon defenses by hacking into the computers that control logistics supply. Using a home computer from overseas, the company's in-house hacker simply swapped two order numbers in the files. When a motor pool at one Air Force base ordered new headlights, they got missiles. When a fighter wing ordered missiles, they got headlights. The threat to security is bad enough. Financially it was a serious drain on budgets. Brock's report said that the Rome Labs affair alone cost $500,000 in personnel and system time. The other attacks cost by very rough estimate anywhere from tens to hundreds of millions of dollars a year.

Hackers are like malevolent imps playing at will, light on their feet, destructive, malicious or sometimes just curious, but often a serious risk and always a nuisance. Their playground was wide open, and in serious danger. As to whether any lasting damage had been done, who knew? The GAO report warned that sophisticated hackers with the intent to do lasting harm could plant all sorts of time- or logic-based programs that would activate in future when a specific set of instructions was run on a system. These "bombs" could lurk unseen for years, then spring to life at the critical moment when a mission vital to national security is put into operation. The entire system could be infected, and no one would know.

The immediate reaction to learning of conclusions like these is that the military put up the shutters, triple-bolted the doors and ensured nobody else ever got in again. Sadly, as the GAO report concluded, it would take a great deal of time, money and education to achieve that kind of security. "Defense's policies on information security are outdated and incomplete," it stated. "Defense personnel lack sufficient awareness and technical training. Technical solutions show promise but cannot alone provide adequate protection," it concluded. The report painted a picture of military bases whose commanders had no idea about the risks involved in hacker attack and therefore paid the issue little attention, and of staffing systems that did not include positions for information security professionals, which meant the job was left to part-time, inadequately trained personnel. On top of all that, the cost of installing the various technological safeguards to stop the hackers getting in, such as network firewalls and proper encryption systems, meant it would be a long time before the barriers could be put in place.

It is easy to conclude that the complacency, ignorance and narrow-mindedness that had led to the point where hackers posed a threat to national security were endemic throughout the military, but that would not be fair, nor is it the case. Just as a certain part of the armed services responded with great speed to challenges in the field of battle, so did elements of the military see the danger and try hard to do something about it. That the kids were looting the candy store with such impunity was not entirely their fault.

Kelly Air Force Base lies outside San Antonio, Texas. Having driven past security, one proceeds through an immaculately maintained eighteen-hole golf course and up a small bump in the flat countryside, known as Security Hill, on top of which sits a low brick-built three-story building. There is not an aerial in sight, nothing that demonstrates this is one of the most intensively wired sites in the whole of the military, the Air Force Information Warfare Center (AFIWC). The battles fought here are along the copper-wire and fiber-optic networks that create the Internet.

I was the first journalist allowed into this nerve center, the front line in the Air Force's information war. The heart of AFIWC is the Operations Center. Twelve computer screens are stacked three high around a circular pillar. Surrounding that are a series of consoles which are manned by a minimum of eight people twenty-four hours a day. It looks not unlike the bridge of the *Enterprise* in *Star Trek*. Computers display the center's logo, a cyberknight that controls a system called Cyberwatch and looks out for threats and warnings of hacker attack around the world. The system is wired to Intelink, which accesses 130 individual servers in all the key intelligence, defense and law enforcement agencies in America, including the Department of Justice, the State Department, the DIA, the CIA, the NSA and the other military services.

In the same Ops Center is the Combat Intelligence System, a fully mobile battle management–intelligence management system. Each unit is a $70,000 Sun computer with two 18GB hard drives. The system runs mission-specific Lockheed Martin software that can be made mission-ready in just twenty minutes.

Between them these two rooms represent the spearhead of the

THE NEXT WORLD WAR 2 0 2

USAF's defensive and offensive information warfare capability. It was created in 1993 in response to the growing buzz in the military about information warfare, the warfare of the future. When America next goes to war, it is from AFIWC that the attempt to attack the enemy in the heart of his information systems will emanate, degrading his supply of information, forcing him through deception to trust that information, crashing his computers, planting false data. While the USAF's planes attack his body, AFIWC will be confusing him with weapons such as "packet forge spoofing," in which the computer file detailing an enemy's order of battle is corrupted by U.S. cyberwarriors to create chaos in the enemy lines. Commanders will try to deploy fighter squadrons that do not exist; when they try to move a tank brigade from its location, they will learn it is actually sitting somewhere completely different. Planting viruses and worms in enemy computers has been under discussion since the Air Staff's Revolutionary Planning Office made the first serious proposal in 1995.

> These viruses would lie dormant until activated by satellite, aircraft radar, jamming equipment etc. When activated the virus would render the weapon useless. This could be done in a very destructive manner or, better yet, this could be a very subtle change for a finite period of time.

The man responsible for the center is Maj. Gen. Michael Hayden, the Commander of the Air Intelligence Agency, who was previously responsible for IW operations in Bosnia. Hayden brims with enthusiasm for his discipline, an infectious smile never far from his lips but his eyes always focused and intense. This is the Air Force's top information warrior, a man who understands information to be more than something passed along lines of communication to help fight the wars of old. Information is a whole new dimension.

"Information is either a place or a mission," he believes. "If you think of it as a mission, you tend to think about how to use information to do things you are already doing. In other words, information is a servant. But if you think of it as a place, then you think about it as an alternative. Here we believe information is a place. It has its own terrain and it is where we intend to conduct military operations and win. War has evolved through land, sea, air, space and now information. We believe that information is just another battle space.

"Information warfare is in two parts. There is information in support of heat, blast and fragmentation. We are very good at that. There is also information in support of IW only. With that there are other ways to accomplish your goals."

If information is a battle space, it is one where America is losing the fight against the world's hacker armies, and it is fending off that threat that most occupies AFIWC's attention. One of the key strategists in dealing with attacks is Larry Merritt, AFIWC's technical director. Merritt is now a civilian but was an officer in the Air Force, serving at the Cryptologic Center at Kelly, which was folded with the Electronic Warfare Center to form AFIWC in 1993.

"AFIWC has 1,000 people. It's a new realm of battle. Five years ago we needed to find out what was going on on the networks," he said. "So we built ASIM." This is probably the most potent weapon yet in the fight against the hackers. ASIM stands for Automated Security Incident Measurement and has become the most powerful tool in the hands of the AFIWC Computer Emergency Response Team, which leads the hunt for illicit intruders. Its operation, which has never been described before, is an illustration of how hacker wars will be fought in the future.

"We are at war every day trying to detect and defend Air Force networked systems against those unwanted intruders that are costing us millions each year and placing billions more in resources they represent at risk," warned Lt. Col. David Srulowitz, the Commander of AFCERT. "ASIM is sniffer technology in reverse," he said. "It is programmed to look for strings of data and monitor patterns and then allocate a suspicion value from 0–10 and capture the data down to the keystroke. The information is processed every twenty-four hours but a newer version will deliver data in real time."

ASIM is the digital bloodhound on constant patrol that automatically chases hackers back up the line when they break in. Without the intruder's knowledge, the system splashes him electronically with an identifying marker so that wherever he goes through the Internet he can be traced. But Srulowitz said the same decision that faced investigators from Clifford Stoll to the present day still has to be made. "We have two choices: Fishbowl, which allows the hacker new access so that we can track him. Or button up the system and make sure that no worms are left behind.

"ASIM is both offensive and defensive," Srulowitz added. "We are seeing 7 [million] connections a month that seem suspicious. ASIM checks 2 billion connections a year on 111 bases."

Bases are increasingly anxious to run their own protections, and AFIWC is now busy loading ASIM on the systems at Air Force bases throughout America and the world. One hundred seven had been installed by April 1997, with many more scheduled to come online soon after.

Although the officers and men at Kelly are reluctant to talk about it, there is both a defensive and offensive capability housed at AFIWC. It is here, thousands of miles from any likely front line, that the new wars can be waged and won. Where ASIM allows for the tracking of attempted hacks, other systems can allow for more aggressive action, including the insertion of viruses and worms to destroy an enemy's ability to wage modern war.

It sounds optimistic, but life in government is not simple. There are wheels within wheels, spheres of influence and territories that overlap, turf to defend. Larry Merritt makes it clear that anyone looking for the root cause of the government's slow response to the hacker threat should look at government itself.

"The technology was not difficult but it took two years to get approval to use it because there were so many legal problems that needed Department of Justice approval," he said. "Then we found we were actually catching intruders and so the criminal investigation guys wanted to get involved so we had to go back to DOJ to work that. Then we wanted to put it worldwide because we realized it was a great indications and warnings system. So then we had to think who to tell, how to analyze the material and how to respond to what we learned. The technical issues have not been the hard part. Getting people to understand and getting people to respond, that has not been easy to do."

The Rome Labs incident, one of the first big hacks that AFIWC was called in to help solve, was a classic case in point.

"When he [Datastream Cowboy] came into Rome Labs, all we knew was that he was coming in from multiple points and from the keystrokes it looked like it was the same guy. We had to find out where the electronic signals originated," Merritt recalled. "It was taking us too long to go back through the routings. Yet it took us three days to persuade the DOJ to use hack back. [Once we did], thirty seconds took us to an Internet service provider in Seattle which was his access to the Internet. From there we were able to go back to find his original user ID [Datastream]. Despite that one-off deal with the DOJ we have not been able to use hack back since. We have tried three times and each time DOJ argues that it is not effective so why use it?"

The legal dilemma that caused the Department of Justice such heartburn is that chasing a hacker back through the Internet means the hunters are breaking the same rules as the hacker with unwarranted intrusion on private systems. Yet, these rules, which were designed for a simpler world where cyberspace did not exist, are ill-suited to the information revolution.

And it is not just other areas of government where it is hard to get quick action. Inside the Air Force itself response times verge on the glacial. "In January 1994 we sent 150 questions in a seventeen-page memo to the Air Force General Counsel to try and get some legal foundation for moving forward in the future. There has not been a single answer to a single question," Merritt said.

David Srulowitz sees it as a matter of attitude. People not familiar with the world of networks bring the wrong mind-set to the issues involved in hacking. "I am not an intelligence officer," he said. "I am a computer guy. The intelligence people will see an intrusion from a foreign country and say 'My God!' Their response is to believe automatically that the attack came from there. I will say it's just the last stop that we can see. A lot of the geography and the old rules have gone out of the window." The fact that hackers like Julio Cesar Ardita and Richard Pryce hopped their way through telephone systems and networks all over the world to disguise their origins proves the point.

In counterpoint to this reported inability at the highest levels of the Air Force to comprehend the nature of the challenge, and respond to people in the field who did, the service's top brass was putting its own stamp on IW and defining the way the Air Force would fight it. The document *Cornerstones of Information Warfare* issued by Air Force Chief of Staff Ronald Fogelman and Air Force Secretary Sheila Widnall defined their view of IW with a series of examples:

- Bombing a telephone switching facility is information warfare. So is destroying the switching facility's software.

- Hardening and defending the switching facility against air attack is information warfare. So is using an antivirus program to protect the facility's software.

- We may use information warfare as a means to conduct strategic attack and interdiction, just as we may use air warfare to conduct strategic attack and interdiction.

The paper also provided a definition of IW for the benefit of those who claim the term is just a new way of describing existing capabilities such as deception and electronic warfare. It made the point that these were forms of indirect attack, whereas IW is direct.

"Indirect Information Warfare [is] changing the adversary's information by creating phenomena that the adversary must then observe and analyze. Electronic warfare, when used to degrade the adversary's ability to accurately perceive phenomena, is conducting indirect information warfare.

"Direct Information Warfare [is] changing the adversary's information without involving the intervening perceptive and analytical functions. Replacing bar code labels with bogus ones on pallets in a deployment staging area is a form of direct information warfare."

Lt. Andy Proud is one of the Air Force's Young Turks in the fight for the new frontier. Blond and in his twenties, Proud sits hunched over his computer screen at AFIWC chugging can after can of Mountain Dew. He uses probing software to find the location of the hacker or to test the vulnerabilities of different computer systems around the world. On his

screen, small lightning bolts appear next to systems that seem vulnerable. He can test 500 systems throughout the defense establishment in
thirty minutes. If his probes get inside, he leaves eleven small brown
and black footprints that walk across the screen. He recalls a famous
incident that put the ASIM, and the team, to the test.

"On March 25, 1996, at Kadena AFB, ASIM detected an attempted
penetration," he said. "The report and data strings were transferred
down to Kelly and ASIM was programmed to check for past penetrations
by this intruder and report future ones. The news was not good. The
same intruder had made 100 attempts at ten bases over a one-week
period and included attacks on Ramstein in Germany, Langley, Gunter,
Yokota, even Kelly itself."

One month later, the hacker tried again and broke through the security screen at Hanscom AFB outside Boston and retrieved unclassified
Defense Department publications and data that were not particularly
valuable in themselves but that contained enough data to give him access to other systems. ASIM tracked the hacker back in a matter of
seconds through various rerouters to the National Center for Scientific
Research at Demokritos in Greece, which links Greece's foremost academic and research institutions. Three years earlier, two Convex supercomputers had been installed at the center with links throughout
Greece, so the hacker could have come from anywhere in the country.
The Air Force Office of Special Investigations (AFOSI) is currently investigating and despite repeated requests has received no cooperation
from the Greek government. Not for the first time the issue of sovereignty stood in the way of apprehending the hacker.

The Air Force (and other services) already have the capability not just
of following a hacker back to his computer but of launching computer
bombs that would destroy the offending computer and any software.
This kind of retaliatory attack, which would be considered standard
tactics in normal warfare, is currently illegal. As hackers gain in sophistication, and governments rather than individuals become the principal
culprits, it is certain that the laws will change to allow for more aggressive action. For the American defense community, there is little logic in
identifying a spy such as the hacker in Greece if nothing can be done
about the person carrying out the attack. This migration of economic
warfare and espionage to the infosphere represents a real challenge to
officials who have to interpret laws that were originally designed to
protect the borders of nation states. In the IW world, the concept of
international borders is largely irrelevant.

But legal issues are only part of the problem. In traditional military
style, when there is food in the trough, every pig wants to get a bite. The
result in the case of IW has been a proliferation of centers, AFIWC and
FIWC, the Navy's Fleet Information Warfare Center, among them, all

of which are designed to scoop up increasingly scarce financial resources in an effort that gets more disorganized by the day. Seven years after the end of the Cold War and the start of the information revolution, there is still no coherent policy in place to handle the challenges and opportunities presented by IW. Instead, every service, every intelligence agency and a multitude of government agencies are going it alone. There is no common standard and even no common definition of what IW is.

But if the United States is wasting resources, other countries such as Britain have not even begun to understand the nature of the challenge let alone attempt to meet it. Where the United States has thousands of men and women working on IW, the British Ministry of Defence has four men working part-time with no budget and a brief that is so narrow as to be almost meaningless.

So nervous is the Ministry of Defence about its IW effort that a request to interview the team running the program takes weeks to work through the system. Eventually, the request lands on the desk of the Director of GCHQ, the British equivalent of the NSA, and he eventually gives permission for the briefing.

"We haven't decided if it's a silver bullet or not," said the head of the IW unit. "Is it just an excuse for employing hundreds of people who would otherwise not have a job and get thousands of billions of dollars? The Americans have made a rod for their own back. It's all overhyped."

Behind these criticisms lie Britain's ingrained distrust of any American innovation. There is a strong perception running through the British intelligence community and defense establishment that IW is just the latest American fad and that once the Americans make all the mistakes and waste all the money, the British can come in with a system that works. In fact, the Ministry of Defence's IW team is only responsible for defending the infrastructure inside the ministry itself. There is no national effort to protect the nation's infrastructure, an extraordinarily narrow view of a society in the midst of the information revolution.

This approach shows a fundamental misunderstanding of the nature of that information revolution. By languishing behind the starting gate, Britain is going to be left behind, unable to catch up with those countries that display more vision.

As well as being the Air Force's first line of defense against hacker attack, AFIWC is also home to one of the six USAF battle labs established by Air Force Chief of Staff Ronald Fogelman. Alongside the Space Battlelab at Falcon AFB in Colorado; the Air Expeditionary Force Battlelab at Mountain Home AFB in Idaho; the Battle Management Battlelab at Hurlburt Field in Florida; the Unmanned Vehicle Battlelab at Eglin AFB in Florida; and the Force Protection Battlelab at Lackland AFB in Texas, the Information Warfare Battlelab at AFIWC is designed to find in the shortest possible time the optimal way for the Air Force to fight its future wars.

On its own AFIWC would not be enough for the Air Force to fight an information war. It needs a portable version of AFIWC, the laptop IW capability versus the PC. The USAF's road warriors are the 609th Information Warfare Squadron (IWS) based at Shaw AFB in South Carolina. Established in the fall of 1996, the 609th is a deployable counter-IW unit that will move wherever the 9th Air Force goes. "We're here for the protection of 9th Air Force assets against computer intrusions . . . and to affect the enemy," said 609th IWS Commander Lt. Col. Walter E. "Dusty" Rhoads.

When up to speed, the squadron will be 100-strong, all systems engineers—about forty officers and sixty enlisted soldiers, with perhaps two civilian specialists—who are skilled at "watching the fence" of a computer system, detecting or stopping intrusions, finding out who the intruders are, and preventing them from causing damage. It is a capability that already can "protect three or four bases," with fewer than a dozen people, and "if it provides a benefit . . . we may set up additional units" like the 609th at other numbered air forces, he said.

When called on to deploy, the 609th would take with it "computers, software and monitoring tools, firewalls, and routers," said Dep. Cmdr. Maj. Andrew K. Weaver. "Almost all of it is commercially available," he added. "The military is using almost everything off the shelf," he said, because the hardware and software are changing so rapidly that a custom-built system designed exclusively for the military would probably always be outdated, compared with an opponent's system. Colonel Rhoads refuses to discuss the 609th's capabilities for offensive IW operations except to say that anything an opponent might try to do to disrupt or disable a U.S. system could be met with a comparable response.

Protecting Air Force assets on the battlefield would be an uphill struggle if the hardware at the front line was not already hardened to IW attack. "Everything we do is aimed at insuring our product lines in tactical air," said Charles A. Anderson, vice president for Information Warfare Programs at Lockheed Martin Tactical Aircraft Systems in Fort Worth, Texas. Anderson said his organization, recently set up to mirror the mission areas outlined in *Cornerstones*, is developing the means to make certain that USAF F-16s and F-22s won't be vulnerable to IW attacks, either in the hangar or in flight.

"Suppose you were able to get into the database of a ground or airborne system and change it," Anderson said. The result could be a plane's sensors "recognizing" a friendly aircraft as an enemy or switching the target coordinates for a standoff missile. Such IW attacks could happen in the middle of a dogfight, sending missiles after phantom targets or disabling their ability to fuse. An aircraft's electronic fly-by-wire system might be crippled by electromagnetic pulses or high-power microwaves.

"We would be remiss in believing our systems are invulnerable" to such threats, he went on. "Nobody knows how much of this is feasible," but the company does not want to wait until it happens to start working on countermeasures. Lockheed Martin is also working on all other aspects of IW, from sensors and processors to jammers and knowledge systems that will push collated, reliable information into the cockpit in real time.

For now, counterinformation operations are not going to replace the F-117, or any other combat aircraft, in launching cyberstrikes against an enemy's command and control nodes and power grid. Colonel Rhoads believes such a scenario might be "ten to fifteen years away" at the earliest, though he acknowledges that technology might bring such a capability sooner. In the meantime progress toward any kind of future is impeded by a military bureaucracy that does not always share the same sense of urgency that the people at the front line might hope for.

"More than anything right now, I'm just scrambling to get desks for all my people," Andy Weaver said ruefully as he tried to get the 609th settled in at Shaw.

The issue of time is critical, according to Pat Gallogly, who commands the Battlelab. The Battlelab opened at the beginning of April 1997, and Gallogly says it has had to hit the ground running in developing new systems to assert U.S. dominance over the information battlefield. "We are losing the initiative because of the acquisition cycle," he said. "With IW, if you think of it now, it needs to be on the street tomorrow. There is simply no time to wait five, ten, fifteen years.

"We want to adapt the experience of the Gulf War with the bunker-busting bomb. The requirement was seen because there was nothing to penetrate Saddam's underground bunkers. We used a stainless steel nose cone, old howitzers and existing tail fins. Every single thing was mature technology and it came into service in a month and those bunkers were history.

"We expect 600 ideas a year and will process twelve. The goal of every operation will be to go from the initial plan to operational in eighteen months. Our motto is get it done and get it done quick."

One of the first priorities, not just for AFIWC but for the whole defense establishment, according to Larry Merritt, is to pool information that will increase protection by sharing experiences. "At present there is no common vulnerability database but we are working to produce one with DISA," he said. "Everyone has different standards, different vocabulary and different priorities."

One of the outfits that will be joining that initiative is the Navy's version of AFIWC, called the Fleet Information Warfare Center or FIWC. Started two years after AFIWC, FIWC is chasing the same imperatives and deadlines and fighting the same battles. Pushing the Navy into an aggressive IW stance was Vice Adm. Arthur Cebrowski, who moved from leading the Navy's Space and Electronic Warfare directorate to become director for Command, Control, Communications, Computers and Intelligence for the Joint Chiefs of Staff at the Pentagon. He saw offensive IW, the ability to destroy an enemy's information systems, as pivotal to future deterrence and defense. "The strategic landscape is forever changed," he said. "Like when [nuclear arms] appeared."

While still at Space and Electronic Warfare he had established the Navy Information Warfare Activity to act as a forum for creating the Navy's IW policy and helping the service implement it. He then set in motion the creation of FIWC, based at Norfolk, Virginia.

Inside FIWC, the Naval Computer Incident Report Team reflects the mixture of old-style military tradition with the new world of bits and bytes where their battles are fought. It is a complex filled with new boxes of replacement computers and the quiet urgency of young men and women on a mission to serve the world and have fun while doing it. The screen savers displayed on the rows of computers show Tom Cruise in the scene from *Mission: Impossible* where he is suspended by ropes over a computer at CIA headquarters, with the team's motto "Venimus, Vidimus, Dolavimus" scrolling across it. Loosely translated, it means "We Came, We Saw, We Hacked."

The team uses the Vulnerability Analysis and Assistance Program, which automatically tests for thirteen vulnerabilities inside the Navy's computer systems. Similar in concept to the Air Force's ASIM program, this alerts naval facilities to potential threats. The team then advises on security.

"Since our name has been out there, the phone has been ringing off the hook," said Lt. Grace Stover.

The outfit is run by Bear Barrett, a hardened war fighter who is the product of the Naval Strike Warfare School and served as a fighter pilot on F-14s and F-18s. Slim and silver-haired, with gold-rim spectacles belying his tough background, Barrett went on to specialize in space warfare. Today he commands the Navy's efforts to protect itself from the backdoor bandits. With an annual budget for 1997 of $8.4 million expected to double each year to 2000, it is obvious the Navy recognizes the importance of his mission.

"We are anxious to find out what are today's vulnerabilities or today's opportunities," he said. "What we found at first was that we needed expertise afloat. There are 380 people at Norfolk, 150 at Coronado and forty people at the EW facilities on the Chesapeake."

His frustrations mirror those felt by his AFIWC counterparts.

"We don't know what the bad guy looks like anymore," he said. "The traditional bad guy went away. The target is no longer people but systems. The computer becomes a target because it processes the ones and zeroes." And fighting those who steal those ones and zeroes, the genetic building blocks of the digital universe, becomes a tough job when the law lags so far behind reality. "We have a legal system based on the physical possession of property. Yet we are in a virtual world with virtual property."

As always, the desire of a cop to catch the bad guy has to be tempered by what society wants in terms of privacy and protection from intrusion by the government. Barrett feels the way cops do; anything that stands between him and funding the bad guy is a frustration, such as the hunt for Julio Cesar Ardita aka Griton the Argentine hacker.

"It took a federal wiretap authorization," Barrett said, a note of indignation in his voice. "The Navy Criminal Investigation Service had to convince the judge that we could limit the investigation and not read anyone else's mail."

Barrett waxes equally indignant over the failure of the establishment to understand just how much is at stake in this new realm of battle. "Our challenge is how to visualize the virtual. In the past, Electronic Warfare was relegated to the corner because it could not be seen. The same extremely impractical people are still out there saying 'Show me.' If there are no visual images, it's a very, very hard sell."

FIWC is more than a hacker tracker center. It is also home to the Joint Maritime Command Information System, the Navy Tactical Command System, which allows the Navy to model attacks by F-18s in full 3-D color and that delivers a tactical picture of the air and sea war. All of these systems are designed to improve communication, manage the battle space and improve their ability to work together. The development of FIWC and AFIWC demonstrates that the services have woken up to the threat from the outside world, but the task they face in protecting the military from attack is enormous. Bureaucratic paralysis, absence of vision, tardy decision making all add up to obstructions that stop the cyberwarriors locking the back door to America's military defenses. Currently it yawns wide open, no matter how hard they work. For the other truth is that even if the cyberwarriors operated within a perfect system, one that responded to their every need, the hackers would still be one step ahead of them. America possesses the finest, most deadly weapons systems in the world. In cyberspace it is the hackers who hold the high ground. And it is not just America's military that lies vulnerable.

The extent of that vulnerability is a matter of speculation for now. There have been alarm bells set off by DISA and others who warn that hackers are trying to break through the thin walls of defense. But there has been no electronic Pearl Harbor and until there is one, many in governments around the world will not take the IW threat seriously enough to change laws and allow those with a developed capability to do something about it. Until now even hackers like Ardita have been relatively benign, amateurs who relish the challenge of taking on Ameri-

ca's finest and beating them. But there is ample evidence that governments with a less benign interest have understood the potential and are
moving to exploit it. French intelligence has hired former hackers to
turn their knowledge to national advantage, the Russians have a strong
record in using hacking to obtain economic and other intelligence, and
there have been reports that in the run-up to the Gulf War Dutch
hackers stole details of Allied military dispositions from Pentagon computers and tried to sell them to Saddam Hussein. Most importantly, the
Pentagon is putting a large sign up in cyberspace that reads "Follow
Me." Where America is leading today, the rest of the world will follow
tomorrow. That pioneering role is bringing with it serious challenges
for the intelligence agencies who spent the Cold War years tracking
flesh-and-blood enemies in back alleyways across the globe. Are today's
James (and Jane) Bonds up to the threat from enemies whose stock-in-
trade are the ones and zeroes of cyberspace?

FOURTEEN

Big Ears and Noddies

T H E National Security Agency has the tightest lips in the intelligence business. Not for nothing is it known as Never Say Anything or No Such Agency.

But it also has the biggest ears in the world. To understand quite how much the NSA can listen to in terms of the world's daily life, assume it can hear everything and work backward. A drug dealer with a cell phone in Caracas, Venezuela; a terrorist sending a fax in Rome; a rogue dictator speaking to military commanders by radio link; an allied politician sitting in his office discussing a secret trade deal with his close advisers; a money launderer transmitting cash to a bank in the Cayman Islands; two friends gossiping on the phone; people talking at a sidewalk cafe— in other words, any communication in the world can be listened to by the NSA at its headquarters at Fort Meade, just north of Washington, D.C., on the Baltimore-Washington Parkway in Maryland.

The NSA is simply the best in the world at planting listening devices in the innermost sanctums of the world's leaders, and unequalled in its ability to suck out of the air vast quantities of information and analyze it.

But of all the brilliant successes it has had over the years in eavesdropping on the world's most private conversations, nothing can match the sheer daring and cunning of the Crypto Caper, and one can only guess what effect it had on the course of world events. For the NSA does not advertise its coups, nor does it boast of its conquests.

Crypto AG is the name of one of the world's foremost private cryptography companies. Its business is selling machines that will secure the communications of politicians, diplomats, corporations, anyone who wants to keep their business safe from prying ears, like those of the NSA. The company grew from the success of Boris Hagelin, a Russian-born Swede, who made encryption equipment for the U.S. military in

World War II. He invested the large sums he made from a grateful Uncle Sam into his business in Switzerland. During his time in the United States, Hagelin made firm friends with cryptologist William Friedman, who went on to become a senior officer at the NSA. It was Friedman who in 1957 approached Boris Hagelin with an amazing proposal. The NSA wanted Hagelin to create a back door in every cryptographic machine he supplied that would allow the agency to read all the traffic that went through it. Details of the arrangement have never been made public, so it is impossible to tell whether Friedman appealed to Hagelin's loyalty or his bank account, but several former employees have admitted that Crypto did rig all the machines it sold from then on. For thirty-five years, the NSA did not even have to make any serious effort to listen to the secret traffic of any nation on Crypto's client list, and that included Iraq, Iran and the former Yugoslavia.

The plan started to unravel in 1992 when a Crypto salesman, Hans Buehler, was arrested by Iranian police and accused of being a spy for the United States and Germany. Believing in his innocence, he protested every day for nine months, enduring endless interrogations before his company bought his freedom for $1 million. He left Iran, still wondering why they insisted Crypto was a spy center and that he worked for the Americans and Germans. Soon after he returned to Switzerland, Buehler was fired by Crypto. Hurt and confused, he discussed his plight with other employees, some of whom had learned of the Friedman deal over the years. The reason the Germans had entered the frame is that when he retired in 1970, Hagelin passed control of the company to the German electronics giant Siemens AG, which has strong contacts with German intelligence. Former employees say that the algorithms devised for use in the machines had to be approved by the Germans, presumably with the full knowledge of the NSA.

Buehler's firing broke the dam of silence around Crypto's secret NSA deal. Stories began to appear in the press, and Crypto's reputation in the marketplace suffered disastrously. The company defended its products to its clients and underlined its commitment by filing suit in 1994 against Buehler. The suit was settled out of court when it became obvious a series of former employees were lining up to tell all in the witness box.

It was an extraordinary story, and testament to the ingenuity of the NSA, which never misses a trick when it comes to being the first to know the world's secrets. Another classic operation was the bugging of the Chinese embassy in Canberra, Australia, which for close to four years poured all communications with Beijing straight into the NSA's lap. During construction of the splendid new embassy, the NSA had planted tiny fiber-optic sensors in every office, reception room and hallway, capable of picking up everything said in the building, and everything typed into computers, and transmitted it all back to Fort Meade. It was an intelligence gold mine that was killed by a press leak alerting the Chinese in 1994.

When it did not have backdoor access to other countries' code machines, the NSA could usually crack any code that came its way in very short order. The agency became the world's leading authority on cryptology and code breaking, using seven acres of supercomputers in an underground bunker at Fort Meade to smash in seconds codes that would take a desktop PC, for instance, over 20,000 years.

For most of history, cryptography was the exclusive domain of military and intelligence organizations, who built and maintained their own coding hardware out of sight of the general public. A few small companies (Crypto AG among them) developed and marketed cryptographic equipment, usually for sale to foreign governments. But the value of these systems was limited by the lack of interoperability and doubts about their strength. So as the need for more efficient encryption increased, the call went out for ideas about how to create a system that could be approved by the government for use by banks and companies. Once this was done, the algorithm—the key formula that underpins the encryption —could be built into a wide range of hardware, allowing for interoperability across different platforms.

After an initially disappointing response, the National Bureau of Standards received a submission based on an algorithm called Lucifer, developed by a team working at IBM in the early 1970s. This was the basis for the encryption system known as DES—Data Encryption Standard. But what emerged as DES was not what the IBM team had anticipated.

For help in determining the strength of the Lucifer algorithm, the National Bureau of Standards had turned to the acknowledged experts in the field: the National Security Agency. The NSA not only evaluated the algorithm, they also made modifications to its operations before returning it to IBM. These modifications have spawned speculation that the NSA installed a trapdoor in DES, or that it weakened the algorithm in some way not apparent to outsiders. Most significant among the changes made by the NSA was the shortening of the key length from 128 bits in Lucifer to 56 bits in DES. The number of bits in a key determines its strength. Each extra bit increases strength exponentially, which means that for every bit increase in length, the difficulty in cracking it doubles, meaning the 128-bit key would have been several million times more powerful than the 56-bit. Other changes were made to the S-boxes, portions of the algorithm that repeatedly substitute one group of letters for another in order to encrypt a message. Alan Konheim, a member of the IBM team, stated, "We sent the S-boxes off to Washington. They came back and were all different. We ran our tests and they passed."

Continued speculation that the NSA had tampered with DES led to

Senate Intelligence Committee hearings on the subject in 1978, with the top-secret-cleared committee fully exonerating the NSA of all charges of imposing weaknesses on DES. When questioned on this topic, NSA refers to the committee findings:

> Regarding the DES, we believe that the public record from the Senate Committee for Intelligence's investigation in 1978 into NSA's role in the development of the DES is responsive to your question. That committee report indicated that the NSA did not tamper with the design of the algorithm in any way and that the security afforded by the DES was more than adequate for at least a 5–10 year time span for the unclassified data for which it was intended. In short, NSA did not impose or attempt to impose any weakness on the DES.

Despite continued (and continuing) worries regarding the integrity of DES, it was adopted as a federal standard on November 23, 1976, and authorized for use for all unclassified government communications.

What happened next may have resulted from a misunderstanding between the National Bureau of Standards and the NSA, but it changed the face of cryptography forever. The NSA had assumed that DES would be implemented in hardware form only, as outlined in the standards proposal; in other words, it would be hardwired by a few manufacturers into specially built chips. Instead, the National Bureau of Standards let the genie out of the bottle and published the entire algorithm, along with associated notes and technical data, in order to ease the implementation of the standard. With the algorithm and its attendant literature available to everyone, the writing of DES software by people outside the government became a possibility. This was something the NSA never agreed to, and they acknowledge that had they known the details of the algorithm would be released, they would not have signed off on the standard.

The release of the DES standard represents the beginning of cryptography outside the purview of major governments. For the first time, computer programmers could study an algorithm the NSA had certified as secure, and begin to determine just what made it so.

The release of DES was revolutionary, but at the end of the day it is a code system for big organizations. Like all the codes in history it was a secret key code that depended on both sender and receiver knowing the key. This was always open to attack by people who got hold of the key, as the British did with the Nazis' Enigma machine in World War II. DES is also cumbersome and impractical for sending small amounts of data. The need for commonly held secret keys, the need to keep changing them and the demands on hardware make it too much of a burden.

In 1976 a new kind of cryptography was invented, called public key. The system works by the sender and receiver of a message each having a private and a public key. The public key is available to anyone in the world, whereas the private key is known only to the individual who generated it. The sender encrypts the message using the receiver's public key; this message can only be decoded by the receiver's private key. This opens a new world of cryptography to individuals as well as corporations.

Very soon after that, in 1977 a team of scientists, Ron Rivest, Adi Shamir and Leonard Adelman, created RSA, the first cryptosystem to use a public key. The creation of the RSA system opened a Pandora's box of problems for the government, particularly when Philip Zimmerman, a computer scientist from Boulder, Colorado, used RSA to create an extremely strong encryption program called Pretty Good Privacy, or PGP —and then in June 1991 gave it away via the Internet. To everybody. Actually, he did not personally place it on the Internet; someone else did, but the deed had been done.

One man had opened the door to secret computer communication to the whole world, and it frightened the daylights out of the American government, which resorted to a law designed to control the flow of guns, bombs and missiles overseas to stop PGP from spreading beyond U.S. borders. The federal government said encryption over a certain strength—forty bits—qualifies as munitions under the Arms Export Control Act. PGP is a 128-bit system, impervious to assault by all but the most powerful of computers. By allowing it to be placed on the Internet, the government contended that Zimmerman had exported it, and for three years he was pursued through the courts. In January 1996, the U.S. Attorney's office in San Jose dropped the case without specifying a reason.

The government's fear was that by publishing such a strong encryption program that could be used by terrorists, organized crime and national enemies, Zimmerman had threatened the security of the nation. Zimmerman disagreed. Citing the government in exile of Tibet, the dissidents in Burma and elsewhere, he said encryption is a force for good. For instance, human rights activists in countries where dissent could lead to a firing squad were using PGP to keep vital information secret from their oppressors, and it had actually saved lives.

While the issue in America is nothing so dramatic, Zimmerman firmly believes people still have a right to privacy from government intrusion.

"The common person needs encryption to function effectively in the Information Age," he says, "so it's time for encryption to step out of the shadows of spies and military stuff and step out into the sunshine and be embraced by the rest of us. Sometimes in a democracy bad people can be elected. . . . If a future government inherits a technology infrastructure that's optimized for surveillance, where they can watch the movements of their political opposition, they can see every bit of travel they could do, every financial transaction, every communication, every bit of e-mail, every phone call, everything could be filtered and scanned and automatically reorganized by voice recognition technology and transcribed. As we extrapolate our technologies into the future, if the incumbency has that political advantage over the opposition, then if a bad government ever comes to power, it may be the last government we ever elect."

Yet again, the spread of technology downward from the traditional controlling elite to the private citizen was turning the paradigm on its head. Encryption for the masses became a profitable line of business for software makers, but the export controls put a serious crimp in their ability to exploit it. The government gradually moved the restriction from forty bits to fifty-six bits, but the concession was meaningless, as foreign companies were already marketing 128-bit systems, which made a mockery of the original export control. Any criminal who wanted that much power could buy it abroad. As if to demonstrate the idiocy of the law, Jim Bidzos, the CEO of RSA, struck a deal with Japan's Nippon Telegraph and Telephone Corporation to import NT&T's fifty-six-bit key chip set for encrypting data in LANs (local area networks), public networks and the Internet, into America. NT&T would still have preferred RSA's software, as it was recognized the world over as the market leader, but the ban prevented that. So by default a foreign company was being given a free run in America and on the world market where American companies like RSA should have been competing. As well as granting Jim Bidzos a degree of satisfaction in demonstrating how wrongheaded the U.S. government's law was, he also managed to get a slice of NT&T's action.

Cracking codes has always been something of a Holy Grail, a very elusive one, for researchers over the years. One of the computer industry's most noted scientists, Matt Blaze of AT&T Bell Labs, posted a proposal on the Internet soliciting donations for building a DES-cracking machine. He said it would cost $50,000 to build a computer that could exhaust DES in eight months through "brute force" decryption. Brute force means running through every single possible combination of codes to arrive at one that works. M. J. Wiener of Carleton University in Ottawa published a proposal that would bring DES to its knees in three and a half hours. He wanted $1 million to build that machine, however, although he had already designed some specialized chips and boards to do the job. These two cracking theories rested on building a single machine. It went without saying that machines capable of cracking DES already existed. Rumors circulated that the NSA could do it in a matter of minutes with the seven acres of computer power sited in the giant underground bunker at Fort Meade.

If it would take a powerful, expensive machine to do the job, why not assemble that power by looping together a whole series of smaller computers to achieve the same end? This was the "Chinese Lottery" system devised by J. J. Quisquater and Y. G. Desmedt in 1991. Imagine, they said, that the Chinese government mandates that every TV and radio sold in China contain a code-breaking chip capable of being linked into a giant national network. When the government needed a code cracked, it would broadcast a signal that tripped every chip into action. The chip that came up with the result would win its owner a prize. Hence the lottery. If there are 257 million radios and TVs in China, DES could be cracked in four minutes and forty seconds.

The Chinese Lottery system sounds a trifle preposterous until you realize this was how a French graduate student broke the encryption embedded in the export version of the Netscape Web browser. This forty-bit code would have taken a single PC 8,000 years to crack. The French student hooked together all the machines at his university computer lab during their downtime late at night and did the job in eight days. Netscape was quick to point out, and rightly so, that the student's feat did not crack the entire code system, just the message in question. Cracking any other message would require a similar exercise. Also, the domestic Netscape browser contains encryption at the 128-bit level, vastly more secure. For now.

The search for ever-stronger encryption is driven by an understanding of Moore's law, the computing axiom that states that computing power doubles every eighteen months. Following this trend means that ten years from now, the average PC could crack DES in forty-five seconds. For executives faced with the cost and difficulty of implementing a companywide data-security regime, Moore's law is serious trouble. Therefore encryption standards have to be set against what will be un-crackable in ten years' time, not today.

A forty-bit key is significantly easier to crack than even the fifty-six-bit key employed by the domestic versions of DES. Assuming the algorithms are of comparable quality, every bit added to the key length doubles the difficulty of finding the key. Therefore, cracking fifty-six-bit DES is more than 250,000 times more difficult than breaking its forty-bit cousin. To take it to its extreme, it would take a Cray supercomputer 40 billion years to crack a 128-bit key. "If we do not address this matter urgently, law enforcement will be out of business," a congressional staffer working on these issues told me.

In the United States, the debate over encryption is essentially about the government's need to access a citizen's conversations, mail and business dealings in the name of law enforcement and national security versus a citizen's right to a life completely free of such potential. It is a debate between those who see the potential for evil around every corner and those who prefer to see humanity as basically good with only a tiny percentage breaking the law. Why should 99 percent of the population be open to government surveillance for the sake of 1 percent? It is a measure of how far technology has liberated and liberalized the individual that people can talk in such terms. In the days when government controlled the technology, it went without saying that it controlled the citizenry. The government's inability to understand that some people perceive as much of a threat to their welfare from too much government as from criminals shows how stuck it is in that historic mind-set.

The argument was given further impetus and shape by the development of secure cellular telephony. Cell phones are among the easiest devices for a hacker to listen to, as the voices are carried by radio waves capable of being picked up by an ordinary scanner. From Britain's Prince Charles and Princess Diana, people had learned that cell phones were not secure. But when AT&T announced in 1992 it had developed a secure phone system using its own encryption algorithm based on fifty-six-bit DES, the government panicked. The thought of virtually unbreakable encrypted conversation being possible was too much for the FBI and the NSA. Under severe pressure from the government's lawyers, AT&T's CEO Robert Allen agreed to incorporate into the new phones a chip developed by the NSA called Clipper that used a different algorithm, the Skipjack. Thus the government, if it wanted to, could decrypt any conversation carried on over these phones. It was an article of faith with the government; whatever form of encryption was used, in phones or in computer communications, the government *had* to have a master key that could unlock it. So they told industry that all private encryption keys would have to be held in escrow so the government could access them whenever circumstances, and a judge's warrant, required.

The debate over the Clipper chip and key escrow was a PR disaster for the government. According to Dr. Dorothy Denning, the head of the computer science department at Georgetown University, it was a failure of presentation by the government rather than intrinsic heavy-handedness that created the Clipper fiasco. She says the government was bypassed by the speed of technological development and did not have a policy strategy in place to deal with the backlash aroused by the Clipper chip. It had been pushed into a tough-appearing stance by the FBI without any thought to the reaction.

"If they had said, in every other sentence, 'This is just an idea,' or 'We're just testing this idea out,' maybe the attacks would not have been so severe, but they didn't," Denning said.

And so the government found itself denounced as Big Brother, but it was prepared to fight for its encryption proposal.

Denning was one of the few in the computer science industry who thinks the argument has gotten way out of hand. While pro-privacy proponents argue that since we do not give the government the keys to our house, why should we give them the key to our privacy; that we can strengthen our locks to any degree we like, so why can't we strengthen our encryption, she counters by saying, "No sane person would build an impenetrable house as they could lock themselves out and never get back in again. No, if they get locked out, they ask a locksmith to open the door. And anyway, the government can *always* break into your house, no matter how hard you make it for them. But if your data is encrypted to a high level of security, they will *never*, in real terms, be able to break in."

Cool heads were hard to find, between the absolutes of the government's paranoia and the libertarianism of the privacy advocates. One voice of reason belonged to Ed Appel, a twenty-seven-year FBI veteran who was director of counterintelligence at the National Security Council and counted a spell as head of counterintelligence and counterterrorism at the FBI field office in San Francisco and a two-year attachment to the NSA. He believed much of the debate should be viewed through the historical perspective of what happened when the 1968 Omnibus Crime Bill gave law enforcement Title 3 powers to tap phones. There was a major hue and cry from those who believed criminals would stop using the phones and find other means to communicate that the government could not listen to. Yet the prisons are full of people who did use the phone system, and did get caught. The FBI took out twenty-nine different warrants to tap phones in Operation Ill Wind, which smashed a price-fixing network at the Pentagon. Almost without exception, each suspect had access to STU III telephones, which had one-time-use encryption keys on them that would have made eavesdropping impossible, as the Pentagon had no way of decrypting its own communications equipment. But while the suspects used the STU III phones, they never pressed the buttons that would have encrypted their calls. They paid the price.

One senior intelligence official points out that just because sophisticated technology is available does not mean criminals will use it. "Most evildoers are like us," he said. "They buy their technology on the open market, they go to Radio Shack like the rest of us." He discounts the argument that the Internet will be a source of aid and comfort to criminals who could download strong encryption programs like PGP. He agrees it will happen, but how often? "Getting stuff off the Net is like seeing a partially smoked favorite cigar lying on the pavement. You may admire its qualities, but would you pick it up and smoke it? What rational sense does it make to use something that has no guarantees?" Of course this is one of the major arguments in favor of universal access to strong encryption—that the Internet will never fulfill its potential until everyone can trust that their communications and transactions are private.

The official cautioned that no matter how blurred the real world is as opposed to the absolutism of the encryption debate, the existence of encryption does make law enforcement's job harder, and relying on human frailty is not always wise. He cited:

• The case of hacker Kevin Poulsen, whose computer hard drive contained over 100,000 encrypted documents. The FBI broke it, eventually, with the help of the NSA, but as they had already enough evidence to convict him he felt no compunction to give them the keys.

• From his base in the Philippines Ramzi Ahmed Yousef, the World Trade Center bomb-maker, created a plan for bombing airliners on his laptop. When he was arrested the data was easily decrypted because he had either bungled the encryption or used a poor program. The data nailed him. But what if the encryption was unbreakable? "After a plane bombing, where is the data to be found? Either in the shattered debris lying on the ground or in the laptop. What a scary world it would be if his data had been fully encrypted. We could end up catching the bad guy and not being able to jail him because we wouldn't get the evidence."

• The Aum Shinrikyo cult in Japan encrypted all its data, but among the floppy disks found by police was one containing all the keys. It may sound careless, but people using encryption will want their keys close at hand so they can access their data. What happens to someone who encrypts all his data, then loses the key?

• A young boy who met a pedophile ring over the Internet and was lured into a meeting with them ended up a murder victim. Police tried to get evidence of the crime from a Maryland pornographer's computer but he had used PGP to encrypt his data. The only way to get access was to use that same computer to download a program from the Internet that helps PGP users who have lost their keys decrypt their data. The program can only be downloaded by the computer that originally downloaded the PGP. So, with a second warrant in hand, they gained surreptitious access to his house, booted up his computer and logged on. With the PGP workaround downloaded, they unlocked the data.

The strange polarized tenor of the debate proved beyond a doubt that the government was stuck in the past, the activists planted in a future that had yet to arrive, and the general populace, as always, was milling around in the middle waiting to hear some proposals it could relate to. And yet, no matter what the merits of each case, it was the fact that the government and its law enforcement and intelligence agencies were demonstrably trying to hold a line that had already been crossed by progressive law-abiding citizens and by the world's criminal classes that boded ill for its ability to combat the changing face of world affairs in general and crime in particular.

The challenge was the most extreme for the intelligence agencies of the United States and the other Western democracies, given that the disappearance of the Cold War as the paradigm that guided their every move left them searching for a raison d'etre. All the evidence suggests that what is required in the field of intelligence, as in the military, is not evolution but revolution. Yet both are proving unequal to understanding the nature of that challenge. Or to be fairer, while there are those in each arena who understand what is needed, the sheer inertia of the system in which they exist dooms them to be constantly behind the curve of change.

It is encryption that has polarized this debate. On the one hand, the public demands the right to control its own security. On the other, the FBI sees its ability to operate effectively in the twenty-first century dissipate if large encryption algorithms reach the open market. In reality, of course, this is a largely sterile argument that shows just how out of touch a traditional law enforcement agency like the FBI has become. Large algorithms are already in use; criminals are using encryption increasingly as they become more computer-savvy; and the United States cannot control the international encryption market.

The NSA has adopted a slightly different and less confrontational approach. At computer security forums across America, the NSA takes a stand, its operatives hand out business cards and prospective customers are given a glossy brochure extolling the virtues of the NSA's own encryption software, known as Fortezza, which is available for sale. As a final piece of marketing, the NSA even hands out a Frisbee engraved with the agency's logo. While this may not solve the encryption problem, there is some chance the NSA may be able to forge a new partnership with industry to control the spread of uncrackable codes.

If the partnership approach fails, American intelligence agencies and their allies are in deep trouble. For years, technology solutions have given the good guys the edge in the fight to keep the peace and win the war against crime. That advantage is now at risk. Unfortunately, that is not the only problem confronting the intelligence community in the Information Age.

FIFTEEN

Puzzles and Mysteries

INFORMATION that was once secret is today openly available, and the possibilities for mining that information to anyone equipped with a modem and computer are endless. Consider the ability the ordinary Internet surfer today has to access libraries of great universities all over the world, read newspapers from just about every country, read manuscripts by some of the great thinkers of the day posted to news groups or Web sites, or ask questions that someone, somewhere in the world can answer in a matter of minutes. Five years ago, intelligence agencies would have killed to get their hands on this kind of capacity. Now you can do it from your den.

And so can anybody, whether their intentions be good or bad. Criminals can go out into the marketplace and buy the biggest computers, the fastest modems and the strongest encryption. So now everybody can access the kind of information that was once the preserve of the agencies charged with defending the nation. What is left for them to do? The soldier on the ground who used to rely on spies to tell him what was over the other side of the hill will now have data streaming at him from satellites, UAVs, flying wings and an array of high-tech sensors. What the soldier does not know, and this is the information gap the intelligence agencies need to fill, is what that enemy is going to do. In the old days, a spy's job was to sniff out the answer to a particular question, say: "What can the new Soviet tank do? How fast is it, how does it defend itself, how powerful is its gun, what are its communications like?"

Nowadays he knows the answer. He knows the capabilities of the missiles China is selling Iran, the capacity of the nuclear reactor being built by the North Koreans. What he does not know is what the bad actors of the world are going to do with their power. Before, that question was always seen through the filter of superpower relationships. Now that countries like Iran, Syria, Iraq and North Korea are free to operate outside a system of patronage that forced them to cleave to whatever line the Soviets or Americans wanted, the range of possibilities has multiplied exponentially. So the world has fewer puzzles and many more mysteries. In solving those mysteries, the spies' job is as much one of panning the gold from the great flood of information sloshing around the world as it is trying to learn a secret that nobody else knows. And all the while the policymakers are yelling for faster service, more timely information. It can be a headache, as Britain's Foreign Office found during the Falklands War. Prime Minister Margaret Thatcher demanded a briefing at noon each day, and the civil servants found it impossible to process all the telegrams and the intelligence pouring onto their desks. So they listened to the BBC World Service news bulletin, transcribed the contents, framed it in their own inimitable civil-servant-speak, and rushed it to Number 10. When the Prime Minister listened to the news an hour later, she would marvel at her advisers' prescience and skill. So much so that she dropped them a note after the war ended, thanking them for their excellent service.

During the Gulf War, politicians learned more about what was going on tactically from watching CNN than from listening to their own spy agencies, a deeply worrying thought given the nature of journalism in general and of CNN in particular. While the CNN coverage was a logistics marvel, one to rank alongside the military's skill in deploying 500,000 troops to a far-off land, the overwhelming reaction among other journalists to the plaudits thrown CNN's way was much the same as Dr. Johnson's reaction on seeing a dancing dog. "It is not done well; but you are surprised to find it done at all." It is hardly the way modern wars should be fought.

Steps have been taken on both sides of the Atlantic to make more intelligence more timely and easier to process but everyone agrees that CNN, or one of the other global news organizations springing up, is usually there first. Often the best the intelligence community can hope for is to gain some sense of strategic intentions before the event and some understanding of what happened after the crisis has passed.

The response of the policymakers and the intelligence communities to this has been disappointing. In the United States and Europe their reaction in appointing new people to senior positions has been to see the post–Cold War period and the birth of the Information Age as a time of instability that needs to be managed, rather than a time of revolutionary change. Given the nature of bureaucracies, this is entirely natural, but also precisely the wrong way to deal with what lies ahead. Even within the intelligence communities themselves there appears to be an ignorance about the threats and opportunities represented by this ongoing revolution. Too often, experienced intelligence officials declare themselves ignorant of, or uninterested in, the information revolution. Yet it is fundamental to the future of intelligence.

The official congressional report into the Rome Labs attack, written by Dan Gelber and Special Agent Jim Christy, related stories that would be hilarious if not so serious. They quoted an anonymous intelligence officer who said the CIA's reaction to the challenge of information warfare was similar to "a toddler soccer game, where everyone just runs around trying to kick the ball somewhere" and that the threat was "not presently a priority of our nation's intelligence and enforcement communities." Lip service was paid to the problem by the creation of such entities as the CIA's Information Warfare Center, but Gelber and Christy reported that "at no time was any agency able to present a national threat assessment of the risk posed to our information infrastructure. [Briefings] consisted of extremely limited anecdotal information." In other words, the spymasters were reading the newspapers and watching CNN like everyone else. The vacuum was often filled by rhetoric that had more to do with increasing one's own self-importance than increasing understanding. Former CIA chief John Deutch's alarmist testimony to Congress on the threat from Libya and Iran was not based on hard intelligence but on extrapolation, rumor and surmise. This does not deny that such a threat may exist at some stage, but it is an example of how feeble intelligence agencies can become when stranded between two epochs.

It was a particular judgment reached by the Consortium for the Study of Intelligence in its report "The Future of U.S. Intelligence" prepared for the Working Group on Intelligence Reform in 1996. Its judgments were profound and important. They included:

• The Intelligence Community (IC) developed during the Cold War and has not existed under "normal" international circumstances. It is a mistake to extend automatically the dominant Cold War reform agenda—centralization of the IC under the Director of Central Intelligence (DCI) authority in the name of efficiency—to the new circumstances.

• The "Information Age" makes many new sources of information available to the government such as commercial satellite photo reconnaissance. The IC cannot hope to encompass all information relevant to foreign and national security policy, and should not even try. The issue is one of "comparative advantage" of the IC over other information sources.

• The IC's major "comparative advantage" lies in obtaining information others try to keep secret and penetrating below the "surface" impression created by publicly available information to determine whether an adversary is deceiving us or denying us key information.

• Counterintelligence (CI) therefore has to be more than just catching spies (Counterespionage); it has to guard the integrity of the government's information collection and analysis process by penetrating, understanding and possibly manipulating an adversary's intelligence efforts against us.

• The demise of the Soviet Union implies not that human intelligence (Humint) collection is no longer needed but rather that there are new opportunities to use it more effectively. This will require greater use of non-official cover (NOC) arrangements (e.g. U.S. officials and assets who do not operate from U.S. embassies) for those involved in clandestine collection and should include a strong DOD Humint capability.

• Covert action must be seen as one foreign policy tool among many, and integrated into the foreign policy process more fully. It has an important role to play under current conditions, including fighting "transnational" threats such as international organized crime, terrorism, and especially proliferation of weapons of mass destruction. . . .

There were other recommendations but these demonstrate that at least the government was receiving advice from people who understood the challenge. Telling spies to go beyond "penetrating below the 'surface' impression created by publicly available information" is a delightfully tactful way of telling them to stop getting their intelligence from CNN or BBC World Service.

The recommendation to develop a strong Defense Department Humint capability was already in effect. The Defense Human Intelligence Agency had been put through its paces in Bosnia and was generally seen as doing a good job. Also, the suggestion that the CIA use more NOCs —unofficial covers—had been overtaken by events. NOCs had long been a standard way of buttressing the spies that already worked under embassy cover, but only when distinct circumstances demanded it. Non-official cover means that when caught such agents lack the diplomatic immunity accorded those with embassy cover. NOCs are actually at the sharp end of the spying game and far more resemble the heroes of spy novels than the embassy types. The dangers they face are much greater, too. The terrorists or criminals they are trying to counter are not in the business of trading spies at Checkpoint Charlie. The penalty for capture is a bullet in the head. CIA NOCs in Colombia who have tried to subvert the cartels' drug business have been wounded or killed in gunfights. The shift in emphasis from Cold War subversion of Soviet or Eastern bloc personnel or stealing the latest weapons plans to combating global terror, drug smuggling and money laundering demanded agents who could swim in the mainstream of a target nation's society rather than in its diplomatic pools. The CIA has accordingly stepped up its recruitment of NOCs, employing people whose education and experience suit the activities that will be their cover, such as executives in banks or overseas branches of big corporations. Ironically, one of the ways in which the CIA loses its NOCs, apart from the regrettable attrition of life in the dangerous world of covert operations, is that the NOCs discover that the jobs they are supposed to be doing, according to their cover, pay so much better than the agency salary that they move out of the shadows into the real world.

On the intelligence/information front, the DIA was demonstrating initiative with the development of the Joint Intelligence Virtual Architecture (JIVA) system, which would, by 2001, provide to 150 U.S. intelligence sites around the world the wealth of intelligence available. Different U.S. intelligence agencies, embassies or military units would be able to log onto a central source that would provide a single database of knowledge combining inputs from all agencies to create a formidable intelligence tool. Eventually full-motion video, 3-D representations of the battle space and a network of powerful computers will create a system of what the DIA calls "virtual intelligence." The DIA also created an Information Warfare Support Office as part of its Directorate for Intelligence Production. TWI, as the office is coded, embraces everything from traditional intelligence gathering to deception operations against the enemy's information base. Its existence was trumpeted in a marketing brochure for the DIA called *Information Warfare Activities*, in which the agency's capabilities were laid out in bold, colorful graphics, for all the world like a commercial company selling manufactured goods. Indeed the brochure goes so far as to refer to "DIA products."

For all its rather crass attempts to sell itself as meaningful in a time of tightening budgets, the DIA was at least trying to mold itself to the demands of the future. At a time when society at large is becoming more open and information is more accessible, most intelligence agencies seem to be moving against the tide. In Britain, for example, MI5 and MI6 are less accessible now than they were four years ago, when at last the government acknowledged openly what virtually everybody in the land had known for decades, namely that the intelligence services actually existed. In many respects, the British agencies have reverted to the defensive posture that did so much damage to their reputations in the 1970s and 1980s. This withdrawal was rooted in the conservative, risk-averse leadership of the Conservative government of John Major. At least being an organization that officially exists allowed MI5, or the Security Service as it prefers to be known, to actually advertise for staff in the newspapers. The advertisement made it clear, for the first time, that applications from homosexuals were welcome, a highly important point, as Britain had been embarrassed by the discovery in its ranks of a succession of Soviet spies over the years who were also homosexuals, most notably Guy Burgess and Anthony Blunt. Barring homosexuals from the security services forced applicants of old to bury their predilections deep, making them easy targets for blackmail when, inevitably, their guard was down, usually and most embarrassingly with the full knowledge of enemy agents. The 20,000 responses received were ten times as many staff as were in the entire service. (MI5 proved it had yet to learn anything about hackers and what they were capable of when the answering machine responding to applicants was hacked, and the message replaced with a guttural Russian-sounding voice that said, "Hello, my name is Colonel Botch. I am calling on behalf of the KGB. We have taken over MI5 because they are not secret anymore and they are a very crap organization. All the details left will be forwarded to the KGB. Thank you.")

The hint of a progressive attitude emerging at MI5 did not, however, signal that anyone in British military or intelligence circles worried about the development of new and dangerous threats from cyberspace.

In the United States, the political will to kick the intelligence community into the twenty-first century was noticeably lacking, too. President Clinton signed legislation that was a pale shadow of what many reformers inside the agencies knew was important after the end of the Cold War and the debacle of the Aldrich Ames case. Within the CIA, there was a real constituency for change that would have made one of the most bureaucratic and inefficient intelligence services leaner, fitter and more responsive to the needs of the policymakers.

Opportunities missed rarely come around again, at least not for some time. The official acknowledgment in 1993 that MI6 and MI5 actually exist achieved much less than had been hoped for from professionals in the two services and GCHQ, Britain's NSA-style listening post. The agencies became, if anything, more bureaucratic than before in an atmosphere made possible by the total absence of debate over how intelligence gathering should be done in the Information Age.

It is understandable that the intelligence community should be deeply resistant to change; spy agencies are the most conservative of bodies and to a large extent they have provided continuity of understanding and thought over several decades. But revolutionizing the gathering and processing of intelligence is not a threat to intelligence itself. On the contrary, it is a way of ensuring there is a future. The British are pinning their hopes on the New Labour government of Tony Blair, whose members have been talking of basic reform for some time. The task is a difficult one, just as it is in the United States. The politicians have to prove that in the new order of things it is important to spend large sums of public money on the gathering and analysis of intelligence. The agencies, who for so long have dealt in suspicion and mistrust, as much of the nation's citizens as of enemies, have to prove they deserve to exist.

Those who argue that we should do nothing more than has already been done significantly misjudge both the problems and the solutions. In the real world, the pace of change is dramatic and accelerating. In the intelligence world, the pace of change is slow and getting slower. If the gap is not closed, then the intelligence community will fail in its duty to act as the front line of defense in guarding national security.

Part 3

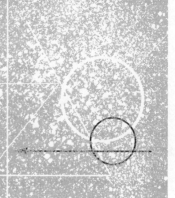

SIXTEEN

The New Arms Race

I F Western intelligence agencies need any evidence of their failure to understand the full import of the information revolution, they should look to Russia. Where America sees information as a liberating influence that frees people to control their own destinies, Russia sees information as subversive, corrupting and dangerous. Where America sees information warfare as a means of waging and winning short, sharp, limited battles in areas of the globe where the nation's national interests are at stake, Russia sees information warfare as a threat to its very existence. A trio of noted Russian scientists wrote: "At the end of the 20th century, more and more problems will arise as a new global threat to security appears—information warfare."

In the spring of 1997, after months of negotiation, I visited Moscow specifically to discuss information warfare. In what proved to be an extraordinary trip, I was given access to the most senior officials in the SVR, Russia's equivalent of the CIA; FAPSI, their equivalent of the NSA or Britain's GCHQ; strategic studies institutes and the President's council on information security. Everywhere, the message was the same: the world is in the midst of a new arms race involving information warfare; Russia is losing the race and its government sees this as a fundamental threat to national security. Russia's perception of this new world was a revelation to me and to Western intelligence officers who had not seen many of the documents I was given and had little idea how seriously Russia is taking this threat.

My visit was in itself a classic IW operation. I had first approached General Yuri Koboladze, the chief of public affairs for the SVR. We had met while I was researching *New Spies,* a book on intelligence after the Cold War. An amiable Georgian with a deep laugh and a love of opera, Koboladze is a very experienced intelligence officer. He served several years in Britain working undercover as a journalist and he developed a love for soccer, the soft British countryside and an affection for the West which never overtook his loyalty to mother Russia. His desk has a glass top under which he keeps mementos of his time in Britain. Center stage is a label from a bottle of House of Commons whiskey which is inscribed "With best wishes, Dennis Canavan MP" (Canavan is a long-serving member of Parliament for the Labour party). "We used to have great drinking sessions," Koboladze said with a laugh.

To my surprise, Koboladze immediately agreed that a trip to Moscow would be a good idea and that officials would be willing to talk. As the official response to similar requests I had made to the CIA, FBI, NSA, MI6 and others had all been entirely unsuccessful, I was surprised. But over the next few months it became apparent that the more details I requested, the more elusive appointments became. It was obvious that the Russians had something to say but they could not decide who should say it or what exactly the message should be. Finally, I got the reassurance I wanted, that if I arrived in Moscow, the right people had agreed to see me. That in itself was an indication of how seriously the Russians view IW. Clearly, they assumed that I had been fully briefed by foreign governments and they wanted to make sure that their message, whatever it might be, would be heard in the West.

Moscow nearly seven years after the collapse of communism was thriving. Everywhere new buildings were springing up and the shops were filled with expensive Western clothes and food. Gangsters, marked by their smart Italian suits, hulking bodyguards and gorgeous courtesans, were everywhere. But scratch that superficial image and there was a darker side to the city that recalled the Cold War and the paranoia that ran through every building and government department. Partly that is because those in the employ of the state are largely unable to benefit from the rampant capitalism that is making millionaires of men who would otherwise have remained local hoods in the pay of the KGB. Partly, too, it is because many of those who still serve the state are time-servers, men and women who have been unable to prosper outside the embrace of the state. It is these men and women who are having to come to terms not just with the capitalist revolution but with the information revolution. These officials see that the former has undermined what was once a comfortable and secure way of life while the latter threatens to destroy the security of their country just as surely as the ultimate power of the free market destroyed communism.

These men all wanted to transmit a common message, that Russia is a nation at war. It is an information war that the country is losing both at home and abroad, and the current technology gap is comparable to the perceived missile gap of the 1950s that did so much to fuel the Cold War. This time, the race is not for space, but cyberspace. And the Russians are angry and frustrated that the Americans appear to be winning the war and that victory appears more assured every day.

There is a certain irony in all this in that the Russians have always considered themselves masters at information warfare in its pre–computer age form. In 1959, the Soviets created a new disinformation section within the First Chief Directorate of the KGB that was initially known as Department D and later became Service A. Its staff of fifty officers had the task of undermining the will of government and people in the NATO countries and of influencing potential allies by planting false information. One of the first acts of the new department was to dispatch a team to West Germany to daub swastikas and anti-Jewish graffiti in several towns. The campaign was remarkably successful and resulted in a blizzard of anti-German publicity, which portrayed the country as being in the grip of a resurgence of fascism.

Encouraged by this early success, the Soviets embarked on a broad campaign of disinformation that involved forging letters, planting false stories in newspapers and recruiting agents in positions of authority where they could influence both the public and political debates. Perhaps the most successful story created by the Soviets appeared in 1983 in the Indian newspaper *Patriot*, which had a broadly pro-Soviet editorial policy. The article alleged that America had created the AIDS virus as an "ethnic weapon" designed to kill only blacks. The story struck a raw nerve, as the Americans had developed a wide range of viruses at the biological warfare center at Fort Detrick, Maryland, and so, like the best disinformation, there was enough truth in the lie to make it believable. For years after the story first appeared, American officials found blacks all over the world who still believed that a perfidious American government had created AIDS.

The Russians have long had an appreciation of the finer points of IW. For instance what the Russians call radioelectronic warfare or *radioelektronnaya bor'ba,* included standard reconnaissance and intelligence, or *razvedka,* with disinformation and psyops, which in its electronic form they call *maskirovka.* This made their concept of EW, electronic warfare, an integrated discipline that was more subtle and flexible than the American version, in which EW is separate from those other tactics. It also means they are psychologically more likely to win an information war if the technological playing field is level. Like most Europeans they tend to be more devious than the Americans, which explains why they excel at espionage involving people and why they so consistently bested the United States in arms control talks. As evidence of the deviousness of their thinking, they perfected a Byzantine philosophy called Reflexive Control as a means of exerting information dominance over the United States.

Reflexive Control is understood as the process of one of the sides giving reasons to the enemy from which he can logically infer his own decision, predetermined by the first side. The term "reflexive control" should be understood as the reflection by the opposed sides in the thoughts of their discussions with each other.

That rather obscure translation from Tarakanov's *Mathematics and Armed Combat* (1974) was clarified by Fred Giessler, who teaches IW at National Defense University.

[Reflexive Control is] conveying to an opponent specially prepared information to incline him to voluntarily make a predetermined decision or to otherwise act in a way that is favorable to the accomplishment of one's own missions.

But there is a world of difference between planting a story in an Indian newspaper or waging psychological warfare against an enemy in war and confronting the challenges of the information age. Without the rigid rivalry of the Warsaw Pact and NATO, communism and capitalism, the Russians appear to have lost their direction and their ability to fight a war that many of them no longer understand. Without the anchors that held the country and its people together, the new world of information warfare seems impossible for Russia's new masters to grasp and control. For Russia's masters of the information revolution, there is ample evidence that the technological superiority that won America victory in the Cold War is being used actively against them to create a new superiority.

Even during the Cold War, American information technology was hard at work undermining the Soviet Union. In 1979, in an operation code-named TAW, the CIA succeeded in planting a bug on the underground cables at Trpoitsk, twenty-five miles southwest of Moscow. The hub connected the headquarters of the KGB's First Chief Directorate at Yasenovo outside Moscow with all the apparatus of the communist party in Moscow itself. As the new headquarters was nearing completion, the CIA managed to infiltrate one of its own men in the construction crew to plant a sophisticated bug that recorded all the most important communications between the KGB headquarters, its stations at home and abroad and its political masters. It was a stunning intelligence coup that folded in 1985 when CIA traitor Edward Lee Howard defected to the Soviet Union. Former KGB general Oleg Kalugin called it "the CIA's greatest coup. They heard every conversation. Everything."

In another operation, when the agency learned that a major order from Russia for 100 IBM and Siemens mainframe computers notionally destined for civilian use was in reality headed for the Defense Ministry and the KGB, they seeded each computer with bugs, logic bombs and viruses that would have funneled secret material back to America or would have crashed the computers and any other machine linked with them on an electronic order from CIA headquarters. The Russian federal agency for government communications, FAPSI, their NSA, examined the computers minutely and discovered what they hoped was all the bugs. FAPSI reports also say that a calculator presented to an eminent Russian scientist as a gift by a foreign national turned out to contain a listening device. How many hundreds or thousands more devices must have been planted in items given or sold to the Russians over the years?

It all added up to a chastening experience for the Russians that colored their attitude in the late 1990s. The U.S. edge in information technology was just not to be trusted: look what they have done with it in the past. But this also put the Russians in a very tight spot. Compared with the United States, Russia is in the Stone Age when it comes to the application of information technologies to daily life. Americans are accustomed to the presence of an electronic or communications component to everything they do, from paying for their groceries by credit card, to satellite and cable TV, the Internet and at least one phone in every home. With their domestic computer industry never fully recovered from the shambles of the Soviet era, and a desperately immature telecommunications system, the Russian government needed to react quickly if it was to drag the country into the Information Age. The Committee on Information Support (Roskominform) was created to oversee and regulate the activities of state and private enterprises in the communications sector. But a sclerotic bureaucracy could not possibly get up to speed quickly enough to satisfy the government's desire to modernize the country's information base. This left a serious technological vacuum that could only be filled by Western computer companies, whose products, inevitably, would be instantly suspect. Throw in the customary paranoia of the Russian people and it becomes easy to understand the following extraordinary extract from an interview given to *Pravda* in December 1996 by Maj. Gen. Vladimir Ivanovich Denisov, a former member of the Chief Political Department at the Ministry of Defense and now part of the Information Security Department at the Chamber of Trade and Commerce. With a Ph.D. in psychology and a military background, he is in all likelihood an expert in propaganda and psychological operations. He certainly believes the Soviet Union was the victim of such a campaign.

> I agree that we ourselves ruined the USSR, that it is not the CIA that did it. But it was a result of the psychological influence which was wielded by the enemy.* And now the same thing is happening again and we are ruining Russia. . . . Socialism is liquidated in our country

---

* Note the use of the word "enemy," a concept Denisov still believes in.

but the Cold War is still going on. It will continue until Russia is cut into small segments and the possibility of reviving an independent state is absolutely lost.

That is why the West is aiming at the privatization of Russian communication companies. In fact the Austrian-American bank CreditanstalBankverein which is controlled by the Americans has already bought up to 20% of shares of Svyazinvest [the biggest telecommunications company in Russia]. Next year Svyazinvest will sell another 25% of its shares. After the sale we can say good-bye to the information security of our state. Many people are happy that they got access to the Internet Web, but the owners are American, not us. Now in Russia lots of American servers have been set up, and they supply their equipment for low prices. But in return, the Americans get huge volumes of information worked out in our scientific centers. Now the West can easily follow all the themes worked out in the centers. We must remember about the "logical bombs" (programmnyye zakladki) inlaid in their programs. Can you imagine what would happen if one day on a special command all the equipment was to be rendered useless? The system of state government will be paralyzed. Who can give the guarantee that the equipment is absolutely secure from that point of view?

Writers like Denisov fear a kind of technological imperialism that threatens to undermine Russian society. The Internet, of course, is the key representation of this. "When the information is . . . controlled by the Americans, it is logical to ask—where do we live? In Russia or in the U.S.?" he asks.

There was some truth behind the Russian paranoia. Both the CIA and the NSA continue to use the importation of computers into Russia for espionage purposes. Where once they bugged mainframes, now they bug PCs exported from the United States. In some cases, this is done with the cooperation of the companies concerned and in other cases the CIA simply intercepts shipments and inserts the devices that have been perfected by its own scientists. The extent of this program is huge and the most successful of the post–Cold War intelligence environment. In espionage terms, it allows for the perfect operation. Intelligence gathering is done at a distance and the bugs are monitored from outside. There are no agents in place, no assets to run and no dead drops to service. The technology does the work, sending back a steady stream of bits and bytes from the computers that have been purchased in good faith on the open market by Russian companies and government departments.

If Russia was vulnerable to this kind of offensive in the 1980s, the situation is much worse today. According to Vitali Tsygichko, a national security expert and vice chairman of the International Committee of the Federation Council of Russia, the country had 1.2 million computers in

January 1995, of which 25 percent were manufactured in Russia. A year later, as the information revolution began to gather speed, there were 4 million computers in Russia, of which none were made in the country. "Now we use all Western equipment for our infrastructure, from telephones to satellites," said Tsygichko. "These come from Western firms but nobody knows what programs might have been hidden inside. Those companies will not give us the specifications. It is a very serious problem and I have tried to explain it to the people but those at the top do not have a technical education and they cannot understand what they can't see. To them information warfare is like a quark."

Although the Russians believe they are under attack, the structure designed to defend the nation is collapsing, mired in a morass of entrenched bureaucracy and assaulted by the temptations of capitalism. FAPSI, the communications agency, is running a major program to scrutinize all technological imports, particularly those targeted at updating the banking and commercial sectors, where a leap of several generations of capability in automation is required, to insure against unwanted intrusion from bugs and viruses. "Until the introduction of certifiable means of defense of its banking information, it is quite possible for anyone to break in [to a Russian bank] undetected and steal billions of rubles," FAPSI's then director general Vladimir Matiukhin warned. Yet, as with so much in the financially crippled country, FAPSI's efforts were stretched to the limit by the departure of its top scientists to jobs in the private sector, where at least they would get a paycheck.

By 1997, the crisis at FAPSI had worsened. The secretive nature of the organization, understandable when compared with the ultra-secretive NSA and GCHQ, had been painted by elements opposed to President Boris Yeltsin as the trappings of a new secret police force. The rumor spread that FAPSI was a key tool of Yeltsin's power and that he was using it to blackmail opponents with compromising information. Lt. Col. Aleksandr Starovoytov, the new director general, stepped out of the shadows to deny the accusations and to defend the need for a powerful agency whose stock-in-trade is information. He saw the attacks on FAPSI as an attempt to derail the presidency and get access to the information which, he believed, equaled power. "There are those who wish to possess this information and use it when reporting to the president or for their own purposes. And this is precisely what the struggle was over. They do not know that information is our product, our day-to-day work, and our daily bread. . . ." He denied categorically that Yeltsin used FAPSI as an internal personal spy service.

At the same time as trying (and failing) to control the information environment, FAPSI was doing everything possible to maintain control of its own political influence. In 1995, the first draft of legislation designed to create an information security policy for Russia that would set minimum standards and improve communications between departments was submitted to President Yeltsin. FAPSI lobbied hard to prevent the legislation being seriously considered. The organization felt that new legislation would inevitably restrict FAPSI's power. The lobbying was successful. Yet FAPSI had a very clear understanding of the problems confronting Russia. During the Moscow trip, I asked for specific examples of American information attacks against Russia and FAPSI promised to provide them. A month later a sixty-page document reached me in Washington, D.C. It had been prepared by FAPSI officials and was the first detailed account ever seen in the West about how FAPSI perceived the threat of information warfare. It was in many ways a chilling document that showed how beleaguered FAPSI feels and how aggressively America and its allies have been using IW to gather intelligence inside the new Russia.

"Information weapons" signifies penetration of foreign services into a country's information-telecommunication systems (ITS) in order to steal, deform, or destruct information, as well as to disrupt day-to-day functioning of the ITS themselves of automated systems of management (ASU) or critical technologies. Critical technologies encompass process of processing information and/or material resources, leakage of which may lead to damage of Russian national interests (economic, political, technical, ecological, etc.).

Understanding the strategic importance of this direction explains great attention to it on the part of the US's administration. Between $1.7 and $2.1 billion is spent in the US annually to prepare and conduct studies of information struggles. Total spending for realization of the concept of "information warfare" until the year 2005 is budgeted at $18 billion. Basically, nowadays "information weapon" is the primary threat to information security of a state.

For FAPSI the challenge is not just to detect the attacks but to prevent the infiltration of bugs and viruses, which it rightly sees as undermining the security of the state. As the document freely admits, it is a battle that is being waged and lost.

Currently, FAPSI possesses broad information on the so called "covert functional capabilities" of programming support and other methods of interception of information. The agency also studies destructive functions that may be implanted in telecommunication systems, and ways to neutralize them. Analysis of imported programming products is comparable to development of the product itself in terms of its complexity. The process requires considerable financing and attraction of large numbers of highly qualified specialists.

Attempts are also being made to import information protection mechanisms. However, practically all such mechanisms are developed by foreign specialists with their countries' national interests in mind. Moreover, many countries are putting a considerable effort to stop exports of such mechanisms.

Considerable problems are also likely to arise with Russian information network joining international information networks, and Internet in particular. According to the US Congress, in 1995 hackers penetrated into computer networks only of the US Ministry of Defense via Internet over 250,000 times. Informational and technological integration of Russia into world systems brings thus new security concerns. At the contemporary stage of civilization's post-industrial development new crisis may be observed, as information becomes as strategic a resource as oil, gas, diamonds, etc.

President Yeltsin has attempted to respond to the fears of FAPSI and others and determined that the Russian military should add its own aggressive IW capacity. "Along with the maintenance of the nuclear deterrence potential at the proper level, it is necessary to devote more attention to the development of the entire complex of the means of information warfare, high-precision weapons, means of ensuring mobility and the development of the defense infrastructure," President Yeltsin said on the appointment of Gen. Igor Rodionov as the new Defense Minister in July 1996.

Work on upgrading existing IW capabilities was demonstrably made urgent by the ease with which Chechnya rebels ran rings round Soviet forces during the uprising. Gen.-Lt. Lev Rokhlin, commander of the Northern Group, complained that "the technical level [of the Russian Army] is so bad that it is below that of Dudayev's gangs." (Gen. Dzhokhar Dudayev was the president of the breakaway Chechnya Republic.) "We cannot identify the coordinates of VHF, cellular and satellite stations because we do not have the necessary equipment. We cannot work with radio networks because we don't have the basic communications equipment. The Dudayev supporters intercept everything that is being communicated and take relevant measures." It is easy to sense the humiliation a career Army officer felt at being outsmarted by a band of irregulars because his Army was so ill equipped.

During 1996 Russian military and intelligence agencies were changing gears as they saw the impending threats and challenges of IW. In many ways they saw it with a clarity that is sometimes lacking in the United States, where greater autonomy among agencies and service units leads to an uneven application of new doctrine. FSB, the part of the former KGB tasked with internal security, cut a deal with FAPSI to have joint oversight over the flow of information over public networks, in particular the Internet. Arching over that, the Security Council, headed by Aleksandr Lebed, created a new department called Computer Security, a section of which is assigned the role of formulating doctrine on IW.

In the Russian context, IW includes all the elements familiar to the United States, plus parapsychological operations such as telepathy and other forms of "mental manipulation." Adm. Vladimir Semenovich Pirumov, who heads the council, was involved with experiments during the Cold War to delve into the unknown powers of the human mind as a potential weapon, including telekinesis, the movement of objects by will alone. That these experiments continue today indicates that after all these years the Russians believe the research has some validity.

The Russian parliament, the Duma, formed a subcommittee of the Defense Committee to deal with "informational security." This subcommittee was to be the launching pad for a series of attacks on the United States by FAPSI and the FSB. FAPSI deputy managing director Vladimir Markonenko openly accused the Americans of subverting computer imports with viruses and logic bombs. The committee then attacked the Soros Foundation, headed by American multibillionaire financier George Soros, as a CIA front organization to undermine Russian society. The foundation promotes the virtues of the Internet in Russia, which offends the more conservative and communist members of the Duma, whose fears of America's intentions run deep. An article written by Maj. M. Boytsov of the Russian army in 1996 illustrates why the Russians, from Yeltsin down, saw the need to get into a new arms race with the Americans.

> The American concept of intimidation has been undergoing changes over time. Thus, because of its inherent shortcomings (answering nuclear intimidation by an enemy equal in might), nuclear intimidation was supplemented by a more flexible and effective intimidation with conventional weapons, above all precision weapons. Intimidation by conventional weapons in turn began to be supplemented by intimidation by non-lethal weapons, particularly electronic weapons. Now we have seen progress towards war in the intellectual sphere, towards a war of machines and electronic equipment without the direct involvement in combat operations of soldiers, whose losses . . . are taken very painfully.

Major Boytsov, like the rest of the Russian military, had been profoundly affected and depressed by the dazzling display of American weaponry in the Gulf War. Faced with a superior America, his reaction, like that of the rest of the Russian government and military, is to try to catch up; no matter where America moves the technology of the battlefield, Russia will always respond to the challenge. The military and its allies in the political leadership clearly believe that not only have they lost the Cold

War but that America is attempting to destroy what is left of their country using IW.

Some Western intelligence agencies believe that this may account for Russia's obsession with developing ever more powerful strains of genetically engineered biological weapons. Since the defection of Vladimir Pasechnik in 1989 to MI6, the British and Americans had learned the extent of Russia's BW work. Pasechnik was a key scientist involved in what was perhaps the best kept secret of the Cold War: a covert program employing more than 15,000 people, code-named Biopreparat, that was ostensibly making harmless vaccines but which was, in fact, developing some of the most deadly weapons ever known. Among the weapons that have been developed are fourteen variants on the Tularemia plague virus for which there is no known antidote in the West. Immediately after Pasechnik delivered his startling news, President Bush in America and Prime Minister Thatcher in Britain began applying pressure on the Russians to shut down the BW plants and destroy the research. Mikhail Gorbachev and then Boris Yeltsin denied the existence of the BW program, despite additional evidence that had been gathered by both the CIA and MI6. Finally, Yeltsin was persuaded that the program did exist and he promised to shut it down. It was a promise that for four years Yeltsin has been unable to keep and the BW program remains as active today as it was when Pasechnik defected.

"We believe that Yeltsin is telling us the truth," said one senior intelligence official. "The military are telling him they have taken action and they are simply lying. What is not clear to us is why. The program is very expensive and risks Russia's relations with many countries including all those in NATO. After all, BW is only an offensive weapon." One explanation is that with the nuclear arsenal crumbling, the Soviet armed forces critically short of men and materiel, and under repeated attack by information warfare, the Russian generals believe that BW is their last line of defense.

With the Cold War officially over close to a decade ago, it is hard to believe that such a state of imagined siege should exist between the two countries. Yet it is clear to many inside the new Russia that a new and dangerous tension is developing between the old superpower rivals. Those wiser heads are arguing for a different, more understanding world where the tensions might be defused before a cataclysm occurs. While acknowledging that "the United States is consistently and aggressively preparing for the conduct of information warfare," a trio of top Russian scientists offered this suggestion:

Russia can act as the initiator of . . . rational agreements based upon international law that minimize the threats of the employment of information weapons. These agreements, as a real contribution to international cooperation, could only strengthen the national security of the states, preserve their independence before the information expansion of the United States and other developed countries and ensure their equal participation in global monitoring of the political, economic and ecological processes.

This was the central theme of my trip to Moscow. In meeting after meeting the message was clear: IW is a threat to us all and we need to negotiate a new kind of peace. For example, Vjacheslav Anosov, a former FAPSI official and now a member of the Joint Committee on Information Security of the Security Council of the Russian Federation, believes that an IW agreement is essential. "We would like an international agreement to limit IW," he said. "But even if we could get an agreement, we recognize that it would be very difficult to enforce."

The Russian suggestion of an arms control agreement on IW was greeted with derision in Washington and London, where it was seen as a simple ploy to allow the Russians to buy time while they attempted to narrow the technology gap between them and the other two countries. But it is too easy to dismiss the suggestion, which certainly reflects Russian ambitions to improve its position in this new arms race. At the same time, the more America exploits its current advantage, the more this will feed into the growing paranoia in the Russian political and military leadership that the Americans are out to destroy what is left of the motherland.

SEVENTEEN

A Mole in the Oval Office

M O T H E R Jones is America's only mainstream left-wing investigative magazine, part of an increasingly thin intellectual thread running back to the salad days of the radical 1960s. *MoJo,* as it is known, keeps a sharp eye on who pays for America's elections, and publishes lists of prominent individuals funding campaigns so that readers can see who is pulling the politicians' strings.

In 1996, *MoJo's* top 400 contributors to Democratic party races brought a new entry at number 262. The name was Jane Huang, home-maker, of Glendale, California. It is fitting that Mrs. Huang receive such recognition. Between 1989 and 1996 she and her husband, John, contributed close to $280,000 to Democratic party funds, an impressive amount for a couple whose only source of income was John Huang's salary as a bank executive.

John Huang came to the United States in 1969 at the age of twenty-four from his native Taiwan to be a graduate student in business administration at the University of Connecticut. A series of jobs in different American banks led him in 1985 to the Lippo Group, a multibillion-dollar Indonesian conglomerate headed by the shadowy Mochtar Riady, an Indonesian of Chinese origin. (Riady changed his name from the Chinese Lee Mo Tie to fit in more easily with the Indonesian business community.) This was the beginning of Huang's being used as the tool for an elaborate espionage operation by the spymasters of Beijing, the central figure in what one senior U.S. intelligence official described as "an espionage coup of epic proportions."

What caused China to take on such a politically risky operation was its obsession with Taiwan and Taiwan's apparent ability to manipulate successive American administrations at will. The separation of Taiwan from mainland China touches the rawest of raw nerves in the Chinese psyche. Since the flight of the Kuomintang under Gen. Chiang Kai-shek in 1949 and the establishment of the breakaway nation, Beijing has been determined to get Taiwan back. The Beijing leadership realized that its refusal to indulge in the kind of sophisticated political and economic lobbying that Taiwan had excelled at in the United States had cost it loss of face and caused an affront to its national sensibilities.

The Chinese embarked on a multifaceted strategy to restore face and to reassert their leverage in American political circles. It was a strategy that was to give them the keys to the Oval Office, deeply embarrass the American President, expose the sleazy corruption of U.S. politics and alert the United States to the threat of a nation that not only understood the rules of future war, but was prepared to fight it on any and every front. It also resulted in a classic example of information warfare that successfully penetrated and manipulated the American political process. Like all successful operations, the Riady-Huang penetration of the Clinton White House was as much the result of luck as good planning. In 1977, when Mochtar Riady first set up in business in America, using Little Rock, Arkansas, as his unlikely power base, nobody could have predicted that the charismatic young politician he befriended and cultivated would one day become President. But of course Bill Clinton did, and thanks to the Riady family's continued political support, the connection propelled them straight into the heart of power. Bill Clinton could not have known that while he was busy getting elected to the Oval Office, Riady was forging an altogether different alliance, with the communist leadership in Beijing. In October 1992, Riady sold 15 percent of his Hong Kong Chinese Bank to the China Resources Holding Corporation (CRC) for "a very, very big discount." China Resources (Hua Ren Jituan) is a government-owned and -operated company that has on its board a senior member of Chinese Military Intelligence, who acts as a coordinator for other agents working under commercial cover for CRC. Now he could use Hong Kong Chinese Bank as a cover too. Four years later, CRC took a 50 percent stake in the company, and won seats on the board.

The Riady connection with Clinton stepped up a gear once he became President. The lobbying intensified, the deals became ever more elaborate, and whether it was by coincidence or not, the President's policies took dramatic shifts in directions favorable to the Riadys. Once violently opposed to doing business with Vietnam until the issue of American MIAs had been resolved, Clinton went on to lift the trade embargo in February 1994. He said it would help accelerate negotiations over the MIAs. It also helped accelerate the exploitation of a promising new market by Riady's Lippo Bank. Clinton's equally fervent distaste for doing business with China for as long as Beijing refused to play by America's trade rules evaporated and the policy of engagement, which had been nurtured by George Bush, but attacked by Bill Clinton, continued.

In Washington, the strong link between Riady and Clinton led to John Huang crossing over from the lobbying side of the fence to becoming an influential employee of the government department most important to an international businessman like Riady, namely Commerce. Riady bragged to friends that Huang's advancement was a quid pro quo for his helping Clinton. Lippo's fortunes prospered, as did those of the Hong Kong Chinese Bank, which formed an alliance with the Lippo Bank in Los Angeles and the First Union Bank of North Carolina. The Chinese now had direct access to the U.S. banking system's financial networks. Huang meanwhile left the Commerce Department and took the job that was to propel him into the public consciousness, as a fund-raiser for the Democratic National Committee. The bizarre chain of events, including Buddhist temple fund-raisers, Lincoln Bedroom sleepovers and vast contributions made by people of apparently humble means has been well documented since the scandal broke in October 1996, too late to affect the election but in plenty of time perhaps to damage Vice President Al Gore's hopes for 2000.

All of this might be dismissed as yet further evidence that American politics is up for sale to the highest bidder, a perennial issue that seems to elude reform. It moved into the ranks of espionage and IW with some startling intercepts being made by the NSA. In June 1996, the NSA picked up signals traffic between the Chinese embassy in Washington and Beijing that indicated the Chinese government wanted to put some money into the campaign war chests of prominent American politicians, among them Senators Dianne Feinstein and Barbara Boxer of California. Beijing had been embarrassed in 1995 when Bill Clinton approved a visa for a visit to the United States by Taiwan's President, Lee Teng-hui. The Chinese leadership realized it had to step up its influence in Congress, particularly as it was seeking permanent Most Favored Nation trade status from the next congressional vote. The National Security Council was warned of the intercepts but, in a famous bureaucratic failure of communication, the word did not reach the Oval Office. Attorney General Janet Reno was advised by aides to make personal contact with Anthony Lake, the National Security Adviser. She made the call, but left no message.

Six days after that, on June 9, 1996, still in the dark about the NSA intercepts, Bill Clinton attended a fund-raising dinner at the San Francisco home of Senator Feinstein and her husband, Dick Blum, a businessman with major investments in China. On the guest list, alongside a host of Democratic party luminaries, including Senator Boxer, were John Huang and Xiaoming Dai, Chairman and CEO of Asia Securities, a company with strong connections to Lippo and to the government in Beijing. A postmortem suggested that Xiaoming was at the dinner courtesy of Huang. Failure to warn the President and the senators that a foreign power was actively seeking to affect the course of American politics was an extraordinary blunder.

It later emerged that the Justice Department possessed what a senior official said was "conclusive proof" that $2 million had been laundered through cutouts in America to fund the political campaigns of politicians sympathetic to China, although there was nothing to suggest the politicians knew of this. In yet another example of the kind of incompetence that characterized America's counterintelligence effort in the Huang affair, the evidence for this had been available for nearly two years, but had lain unexamined among a mass of reports. Soon a whole rash of federal and congressional investigations were under way, focused mainly on the fund-raising scandals, although the most important one from China's point of view was by the House Intelligence Committee into the alleged attempt by China to influence the elections and into the security clearances given to John Huang.

As the fund-raising scandal began to unravel, the focus of the investigation by both the media and the Congress was improper contributions to the Democratic party during the presidential election. Under pressure from the media, Attorney General Reno established a fifty-person task force composed of Justice Department lawyers and investigators and FBI agents, who were charged with investigating the affair. In one of the first meetings of this group, one FBI agent recognized some of the names that were central to the probe.

As a matter of routine, the FBI listens to every conversation inside the Chinese embassy in Washington and at the nearby ambassador's residence. Most of these bugs, both inside the buildings and planted nearby, operate automatically, filing conversations into the database. Once there, the data can be scanned for specific information by hitting key words in the dictionary. Until the probe, nobody had bothered to listen to the tapes or to punch in "Huang" into the dictionary. When that was done, a treasure trove of data appeared. Not only had Huang been a frequent visitor to the embassy, but he had also been to the ambassador's residence.

CIA agents occupy a number of offices in the Department of Commerce building and coordinate any economic information they pick up with Commerce to maximize business opportunities for the United States. On a regular basis, John Dickerson, head of the CIA's Office of Intelligence Liaison, briefs key members of the Commerce Department on this sensitive information. John Huang sat in on 146 of these briefings and had access to fifty-two classified documents, many relating to China. Phone records show that he had been in touch with Lippo immediately before and immediately after some of the meetings. More ominous still, according to Representative Dan Burton of Indiana, who ran the House inquiry into the fund-raising scandal, as well as passing classified information to the Riadys, Huang was also feeding the Chinese. "He was at the Chinese embassy many times, and had a security clearance that we didn't think was proper," Burton said.

Burton's colleague Representative Gerald Solomon of New York, chairman of the House Rules Committee, went even further. Calling the fact that Huang had received these briefings "extremely serious and dangerous," he alleged that Huang's access to this secret intelligence could have resulted in the death of a CIA agent.

Both the congressional investigators and the FBI team began to try to connect the known visits Huang made to the embassy, the telephone calls and the CIA briefings. In many cases, there was a perfect match. A briefing would be followed by a phone call and then a visit. That in turn was followed by visits to the Chinese embassy. "Was he a spy?" asked one congressional investigator. "Sure he was. Was he a willing agent or an unwitting dupe? Who the hell knows. But what we do know is that he had access to the best stuff and we are certain he passed it on."

Lippo's castle of influence crumbled under the weight of the investigations but not before Mochtar Riady had enjoyed a full presidential term of access to the Oval Office and had seen policies that would benefit his company pursued in Indonesia, Vietnam and China. And despite the scandal that had his name at its epicenter, his reputation in Asia soared because of the evidence that he had been a player in U.S. politics.

Standing in the shadows behind Riady, the spymasters of Beijing had been enjoying a bridgehead into the banking and political systems of the United States. What had been a good investment for Mochtar Riady was a good investment for his partners in the Hong Kong Chinese Bank, the Politburo in Beijing. Chinese interests had admittedly suffered a setback, but in the grand scheme of things, the inconvenience was minor. Whatever the result of the Justice Department probe and the congressional investigations it is unlikely that Huang will ever be prosecuted, even if the intelligence community gathers enough evidence to stand up in court. The NSA has told the FBI that it will not allow the transcripts of the intercepts of communications inside the Chinese embassy or between Washington and Beijing to be used in evidence, and the FBI will not reveal its own data either. Both are concerned about revealing the extent of their eavesdropping capability.

The final report of the Senate Governmental Affairs Committee into the fund-raising scandal was bedeviled by precisely those concerns. The first draft of the report laid out in considerable detail the extent of Chinese attempts to influence the American democratic process. However, objections by both the CIA and the FBI ensured that the unclassified version had most of the contentious material excised. Both agencies were concerned that publication of the initial draft would have seriously compromised sensitive sources and methods both in Washington and China.

After much haggling, the Committee and the intelligence community were able to agree on a single statement concerning China: "There are indications that Chinese efforts in connection with the 1996 elections were undertaken or orchestrated, at least in part, by People's Republic of China intelligence agencies."

What the report did not mention was the enormous counterintelligence effort that had been provoked by the fund-raising investigation. Horrified FBI and CIA officers, supported by intercepts from the NSA, understood for the first time the extent of China's intelligence activities against America. As so often happens in such investigations, one stone unturned led to many more, and at the end of 1997 and for the early part of 1998 an intensive manhunt was under way to uncover what both FBI officials and CIA officers say is an extensive network of Chinese agents buried in the American establishment. At the time of writing, one focus of the investigation is on the Pentagon, where sources say "a high level penetration" has been uncovered.

In the United States, the cycles of the nation's life are ruled by daily reports from Wall Street, opinion polls, monthly economic indicators, biennial elections to Congress and quadrennial presidential battles, but China measures time in generations.

The Chinese understand time. While Westerners are always fighting it, the Chinese embrace and use it. Their political system is geared not to election cycles but to the lifetimes of China's leaders. Constancy, consistency and single-mindedness of purpose characterize the mindset, underscored by patience. It is little wonder that China not only understands the concept of information warfare but wages it with a sophistication and understanding that is quantum leaps ahead of Western thinking.

In 1985, Shen Weiguang, a twenty-five-year-old soldier serving with the ground forces of the People's Liberation Army, the PLA, wrote a monograph entitled *Information Warfare*. This was years before the phrase caught on in American military circles. His work foresaw information as an overarching feature of society, from military through to cultural affairs. He talked of the "information frontier," "information factory," "informationized army" (what the U.S. military refers to as the digitized battlefield), the "information police" and the "take-home battle."

Given the soldier's age and the relative immaturity of IW concepts in the West at the time, this was a remarkable paper. The Chinese leadership was quick to grasp from this and subsequent works that the rapidly spreading technology of computers, networks and digital communications would greatly enhance its ability to project its agenda on the rest of the world and leverage existing military assets. Writings on the subject were a regular feature in the Chinese military newspaper *Jiefangjun Bao,* often reflecting much the same thinking as was being expressed in the West, but with an occasional, very Chinese, twist.

> The sanguinary type of war will increasingly be replaced by conten-
> tion for, and confrontations of, information. . . . The most effective
> weapon is information itself. Information can be used to attack the
> enemy's recognition system and information systems either proactively
> or reactively, can remain effective either within a short time or over
> an extended period, and can be used to attack the enemy right away
> or after a period of incubation.

The Chinese military see the Internet as a powerful means of attacking the West, "so we can drown our enemies in the ocean of an information offensive." This will include the use of viruses to destroy the enemy's ability to wage war by crashing computers, logistics systems, electrical power and water supplies. The writer went on to say that computeriza-tion of production meant that production centers were moving away from the big cities and into rural areas, and that networks would be used to help mobilize the people, "sending patriotic e-mail messages and setting up databases for traditional education." It sounds curiously quaint, yet the writer is adamant about one thing: that information in the civil arena is no less secret or important than in the military sector. In stark contrast to the freewheeling, jealously guarded privacy of the ultra-democratic cyberspace in the United States, Chinese cyberspace should be fenced in and controlled. "If no security measures are taken to protect computers and networks, information may be lost."

What is interesting about China's foray into the Internet is that it is using an American online service, Prodigy, as its guide. Prodigy signed a joint venture with Norinco, China's government-owned arms-trading conglomerate, to set up such a sealed enclave inside cyberspace.

In 1996 China established a top secret information warfare center to bring under one roof all the operations seen as IW, including the projec-tion of positive propaganda, the theft of economic secrets and offensive computer attack. The new center was established under the direct con-trol of the central committee and with representatives of every important department in Beijing. "The Chinese understand the importance of in-formation warfare and are ready to exploit the opportunities presented by the information age," said one intelligence source.

Articles appearing in Chinese military journals talk openly of using doctored chips and software to plant viruses in an enemy's information systems, which would be done by "paying off" chip producers to make the necessary adjustments. The articles are not specific about which producers, whether Chinese or in other Asian countries, would be paid off, but the threat should be regarded with some apprehension. And if these producers are making chips destined for American markets, then apprehension should become alarm. It is doubtful however that circuitry in the most sensitive systems in America would include chips and software created outside the United States, in which case this threat could apply more directly to China's neighbors in the Pacific region, who would be far more likely to buy low-priced electronics from the rapidly emerging regional economic superpower.

The nation's military commanders understand that information provides a tactical and strategic edge that when employed properly can tilt victory toward the weaker but smarter combatant or can give the numerically larger combatant the means to win wars more cheaply. Maj. Gen. Wang Pufeng of the Institute of Military Science maintained that in modernizing the nation's military capability the guiding principle should be that firepower superiority depends on "information superiority."

> When we engage in war with strong enemies in the future, we will face comprehensive and powerful information suppression. There is a question of how to use weakness to defeat strength and how to conduct war against weak enemies in order to use information superiority to achieve greater victories at smaller cost. It must be confirmed that information and weapons are all controlled by people. People are the main factor in combat power. However, it must also be confirmed that the functions of people and weapons will primarily be determined by the flow of information.

General Wang advocated the upgrading of military capabilities across the whole spectrum of warfare, from space-based laser weaponry to ballistic missiles, underpinned all the while by information technology. But who are the "strong enemies" he spoke of? Without a doubt, the United States is the only country that fits the description, and the clarity with which Wang sees the challenge is a strong caution to those who see America as invincible.

> In wars of the future, China will face the enemy's more complete information technology with incomplete information technology. Based on the fact that sometimes superior tactics can make up for inferior technology, China will carry out its traditional warfare method of "you fight your way, I'll fight my way, and use my strengths to attack the enemy's weakness. . . ."

Sabotage, psyops and special forces would be used to whittle away the enemy's will and stamina, the general said, with growing IW capabilities coming onstream as they are developed. This is already happening. China is one of the top ten countries on the FBI's list of nations using information warfare for economic and industrial espionage. Chinese government officials routinely target American companies to steal their secrets. In December 1996, Eric Jenott, a twenty-year-old Army private based at Fort Bragg, North Carolina, was charged with using his skills as a hacker to break into the Pentagon's computer and telecommunications system. He was charged with spying after he passed a top secret code word to a Chinese national, a code that would have given Chinese intelligence access to a secure telephone system. The arrest of Jenott was just one visible sign of a new and aggressive China prepared to do battle on a number of fronts.

In November, China attacked on the cultural front by trying to stop Disney from producing *Kundun*, a film by renowned director Martin Scorsese about the life of the Dalai Lama, the exiled ruler of Tibet. China conquered Tibet in 1951 and ruthlessly suppressed Buddhism while destroying hundreds of temples and jailing thousands. Remarkably, considering the company is building a theme park in China, Disney refused to be intimidated, displaying the kind of strength that China actually respects. Beijing also went after the makers of a movie starring Brad Pitt called *Seven Years in Tibet*, about the celebrated mountaineer Heinrich Harrer, who escaped a British internment camp in India by scaling the Himalayas, ending up in Tibet, where he became a teacher of the young Dalai Lama. The filmmakers knew they would not be able to film in Tibet, so they set their locations in northern India. Six months and over $1 million was wasted when the Indian Prime Minister caved in to Chinese pressure and ordered a halt to the project. Director Jean-Jacques Annaud went on to make the film in Argentina. The Chinese even tried to stop it there, but the Argentines, with commendable resilience, greatly encouraged the project.

Soon thereafter China posted a Hollywood blacklist at the state-run travel agency in Lhasa, the Tibetan capital. The list included more than fifty Westerners, including Harrison Ford and Brad Pitt. According to the Campaign for Tibet, everyone on the list had been involved in the effort to establish democracy in the Himalayan country. The blacklist represented China's second run-in with Hollywood in as many months. In December 1996, forty-one Hollywood celebrities wrote to the Chinese government criticizing it for trying to block the Disney movie. Again, it would appear China lost the battle, yet it soon transpired that in China losing one battle is acceptable if in the long run a war can be won. Film companies and television stations interested in doing business in China started scrapping scripts with an anti-China theme.

This opposition to films critical of the country appears overly sensitive to Western thinking, but actually it was part of a broader information warfare strategy aimed at projecting a positive image of the nation and deterring criticism. Some American businesses with contracts in China were quietly persuaded to produce a steady stream of videos and books for circulation to schools across America about the projects they were engaged in, all with a positive spin on China. Influencing a new generation in America made good strategic sense, and for the companies who stood to make millions from their deals, a few thousand dollars spent on a video or two was meaningless.

It was a subtle tactic. Even more subtle—indeed, insidious—was the way in which China used as its advocates some of the most prestigious names in the field of American foreign policy. A *Washington Post* article on March 16, 1994, reported:

> Much of the American foreign policy establishment, including three former secretaries of state and other former senior officials of both parties, turned a collective thumbs down yesterday on the Clinton administration's policy of linking trade with China to Beijing's human rights performance.

The three former Secretaries of State at that meeting of the Council on Foreign Relations were Henry Kissinger, Cyrus Vance and Lawrence Eagleburger, respected public servants all—in their time. Once out of public service they went to work for companies that wanted to do business in China, so they had a direct financial interest in the degree of openness of trade between America and China. They were not alone. Other former high officers of state such as Dick Cheney, who was Defense Secretary under George Bush, and Brent Scowcroft, who was Bush's National Security Adviser, also played the game from both sides of the net.

Given the esteem accorded the pronouncements of former officials of such stature, it was an astonishing betrayal of the public trust by these men and an equally huge propaganda coup for China. Clinton actually did de-couple trade and human rights issues shortly after that meeting. These unofficial lobbyists for the communist government in Beijing continue to maintain a steady drumbeat of disapproval at any mention of sanctions for human rights abuses that might slow down the pace of the business that pays their way.

While propaganda subverts the American attitude toward China, covert operations and economic espionage raid the national armory, and ruthless trade practices distort the economic balance between the two countries.

While remaining contemptuous of American society and character, the Chinese covet its technology. Companies were established across America whose sole purpose was to take back to China "dual-use" technology, that is, technology which originated in the military sector and as defense companies closed down after the end of the Cold War migrated into the civil sector. One such was China Yuchai International, whose Cleveland subsidiary Yuchai America bought two highly sophisticated five-axis milling machines at an auction held by the Heinz Corporation, which made jet engines for the Pentagon. U.S. Customs impounded the machines during transit back to China in May 1994, and they languished there until late 1995 when the Commerce Department dismissed Customs' concerns about the possible military use these machines could be put to, and let them go. The most celebrated incident was the sale by McDonnell Douglas to CATIC, the China Aero-Technology Import-Export Corporation, of thirty-one high-tech machine tools from a former B-1 bomber plant that CATIC promised would be used for civilian purposes. The Pentagon objected to the sale, but Commerce Secretary Ron Brown approved it. As soon as they landed in China, the tools were diverted from the intended civil aircraft production plant to a factory making the SU-27 fighter-bomber. The outrage provoked a grand jury investigation, but despite demands the tools be returned to the United States, they remained in China.

There is little doubt the Chinese technology gap is closing, as China plunders the United States, with the willing assistance of American business. The Chinese can be stunningly ruthless. In 1988 a Fort Lauderdale company, Revpower, which made specialized batteries, was recommended by McDonnell Douglas to enter into a joint venture with Shanghai Aviation Industries, making batteries at a price that made sense to the American company's boss, Robert Aronson, who put up $5 million and his company's technology. As soon as manufacturing began, the Chinese partners charged him 40 percent more than the agreed price. Aronson won an international tribunal ruling and was awarded $4.6 million against the Chinese, who refused to pay, stole the batteries, stole the technology and stole the plant. Aronson lost $8.5 million in total, while China suffered not one bit for its outright piracy. A startling new book written by two Chinese military strategists, *America, Russia and the Revolution in Military Affairs* by PLA officers Zhu Xiaoli and Zhao Xiaozhu, predicts that the gap will actually close by 2007, at which point America's much vaunted dominance information technologies will be over, they say, thanks to its own complacency and overconfidence: "Those who believe that the current revolution in military affairs will be under the control of the United States or can develop only according to the speed and directions set by the United States are extremely wrong and quite dangerous."

The book was brought to the Pentagon's attention by Andy Marshall, the Pentagon's Director of Net Assessments and one of the most prominent strategic thinkers of the age, in a pamphlet entitled *Chinese Views of Future Warfare*. It was part of a movement among people alarmed at China's growing might to alert the administration to the dangers of underestimating China's capacity. One congressional aide who read the pamphlet and who had conducted similar research noted: "The disposition of the Chinese military character is changing and the leadership is contemplating a form of warfare never undertaken before. It appears that our national security strategy on China is to keep our fingers crossed and hope for the best."

The warnings provoked reactions from experts who dubbed them unduly alarmist. "Anybody who makes the case that the United States is falling behind, or will fall behind absolutely has no knowledge of China," Al Wilhelm, a former U.S. defense attache to Beijing and executive vice president of the Atlantic Council, retorted. Harold Leach from the U.S. Navy's Office of Naval Intelligence added, "They want to do what the big boys do, but the bottom line is that the Chinese Navy today can be compared to the U.S. Navy 30 to 40 years ago. But sometimes we need to be reminded that even slingshots can kill."

The strong reaction to Marshall's pamphlet either was symptomatic of the overconfidence the Chinese had noted or was a reminder that America is too easily swayed by alarming-sounding predictions that may, in themselves, be part of an IW offensive.

Whether China is in the technological Dark Ages or not, the leadership in Beijing is pursuing the modernization of its economy and military with a vengeance, all the while cleaving to a communist hard-line grip on government. It is no accident that while the Chinese are embracing the new technologies of the Information Age, President Jiang Zemin has taken to wearing Mao-style suits and military fatigues or that the people are ordered to display patriotic flags and emblems while getting rid of foreign symbols and logos. In other words, the Chinese leadership determinedly avoid the dilution of its rigid political structure that the West is hoping market forces will produce. Chinese culture and history demonstrate why it would be foolish to think they would do otherwise.

The military philosopher Sun Tzu wrote the most complete manual of martial psychology and tactics 2,500 years ago, called *The Art of War*. His book is required reading in Western military academies for the simple yet devastatingly accurate truths he wrote about warfare, as applicable today as they were then. Western military officers writing strategy papers quote freely from the book to demonstrate their intimate understanding of his philosophy and there is no doubt that applying Sun Tzu's philosophy will make a better officer out of anyone. But Sun Tzu's writings were the external manifestation of a complete and deeply rooted philosophy on life that a Westerner can only dimly comprehend. He was Chinese, and Westerners have been trying and failing to comprehend the Chinese mind for centuries. It goes without saying that a Chinese mind is going to understand more closely, and implement more effectively, the ideas of Sun Tzu.

Take this one, fundamental maxim:

> If you know the enemy and know yourself, you need not fear the result
> of a hundred battles. If you know yourself but not the enemy, for every

victory gained you will also suffer a defeat. If you know neither the enemy nor yourself, you will succumb in every battle.

One thing can be absolutely taken as real: the Chinese leaders know themselves. They are unswerving in their hold on power and in pursuit of their objectives. Can the same be said for Western leaders? Does a president like Bill Clinton stick unswervingly to his principles and philosophies or does he react to daily opinion polls and listen to spin doctors like Dick Morris?

The Chinese also understand the way the West works. They know that simple cash, and not very much of it, can buy access to the heart of power; that America may have more raw military power at its fingertips than any nation on earth but that the will to use it is weak. Does America understand the way China works? Successive U.S. governments have maintained that China will eventually be tamed by the power of the marketplace, that Adam Smith, perhaps in cahoots with Mickey Mouse, will depose Karl Marx and set the people free; that it is wiser to do business with the dictators and treat issues like human rights violations, proliferation of weapons of mass destruction and trade piracy as completely separate. This is what the Bush administration termed the "campaign of peaceful evolution." The Clinton administration talks in terms of "engagement." The policy is seen as the only alternative to Chinese isolationism; better to embrace than to shun. But this is how the Chinese felt about being embraced: in 1992, Chinese intelligence was warned that "hostile forces and reactionary organizations overseas are carrying out mental offensives, sabotage activities, religious infiltration, fund [scholarship] infiltration, and cultural infiltration."

This does not sound like a country ready to succumb to the pleasures of shopping malls and Tarantino movies. Indeed Chinese policy toward the West is characterized by rigidly opposing anything that smacks of undermining or criticizing Chinese policies while simultaneously manipulating the West's endless pursuit of money and business to its own ends.

The Chinese understand both themselves and the West far better than the West understands itself. By Sun Tzu's reckoning, that means the Chinese are winning the information war.

EIGHTEEN

It's the Economy, Stupid

A S with so many things, the end of the Cold War and the advent of the Information Age caused a seismic shift in the world of espionage. Spy agencies needed a reason to be; although the need for intelligence had not lessened, the fact that most required knowledge was rapidly becoming available on the Internet meant that cloak and dagger was beginning to take second place to the drudge of reading and analyzing mountains of online reports. This left a hole in the job description of intelligence agencies. If most military secrets were becoming irrelevant, the spies had to find something else to steal and quickly, before the budget axe started swinging their way.

Sophisticated survivors that they were, agencies realized they could make a big impact on national economies, particularly in those countries which did not have the R&D and manufacturing base to keep up with the revolution in information technology. Thus what had been a secondary activity in the spy business, economic espionage, took center stage.

It is a fair rule of thumb that countries where governments exert direct control over industry are more likely to use their spies for economic espionage than countries where markets are more open and free. Although economic prosperity is a strong component of any country's national security, a government which has direct influence over key areas of major national, and perhaps nationalized, corporations will see nothing wrong with leveraging other governmental assets to improve those companies' edge in the global marketplace. And if that means stealing technology that their own companies are not yet able to develop, the spies are only too happy to oblige.

The early 1990s saw an explosion of economic espionage, particularly against the United States, which had absorbed the pain of changing gear from an economy where defense spending was a real engine of growth to becoming a hothouse in which small, focused, high-tech companies could develop and prosper. Individuals may have lost out as companies downsized, shedding people and costs to enable them to survive and compete. But what emerged was a vibrant economy that set a new pace and agenda in world markets. Europe, with the exception perhaps of Britain, which went through its own painful metamorphosis, languished in its paternalistic, socialistic stasis, unwilling and unable to create the social and economic conditions necessary for the freewheeling ways of the new age. One of the most sclerotic nations was France, and it was no accident that its external secret service, DGSE, had since the 1960s run a separate department for economic espionage known as Service Seven.

According to Pierre Marion, the former head of DGSE, the special unit was set up at the organization's headquarters in Paris, which is known as La Piscine or the Swimming Pool. Marion was perfectly open about the unit's purpose. "I would try to get documents and intercept communications. All in the service of French companies," he said. Marion said that while the military and political secrets of France's allies were off limits "in the economic competition, the technological competition, we are competitors and not allies."

The members of Service Seven were no less aggressive than their more traditional spy colleagues in the DGSE just because their targets were economic rather than political or military. Not for nothing did they earn the memorable sobriquet "the economic KGB of the post–Cold War era." They employed the entire range of espionage techniques, or trade craft, from recruiting agents inside companies they wanted to steal from, to using bugs and surveillance cameras to lift data.

The agency's reputation for economic spying is such that myths and legends have grown up around it, such as the delightful notion that the headrests of first-class seats on Air France are bugged so that conversations among traveling American executives can be recorded. There is no truth to this story, yet its very existence is testament to what might be going on. In fact, intelligence analysts warn businessmen going to France to act as if every conversation might be overheard, for even if the headrest is not bugged, the flight attendant might be an agent. Or the hotel room might be wired. And don't leave a laptop computer in the room, it is just begging to be either stolen or data-raped. The importance of arms exports to the French economy makes it doubly vital for them to steal what they cannot develop for themselves. One of the most crucial technologies they needed in the early 1990s was stealth, as the Dassault aerospace company set about the development of a French stealth fighter. After trying to plunder British aerospace companies for anything they could lay their hands on during a concerted operation in the late 1980s, the DGSE agents turned to American companies that had access to elements of the stealth process, making something of a mockery of Pierre Marion's avowal that allies' military or political secrets were off

limits. The FBI was alerted to their attempted penetration of Dow Corn-
ing in 1991, and followed them as they tried to subvert employees in
twenty-four other companies. The agents were sent home empty-
handed.

The irritation between America and France was catapulted to a new
level in 1993 when a copy of a twenty-one-page memo addressed to
French diplomats in America was delivered in a plain brown envelope to
the CIA and to the Knight-Ridder newspaper chain. This memo,
stamped "Defense Confidential," was the most complete evidence yet of
a systematic effort by France to spy on American business. Analysis of
the document revealed that it was issued by the Department of Commerce, Science and Technology, a front agency for the DGSE. Senior
Clinton administration officials decided to fire a shot across the French
bows and Peter Tarnoff, the Assistant Secretary of State, was told to
confront them. In a heated exchange, officials from the French embassy
forcefully disavowed the document.

"They just lied, lied, lied and lied," said a Treasury Department official.

The memo listed the companies that were being targeted, together
with the information that had to be stolen, including "computers, electronics, telecommunications, aeronautics, nuclear armaments, chemical, space, consumer goods, capital goods, raw materials and major
civilian contracts." It was quite a shopping list, which demonstrated that
it was not just gaps in certain areas of French knowledge that had to be
filled—they were after anything and everything. One U.S. bank called
in investigator Tatiana Gau of the Silver Spring, Maryland, company
Parvus-Jerico, after a suspected attempt at subversion by a French spy.
"This executive was introduced to the Frenchman at a party and they
developed a friendship," she said. "Before long he was regularly passing
information to the spy." The operation was so subtle that the executive
was not aware of what was happening, even though the information was
extremely sensitive. The exposure of the memo sent shock waves
through American industry, and even prompted the Hughes Aircraft
Company to cancel plans to attend the Paris Air Show in June 1993.
The CIA had sent out a specific warning to aerospace companies about
French spies, and given that Hughes had already lost a $258 million
satellite contract in the Middle East to France, and that the Hughes HS
601 satellite had been one of the items on the French spy memo, chairman Michael Armstrong pulled the plug on the trip. Such was American
annoyance over the memo that Secretary of State Warren Christopher
delivered what was described as "an earful" to his French counterpart
during a trip to Paris in May 1993.

Of course, the American outrage was pure hypocrisy, although that
was not apparent at the time. The administration played the grievance

to the hilt and the French were portrayed, with absolute justification, as thieves and rogues. The trouble was, America was playing the same game, too, although, as it turned out, with much less finesse than the French. In February 1995 a group of CIA agents, using the U.S. embassy in Paris as cover, mounted an operation to establish the French position on sensitive trade and technology negotiations involving the entertainment industry. The French government was trying to restrict the number of American-made films and TV shows being shown in France, to stave off the "cultural imperialism" of Hollywood. Given the importance of such exports to the American economy, someone in the CIA thought it would be useful to know how the French were going to play their hand.

It was a complete disaster. One of the operatives, a part-timer of questionable abilities, was exposed early on by the French security services after she tried to bribe a member of the French government. Henri Plagnol was at the time a thirty-four-year-old government researcher who was befriended by an American woman who described herself as a PR director for a Texas foundation interested in preservation of the ecosystem. French counterintelligence warned him she was a CIA agent, and to take care. Shortly thereafter he was recruited to serve in Prime Minister Edouard Balladur's office and when he was once again contacted by his friendly American PR friend, counterintelligence asked him to play out the scenario as a double agent. She took him to a series of meetings with a CIA case officer, who interrogated him about French policy on agricultural and entertainment trade issues, and he fed them data supplied by his counterintelligence handlers in return for hard cash. The French amassed a pile of incriminating evidence along with the identities of four other agents involved in the operation, including the CIA Paris station chief. They had also tried to bribe a senior member of the Ministry of Communications, Thierry Mileo, offering him cash in exchange for details on the French negotiating position at GATT (General Agreement on Tariffs and Trade) talks on telecommunications, and a technician who had extensive knowledge of French domestic and international telephone networks. The operation was rolled up and the spies were asked to leave the country quietly. The Foreign Ministry at the Quai d'Orsay then called in Ambassador Pamela Harriman for a sorely embarrassing dressing-down. It was a total triumph for the French, for not only did they demonstrate quite clearly that the earlier American attacks on them had been hypocritical, they also forced a shell-shocked CIA, under strong attack from Congress for not informing either Ambassador Harriman or Congress about the operation, to virtually wind up all intelligence gathering in France. Given that the French had also briefed the intelligence services of other countries, like Germany and Italy, on what the Americans were up to, it was a severe setback for intelligence gathering in large parts of the rest of Europe, too.

The incident in Paris may have been notable for its incompetence, and the agency could have used some Gallic bluster in weathering the storm instead of slinking away chastened, but it was by no means the first nor the last example of U.S. economic espionage. A much more sophisticated and successful campaign had been waged in early 1994 to stop the French winning a $6 billion order from Saudi Arabia for military equipment and civil airliners. The French had been salivating over the deal, right up to the moment when Prime Minister Edouard Balladur sat down with King Fahd to discuss the final agreement.

At the last minute the Saudi monarch backed away from signing, and Balladur's team went home scratching their heads over the deal-that-got-away. They found out what had really happened when it emerged that the CIA and NSA had employed their combined resources to get the inside track on the terms the French were offering and the bribes they were paying, which then enabled President Clinton to wage a very effective personal lobbying campaign with King Fahd that delivered the airline contracts to Boeing and McDonnell Douglas while the arms contracts went to a variety of different companies. It was a stunning blow, compounded by CIA intercepts alerting the U.S. government to the French negotiations and bribes aimed at securing a $1.4 billion high-tech radar contract in Brazil, thanks to which Raytheon took the deal from under the nose of the French company Thomson CSF. It was little wonder the French set about humiliating the Americans so completely the next year.

If the French called their specialist economic espionage unit Service Seven, the KGB had another numerical name for theirs: Department 8. It was set up in the late 1970s by KGB head Yuri Andropov, who later became General Secretary of the Communist Party, as part of the First Directorate. In the 1990s, with the communist party and the Russian economy in ruins, Department 8 was split off and made into a stand-alone organization. The gap between the Russian economy and that of the developed countries of the West was so huge that the order went out for agents to steal whatever technology or economic intelligence that the country could not otherwise beg, borrow or, at a real pinch, buy. In 1996, President Yeltsin publicly urged his spies to increase their efforts, "to close the technology gap," so vital was it to the national economy. The Russian Security Council, headed by Gen. Aleksandr Lebed, set out a detailed demand for what the SVR, which took over the KGB's external operations, should be doing. Lebed added:

I believe it is necessary to adjust the structure and goals of the intelligence services, directing their efforts to back Russia's economic interests in the first place. I will demand that more effort be applied immediately in the following specific fields: to ensure an uninterrupted monitoring of the situation on the world markets of armaments, aviation and space equipment and to search for information about existing or developing technologies in the design of new armaments; to search for new designs in commercial technologies, both by state-run and private enterprises; to search for critical information on the plans and activities of the leading international financial institutions, major transnational corporations, banks and investment companies of all the countries of the world. . . .

That was quite a shopping list, unequivocal in its exhortation to watch, listen, learn and, above all, steal whatever Russia needed. This upgrade in Russian economic espionage efforts drew a grim response from Louis Freeh, the Director of the FBI. "It's an ominous sign whenever a foreign intelligence service prioritizes in any way theft or stealing by espionage. . . . American corporations aren't equipped to defend against that kind of attack." Freeh might be forgiven for his disingenuousness. He was playing by the rules of the game that say "everybody does it but us," the official line on all espionage matters. Oleg Kalugin, the celebrated defector from the KGB, adopts a much more laid-back approach to his former employer's aggressive behavior. "Don't look at Russia as a complete rogue," he said jovially. "Russia has entered a club wherein Russia will steal secrets. It is a noble process, and an evil we have to live with. Don't get hysterical about it!"

Kalugin's point about there being a "club" at work was well made. In February 1996 the General Accounting Office (GAO), Congress's investigative arm, published a report into the economic espionage activities of countries otherwise regarded as allies of the United States and listed five countries engaged in direct attempts to steal or bribe away America's technology. Unfortunately, the GAO referred to each country as Country A, Country B, etc., through to E, but intelligence sources were quick to help identify them. Country B, for example was France, D was Japan and E Israel.

Thus it should have come as no surprise in May of 1997 when it was learned the FBI had opened an investigation into possible espionage activities by Israel, which had allegedly recruited a senior U.S. government official to pass it "highly sensitive information." For five months, the bureau had been screening top officials in the Clinton administration, looking for the agent called Mega. This agent was a highly placed civil servant, possibly a member of the President's National Security Council, and his, or her, existence came to light when the NSA picked up signals traffic between the Israeli embassy in Washington and Tel Aviv in which agents in Washington reported they would ask Mega for a copy of a letter that detailed American commitments to the Palestinians in the Middle East peace process. "The ambassador wants me to go to Mega to get a copy of this letter," the Mossad agent in Washington said. Back came the reply from Tel Aviv ordering him not to. "This is not something we use Mega for."

When the story broke, the Israelis angrily denied the existence of a spy. "The story is absolutely baseless," Israel's Ambassador to the U.S., Eliahu Ben Elissar, said. "Israel is not involved in any kind of espionage or trying to obtain intelligence from the United States." To American intelligence officials, this protestation of injured innocence fell into the "methinks he doth protest too much" category. At first Israeli officials claimed that Mega was the short form of Megawatt, a communications system used between friendly countries. In fact, Kilowatt is such a system used in counterterrorism but Megawatt does not exist. Then Israeli officials claimed that Mega is a slang description for the CIA officer responsible for liaison with Mossad. If true, that was news to the CIA. In fact, the FBI counterespionage investigation focused in June 1997 on a senior official in the National Security Council who was believed to have been friendly with Israel for some years. Although the FBI was convinced it had uncovered its man, there was little hope that he would ever be prosecuted. "We are dealing with reality here," said one intelligence official. "The political fallout from such an arrest is too terrible to contemplate."

The investigation into Mega was just one of a long list of episodes that had created a cynical view about the United States's supposedly close ally in the Middle East. Any intelligence officer involved with the Israelis will talk privately about the agents in place Israel uses, the information warfare that is routinely practiced in Congress and across the country and the routine stealing of business secrets. But there is a weary recognition that, whatever the evidence, nothing will be done. "I've read a two inch-thick file on Israel's spying against America," one intelligence source told me. "Make no mistake, Mega is just the latest visible sign of a big business."

It was a business that had been going on for some years. In the mid-1980s, according to CIA testimony at a Senate hearing in 1996, a large defense manufacturer who had invited Israelis to visit found later that it was missing some very important, and very exclusive, test equipment involved in the development of a new radar system. Challenging Israel was out of the question as there was no proof the visitors had stolen it, until two years later the manufacturer was contacted by the

Israelis asking for help. The missing equipment had malfunctioned and, with admirable chutzpah, they were seeking help in repairing it.

Then there was the presence in a U.S. jail of Jonathan Jay Pollard, who in 1986 was convicted of selling military secrets to the Israelis, including satellite photographs. On that occasion, Israel also denied it was spying on its closest friend; then, when the evidence became incontrovertible, changed the story to say Pollard was a "rogue agent" and that all such spying would stop. (In the case of Pollard, what particularly infuriated the Americans was that they heard of his spying from a communist source who had received from the Israelis some of the intelligence Pollard had delivered.) Of course, the Israeli spying did not stop. In the years since Pollard, the FBI ran several investigations into Israel's spying, with one case so clear-cut that there was no room for equivocation on Israel's part. A U.S. citizen they tried to recruit went straight to the FBI, who ran him as a double agent. One U.S. intelligence source who saw the file on that operation said, "There was video and audio of the Mossad guy receiving top-secret documents from the trunk of a car. I expected to see the arrest announced, but nothing happened."

The fact that so little was heard of this case can be attributed to a desire on the part of successive administrations not to be too openly hostile toward a close ally and also a desire not to ruffle the feathers of the always-sensitive American Jewish lobby. In December 1995, the Pentagon had been obliged to retract a warning to defense contractors about Israel's aggressive spying, which said that Israel was leveraging "strong ethnic ties" to recruit spies. The Anti-Defamation League regarded that as a racial slur and forced the government to back down.

In all the uproar over the Mega case, few bothered to ask what the NSA was doing eavesdropping on communications between the Israeli embassy and Tel Aviv. Of course, America was spying on Israel. It was deeply embarrassing for the NSA that someone leaked the story that there was a mole hunt under way, because it meant they, too, had been caught with their hands in the cookie jar.

Why would a close ally like Israel spy on America? A better question to ask is why wouldn't they? Everybody else does, and what is more, America spies on Israel. It is the game they all play; no friendship is so close that trust is absolute. Diplomacy and realpolitik have a bottom line that dictates nobody can be trusted, no matter how close the alliance. After all, today's ally might become tomorrow's enemy.

The issue of Country D, Japan, in the GAO list was a ticklish one. Throughout the 1980s and early 1990s, America viewed Japan with fear and anger, as the Japanese seemed to dominate every market they entered, while stopping America from gaining reciprocal access. Yet nobody in government spoke openly of this fear, wishing not to attract the label "racist" and not wishing to give Japan the opportunity to exploit any such argument. So it was extremely embarrassing in 1991 when an academic wrote a report commissioned by the CIA on Japanese intentions toward America that spelled out in no uncertain terms what the U.S. administration no doubt felt but would rather nobody said openly: "[The Japanese are] creatures of an ageless, amoral, manipulative and controlling culture [who are intent on] world economic dominance."

The report added that Japan was a racist and nondemocratic country whose population believes might is right and feels superior to other people. The report was swiftly disowned, one suspects for telling more truth than anyone wanted to hear.

The problem was that Americans, like many other Westerners, deeply resented the Japanese ability to penetrate the U.S. market so comprehensively, making products consumers proved eager to buy. Japan was reaping the benefits of an extremely well thought out, brilliantly executed economic intelligence strategy that they had developed in the postwar years. The country's new constitution denied it the opportunity to operate a conventional intelligence and espionage service. So it directed all those energies, and the money it had thus saved, toward establishing a global economic intelligence operation that scoured the markets of the world for data, trends, ideas and inventions that gave Japan a real edge in understanding what consumers around the world wanted. A classic example is that some of the 10,000 Japanese businessmen sent all over the world from 1956 onward to gather intelligence noticed a growing enthusiasm for amateur photography. Data was collected, trends identified and Japan got a jump on the coming boom in home photography. In his seminal work, *Kempei Tai: A History of the Japanese Secret Service,* author Richard Deacon points out that perhaps only the Japanese could have succeeded in an intelligence undertaking of this scale.

> The sheer range of the Japanese conception of total intelligence is that much greater, so the volume of information collected is greater. The Japanese are tireless, perpetual-motion observers. . . . Then there is the talent for thinking twenty and thirty years ahead. It was this that enabled the Japanese to foresee the markets for electronics, the high-speed train and the boom in cameras.

The operation was run by the government, which had an intimate, almost symbiotic, relationship with business through the interface of the Ministry of International Trade and Industry, or MITI. Over the years individual companies began to operate their own in-house intelligence-gathering operations, all aimed at one thing, prosperity for the company, which in turn meant prosperity for Japan. Any information gained that would have a bearing on national security or economic policies would immediately be passed on to the government. Economic patriotism demanded nothing less. In the main, the Japanese eschewed the cloak-and-dagger style the French loved so much for the more mundane, but probably much more productive, option of exploiting readily available, or "open," sources, which is entirely legal. Only a loser would seek to attack the advantage gained through hard work and enterprise, a role unfortunately suited to America during the time when Japan was so predominant in world markets.

Occasionally, however, the Japanese did move into illegal areas. In 1984, CIA Director William Casey was informed of a top secret FBI investigation into a leak at CIA headquarters. It had been exposed by an NSA intercept of a signal between the Washington office of the Mitsubishi company and its headquarters in Japan. To everyone's alarm, the Mitsubishi signal had contained verbatim extracts from two National Intelligence Daily reports of July 1982, in which the movements of troops on both sides of the Iran-Iraq War were detailed. The NID is a daily digest of the latest intelligence distributed to only the most senior administration officials and so a verbatim leak was a serious breach of security. The extracts were accompanied by top secret CIA analysis that maintained that Saddam Hussein would have to fall before there could be peace between the two countries. Given America's then support for Saddam, this was explosive stuff. The means by which Mitsubishi got hold of this was spelled out in the signal, and demonstrated that a Japanese businessman is not just a businessman. The information had been given to a member of a Washington consulting firm that was also working for Mitsubishi. Perhaps the firm's employee had once been a senior government official who had taken advantage of the always revolving door between politics and commerce. For whatever reason, the information now resided with the Japanese government.

Senator Arlen Specter, Chairman of the Senate Intelligence Committee, came right out and accused the Japanese of economic espionage, saying they were trying to steal America's technological secrets. He cited a number of examples, including the celebrated Hitachi-IBM case of 1982, in which Hitachi's theft of trade secrets from IBM resulted in a lawsuit and total victory for IBM. So crushing was Hitachi's defeat that the out-of-court settlement mandated that IBM have the opportunity to inspect Hitachi's new products for five years.

The accumulated evidence that America's treasure of knowledge and innovation was being plundered by an array of nations led eventually to a realization that the only laws applicable to economic espionage were woefully out of date. The Interstate Transportation of Stolen Property Act was passed in the 1930s to deal with increasingly mobile gangs who were using cars to take stolen property across state lines and beyond the reaches of state law. Modern prosecutors had to convince judges and

juries that stolen ideas and information that a foreign intelligence agent
had removed from an American company amounted to stolen property.
To be guilty of theft one has to intend permanently to deprive the owner
of the stolen goods. Arguably a spy is not intending permanently to
deprive an inventor of his proprietary information, he is merely seeking
to let someone else in on the secret. It is a lot of things, all of them
dishonest, but arguably it is not theft. The picture became even murkier
as computers became commonplace and it became easy to copy a file
and send it via modem to someone else. At Senate hearings designed to
help formulate new laws to govern economic espionage, FBI Director
Louis Freeh bemoaned the fact that he had so few weapons to deal with
data thieves and spies.

> We have approximately 800 pending cases involving 23 foreign coun-
> tries. These are state-sponsored economic espionage—forays and ini-
> tiatives into the United States, using all the various techniques of
> intelligence officers, from compromising individuals, to unlawful wire-
> tapping, to bribery. . . .

A few months later, a high-profile economic espionage case demonstrated that the old law still had teeth, but also exposed the FBI to
possible censure as an aider and abettor of economic spying and demonstrated that when commercial interests become tangled up with national
security America does not have as clear-cut a view of things as does, say,
Japan.

Guillermo "Bill" Gaede was an Argentine communist who in 1979
went to work for the U.S. microchip maker Advanced Micro Devices.
He was in the United States illegally, having outstayed his visa and using
forged residency documents. In 1982 he started stealing the designs for
AMD microchips and passed them on to the Cuban government, in the
hope of kick-starting a Cuban computer industry. Seven years later,
Gaede visited Cuba, an experience which stripped him of his communist
idealism. "The visit to the island . . . served to destroy what little was left
of my socialist dreams," he said. He decided to turn himself in to the
CIA, something the agency will not verify or deny. Gaede then said the
CIA turned him over to the FBI, who asked him to carry on spying so
they could entrap the Cuban agents who were running him. Fearful of
discovery at AMD, Gaede eventually resigned and in 1992 applied for a
job at Intel, the most powerful microchip maker in the world, and as his
references included a supervisor at AMD and several FBI agents, he was
welcomed aboard. In 1993, he was stealing again and still, he says,
passed secrets to the Cubans with the full support and connivance of
the FBI. It was too much strain, so in May 1994 he decided to make
one final raid on Intel and then get out. He downloaded Intel's data and
designs on the 486 and Pentium microchips onto his home computer,

and then used a video camera to record all the screens of information as they appeared. He sold the tapes to China and Iran and made "large sums of money." Intel accused him of selling the data to AMD, too, something he denied, saying the FBI did that to set him up for a criminal charge in the United States. He fled home to Buenos Aires, but eventually came back to America and gave himself up after the stress of being on the run became too much for him. He was sent to jail for thirty-three months on charges of mail fraud and under the Interstate Transportation of Stolen Property Act. If it was a success for law enforcement it was because Gaede was a willing witness against himself. It was otherwise a cautionary tale about patently inadequate internal security at a company of Intel's size and stature and, although they deny Gaede was their informant but admit they met him several times, about the FBI almost causing a disaster with reckless use of an unreliable criminal to further its counterespionage goals.

In the early 1990s, the question of whether the CIA and the NSA should get seriously into the business of economic espionage was raised time and time again, as frustration with the activities of supposed allies grew. As has been made clear, the CIA and the NSA were themselves up to their necks in economic espionage. The NSA was listening to all the signals traffic between foreign companies, embassies and their homelands. And the CIA was aggressively stealing secrets that would help America in trade deals. In 1995 the computers of the European Commission in Brussels and Luxembourg were broken into by CIA hackers to gather information that helped the United States in delicate worldwide trade talks at GATT. There was the entire floor at the Commerce Department where the CIA's Office of Intelligence Liaison filters important strategic economic information for use by trade negotiators and major companies. But the line they were being asked to cross was into a new area that set alarm bells ringing. The suggestion was that they use their talents to spy on foreign companies, in the way the French spied on U.S. corporations. The intelligence community did not like the idea. "Every two or three years when I was in intelligence, some turkey would come up with this idea," said William Odom, former head of the NSA. "I'd quash it."

The reluctance to get involved so closely is understandable. As explained earlier, the government of the United States is very far removed from the day-to-day operations of American business. There is not shared culture, and the days when "What is good for General Motors is good for America" are gone. While still basically true, such a philosophy overstates the identity of purpose between government and industry as compared with the attitude in Japan, for example. The old adage found a sardonic new counterpoint in the apocryphal quote, attributed by former CIA Director Robert Gates to a CIA agent, but used many times by others: "I'll die for my country but not for General Motors."

Robert Gates spelled out his position clearly in an interview with me.

> There are a number of reasons for us not to do it, some practical and some philosophical. One of the practical reasons is that I think it gets us into a welter of legal problems here at home in terms of what happens to the information we gather, who it goes to, how do we avoid advantaging one business over another.

Business was not too thrilled about the idea either. The CIA did not have a brilliant reputation for analysis during the Cold War, arguably costing the American taxpayer hundreds of billions of wasted dollars by overestimating the size and strength of the Soviet economy. The CIA fiasco in Paris put the lid on the notion of proactive economic cloak-and-dagger work.

Then there was another important question in a society as litigious as America. If the CIA did find a nugget of information, say in the defense field, to whom would they give it? Boeing? Lockheed Martin? Northrop Grumman? Stewart Baker, former legal counsel to the NSA, believes this would be an insurmountable obstacle. In a speech to the National Information Systems Security Conference in Baltimore in October 1996, he said that the people who do it most are, almost by definition, the people least capable of exploiting technology on their own. As an example, he mentioned the theft by Bulgaria from Digital Equipment of huge amounts of technology. Digital Equipment was initially horrified, but says that Bulgaria is their "best upgrade market in the world today!" "People who do it need to cheat just to keep up," he added, citing France as an example of a country where industry spends too much time in government ministries and not enough time in the lab.

That economic espionage can, in the long run, be self-defeating was confirmed by former KGB general Oleg Kalugin, who said that no amount of stolen or acquired technology can replace homegrown R&D. Importing it by theft or otherwise simply increases the dependence of a country like Russia on external powers like America.

These arguments appear increasingly clear as societies begin their migration to the infosphere. Industrial espionage has been part of life since commerce began. Strategic economic espionage (Is Russia about to sell gold? What is the Italian negotiating position in the upcoming talks on the General Agreement on Tariffs and Trade?) has also been a central part of intelligence gathering. One NATO intelligence officer cites as the greatest coup of a distinguished career his recruitment of a senior Italian trade official in Rome to spy for an ostensible ally.

What makes the new world different and much more threatening is the destruction of the old boundaries. As borders disappear in cyberspace, so the very concept of economic security becomes more elusive. For governments used to defending the nation structure, the Information Age presents a series of new challenges. To meet those, better defenses will be needed and that has to mean an increased offensive capability as well.

Every Picture Tells a Story

''*T A K E your spears, clubs, guns, swords, stones, everything. Sharpen them, hack them, those enemies, those cockroaches . . .*''

Rarely can the history of human conflict have witnessed such venomous imprecations to slaughter. In the summer of 1994, these violent words drenched the airwaves of Rwanda, urging members of the Hutu tribe to kill the rival Tutsi people, whose Rwandan Patriotic Front had already succeeded in deposing the Hutu regime. In the chaotic days following the coup, before the Tutsis could gain control of the countryside, Hutu extremists used a mobile radio to urge their people and their militias on to a mass killing spree as they themselves fled to refugee camps in Zaire, Tanzania and Burundi. The battle may have been lost but tribal hatred mandated that the killing go on.

The radio station was called Radio Television Libre des Milles Collines—Free Radio-Television of the Thousand Hills—a rather grand title for an outfit eventually operating from the back of a van. When the plane carrying Rwandan President Juvenal Habyarimana crashed on April 6, 1994, it was RTLM that took to the airwaves to accuse the Tutsi-led Rwandan Patriotic Front of shooting down the plane. The effect was electric. The river marking the border between Rwanda and Zaire became choked with Tutsi bodies, women, children, men, it did not matter as long as the tribe was exterminated.

"Hunt out the Tutsi. Who will fill up the empty graves? There is no way the rebels should find alive any of the people they claim as their own . . ."

Little wonder it became dubbed Radio Hate.

Even when the Hutus were in the refugee camps, RTLM kept going, ordering the Hutu refugees to stay out of Rwanda where they would be certain to be killed by the Tutsis. So the refugees stayed, in stinking, cholera-infested camps, pawns in the hands of their leaders whose power rested in the number of people they could keep in the camps.

When it was finally shut down in August, an independent team of journalists and broadcasters from the Swiss group Reporters sans Frontieres understood that just as RTLM had polarized the Hutus with hatred, a new force was needed, one that would reassure instead of terrify them.

The fact that RTLM was allowed to broadcast was a colossal mistake on the part of the international community, which was mobilized to "do something" by the TV pictures of the slaughter and the refugee flight being shown around the world. President Clinton ordered an immediate American operation into central Africa. Maj. Gen. Michael Hayden, now head of the Air Intelligence Agency, was J-2 (Intelligence) U.S. command in Stuttgart, Germany, and part of the planning team that pulled together a U.S. Joint Task Force (JTF) to go into Rwanda and Zaire to relieve the suffering. He remembers Gen. George Joulwan, Supreme Allied Commander Europe, entering the briefing room and writing on a drawing board: "Stop the Dying."

" 'That is your mission,' he told us, 'Stop the Dying,' " Hayden recalled. Crucially, however, the dying they were going in to stop was that which resulted from disease, not the slaughter.

The first part of the mission was to gather intelligence on the scale of the problem, which the JTF did with a U-2 redeployed from Saudi Arabia, a P-3 Orion fitted with a camera that flew low over the columns of fleeing Hutus, and special forces units gathering intelligence on the ground. As refugees started filing into Zaire, Hayden and his teams decided the priority was not getting food to the people but clean water. The second stage was to reduce the death rate. "When we went in," Hayden said, "3000 people were dying every day from waterborne diseases. By the end we had reduced that to 300 per day, less in many cases than if they had stayed at home."

Stage three was to organize the withdrawal of the Hutu refugees from the camps in Zaire and get them to go home. It was then that Hayden came to a crucial understanding. "I was lying in my tent. It was the middle of the night," he said, "and I woke up realizing I had this all wrong. At first light I sat down and typed up a 400-page report that said we had missed something. The migration from Rwanda was an act taken in a political context, not simply the CNN-derived images of two and a half million people fleeing for their lives. The Hutu leadership that told them it's okay to kill the Tutsis was the same as the leadership that told them to walk to Zaire, and the method of communication was Radio Hate."

Hayden's report made its way through the JTF command in Africa, to Europe and then to Washington. Soon after that, the JTF was instructed to ignore the task of getting the Hutus back to Rwanda and instead to hand over responsibility to the relief agencies in the field and then withdraw.

Hayden recognizes now that having belatedly come to the correct conclusion about Radio Hate, the United States had the power to do something about it. "We had psyops [psychological operations] units in the JTF so we could have counterpunched, but we didn't. We are a rich and powerful nation, we had the means to find out where the radio station was, but we did not. To have done so would have been to get involved in a civil war in a way that we had not contemplated."

In fact, Radio Hate had been broadcasting from the back of a Toyota pickup. A single strike by a special forces unit would have eliminated the problem and altered the course of the mission and its outcome. But the American mission was defined as peacekeeping and not peacemaking, a critical difference that helped confine the parameters of what could or could not be done, whatever the real circumstances on the ground. The experience was something of an epiphany for General Hayden, who realized that a new way of thinking about war had to be adopted. "As theater J-2, I had not thought through my function in life to include something I am now much more comfortable with, namely operating in the information domain."

The operation of Radio Hate is a classic example of an information operation that was a crucial part of the theater into which American forces were being deployed with nobody in the U.S. military or body politic, until Hayden's report, understanding how it would impinge on and ultimately alter fundamentally their mission.

The existence of Radio Hate and its role in creating and perpetuating the misery was recognized by the United Nations force in Rwanda, UNAMIR, commanded by Canadian Maj. Gen. R. A. Dallaire. In a strongly worded article published in early 1995 Dallaire blamed lack of interest on the part of member nations of the U.N. for a response to the crisis that was too little, too late, and that lacked the necessary logistical support for the limited number of troops on the ground to keep the peace and protect the weak. He sees the failure to block Radio Hate as a direct result of this tepid response.

> These broadcasts were largely responsible for spreading panic that, in turn, drove large numbers of people to refugee camps in neighboring states, thereby spreading instability throughout the region. . . . It should also be pointed out that the broadcasts discouraged survivors from returning to their homes, and should have been jammed. The United Nations should have aired counter-broadcasts to give the population a clear account of what was actually happening . . . yet . . . no country came forward to offer jamming or broadcasting assets.

The ability to understand the use of information as a weapon of war, and how it impacts the conduct of military and foreign policy, is something very few in government possess. Those that do tend to be several times removed from democratic accountability, and use the media not only to influence events on a global scale but also in their own backyard, perhaps to push wavering politicians down the path of rectitude.

In the course of my twenty-five years in journalism I have developed sources at the highest levels of intelligence agencies in America, Britain and Russia. Those relationships have been very clearly predicated on the assumption that there would be times when I would be a conduit for information that would redound to an agency's benefit, and they knew I would also write stories from time to time that would cause embarrassment. Two examples illustrate how this relationship frequently works.

Western Intelligence knew that the apartheid regime in South Africa had developed a BW capability, and had indeed used it against members of the African National Congress. When Nelson Mandela came to power, Washington and London asked him to scrap the BW program. He did, but under pressure from his military commanders refused to destroy all the research that had gone into it. This posed a major problem for the West, because waiting in the wings to exploit this situation was Libya's Colonel Qaddafi, who was eager to lay his hands on some BW weapons.

Qaddafi sent a secret emissary to South Africa to meet one of the key BW developers in a bid to buy his skills and the research for a Libyan BW program. Western intelligence agents had penetrated Qaddafi's procurement program and knew about the mission before it took place. They were concerned that the Libyans would succeed. The intelligence community had a choice: continue to negotiate with Mandela in secret in an effort to get the BW research destroyed or see if there was some other way to prevent the Libyan procurement effort from succeeding. One individual took the decision to leak the story to me in the hope that publicity would achieve what private negotiations had failed to do.

In the course of one week in February 1995, I began work on trying to fill in the details that had been missing from the lunch that had produced the initial lead. Inevitably, news of my investigation filtered through the political and intelligence communities on both sides of the Atlantic and I was able to track the different meetings that were called to discuss what to do. Under such circumstances, there are two choices: try (and usually fail) to suppress the story or draw the reporter into the problem and see if a deal can be struck.

On Saturday morning, my home phone rang. It was a senior British intelligence officer, who told me he knew I was doing the story but was concerned that running it might compromise sources and methods involved in the operation. It became clear that revealing some of the details I had by now uncovered would put lives at risk, so I made some changes to the planned story that did not alter the truth of what I was writing but would encourage Libyan counterintelligence to look in the wrong direction for the source of the information that had leaked from their procurement program.

The story that eventually ran in *The Sunday Times* was headlined "Gadaffi Lures South Africa's Top Germ Warfare Scientists." Immediately after publication, the article was followed up by the South African media, which found out for the first time about their country's extensive BW program. The research for that program was destroyed and Libya's BW program was put on hold. (Qaddafi correctly found that it had been penetrated.)

From an intelligence and national security perspective, publication of the article could be considered a completely successful operation. But the political fallout from the piece continued for several weeks. President Mandela personally protested to the White House, blaming the Americans for the leaks. The Americans protested to the British, who blamed the Americans and then launched their own investigation, knocking first on the door of MI6. A series of meetings were held in Washington to try to track down the source of the leak. At one such meeting in the Pentagon which involved senior officials, the conversation focused on a single quote sourced to a Pentagon official which used the word "incontestable." Was this word, they earnestly asked each other, a British or American word? After much debate it was decided that the word was British but was sometimes used by Americans who had been in Britain. So who might that be? A Rhodes scholar was one answer and so an official was sent off to check the records to find which Rhodes scholars had access to the sensitive intelligence that formed the basis of the article. The source of the leak was never uncovered.

In the summer of 1996, the British, French and American governments had agreed that no effort would be made to snatch the Bosnian Serb leader Radovan Karadzic and his chief military officer, Gen. Ratko Mladic, and deliver them to the war crimes tribunal in The Hague. The Europeans feared that a kidnapping would provoke a backlash against the allied peacekeeping force in Bosnia that could produce a heavy loss of life and the possible withdrawal of the force. The British and French believed that the operation, while militarily feasible, would have no military purpose and would simply be carried out to satisfy a narrow political constituency in the United States that was pressing for action against the war criminals. Reluctantly the Americans agreed to hold off.

It was frustrating because video images taken by the Predator UAV near Karadzic's home in Pale clearly showed him vulnerable to a snatch operation. As the American presidential election drew near in November 1996, the British and French heard that the idea of a snatch operation had been resurrected and that the Americans might be considering unilateral action. Extra intelligence assets had been placed on the ground to gather real-time intelligence on the location of Karadzic, who was considered the prime target. Special forces had been earmarked for the operation and there was real concern that it would go ahead.

Some intelligence and defense officials in Washington as well as in London and Paris were dismayed by the idea that military lives were going to be put at risk for what they considered would be simple political grandstanding in an election campaign. I heard about the story and interviewed a number of people in the process of checking it out. Inevitably, news of my inquiries leaked.

On paper, Britain possesses a tough piece of censorship legislation known as the Official Secrets Act. In theory the government can prevent the publication of just about anything to do with national security and prosecute anybody who breaches the strict terms of the act. But, Britain being Britain, the law is only very rarely used (never in my fifteen years writing about national security matters and breaching the letter of the act every week). Instead, there is an informal, behind-the-scenes arrangement known as the D Notice Committee. If the secretary to the committee—usually a retired admiral—hears that an article is about to be published or a broadcast made that is considered a danger to national security, he picks up the telephone and calls the editor responsible. There then follows a wonderful conversation where the admiral describes his concerns and politely asks that the offending item be changed or stopped. If the editor agrees, then all is well, and if he does not, nothing happens. It is a discussion between responsible people where both sides understand that the other has a job to do.

On this occasion the D Notice Committee secretary phoned my editor to express his concern about the proposed article on Karadzic. I immediately called the admiral back directly. "We are very concerned that this article may put the lives of special forces at risk," he said. I reassured him there was no intention of betraying sources and methods and the sole intention of writing the article was to reveal a planned military operation that my own government, senior officers in the Pentagon and French officials believed was a very bad idea. The admiral was satisfied and the conversation ended. The effect of his call was to galvanize my editors in London, who judged the official response to the impending story as not only confirmation that it was entirely accurate but confirmation, too, that it was an even better story than they had thought. Instead of it going on an inside page in 1,000 words it started on page one at 800 words, followed with an additional 1,200 words inside, complete with a dramatic illustration of Karadzic being snatched from his house by a group of tough-looking characters in black who descended on ropes from a hovering helicopter.

The story was widely followed up and on Monday dismissed by Mike McCurry, the White House spokesman, as a "complete fabrication."

More importantly, any immediate plans for snatching Karadzic were abandoned (to be resurrected in the spring of 1997 but again not executed). The article had quoted verbatim from a British Joint Intelligence Committee report on the plan, and on the personal order of Prime Minister John Major, a formal leak inquiry was launched. People on both sides of the Atlantic who were known to know me were interrogated. The Foreign Office officials blamed either the CIA or the Pentagon for the leak while MI6 blamed the Foreign Office. The Ministry of Defence refused to cooperate in the inquiry at all. After several weeks of fruitless investigation the inquiry was abandoned.

What these anecdotes illustrate is that information is alive and well in some parts of the intelligence and defense establishments. They show, too, that effective use of the media is compatible with being truthful and achieving clear national security goals. But it is also clear that all too often national security goals are confused with political goals, or rather, political fallout from successful information warfare operations can result in a political backlash. This demonstrates a clear lack of understanding about how information warfare should work in a modern media environment and shows that there is still much ground to make up if IW is to be used effectively.

On Saturday, February 3, 1996, Sgt. 1st Class Donald Dugan became the first American soldier to die during the NATO peacekeeping operation in Bosnia. It was no concerted enemy attack or precision sniper fire that killed Dugan. That might have jeopardized the fragile peace agreement and been a serious cause for concern. Instead, he had picked up a land mine, which then exploded in his hands. In other words, it was an accident and the kind of accident that happens frequently in every army. Yet, that night, the death of Dugan was the lead on every network news broadcast and it was the front-page lead in every major newspaper the following morning. According to the Lexis-Nexis database, in the course of the first twenty-four hours after his death newspapers and magazines devoted more than 50,000 words to this single fatality and accorded Dugan the kind of national stature that might have been given a general fallen in battle in another age.

The difference in the coverage now compared with then is the immediacy of the news and the ease with which it is transmitted around the world. Within minutes of Dugan's death, the information had been relayed from the U.S. army headquarters in Bosnia to the NATO force headquarters and from there via a press conference to the world's media. With CNN leading its half-hourly news shows with the death, the agenda for the rest of the nation's media had been set and they all dutifully followed the lead that had been established for them.

This kind of group feeding is a hallmark of news operations that began with the development of the wire services and has been honed by the revolution in international communications brought about by the growth of the Internet. But it is also a graphic illustration of the collective ignorance of the media and the politicians about how warfare actually works. Conflict today is judged not so much by victory or defeat but by the numbers of casualties on the allied side, where deaths above single figures are routinely seen as a political, and thus a military, defeat.

In Somalia, the media were on hand to watch the first troops ashore and were in a mood to act as national cheerleaders. But as the operation unfolded, an initially compliant media turned hostile when poorly trained and equipped gangsters managed to defeat the United States of America, the last remaining superpower and the strongest military force in the world. The facts of the reversal of American foreign policy are not in dispute. For America's allies, the reversal of U.S. fortune in Somalia caused concern about the strength of American political will in prosecuting foreign policy. These concerns are justified. But underlying such questions is a new reality that exists in the media and in government and that helps explain why peacekeeping, peacemaking and war fighting in the future may be impossible if we apply the criteria that once worked so effectively.

Peacekeeping and warfare today are taking place in a world the likes of which we have never seen. All the old certainties have disappeared; there is no Cold War, no superpower rivalry to provide both tension and stability. Instead there has emerged a series of relatively small, unexpected crises. These have provided some fresh challenges for the policymakers, the politicians and the media that report on them.

But it is not just the end of the Cold War that has transformed the debate. The end of that era has heralded a new generation in the political leadership in many democratic countries, and has consolidated changes that have been under way for some time in the media. Consider this: the President of the United States is fifty; the Prime Minister of Britain is forty-four. The average age of staff working in the White House is around the mid-thirties and not a single member of Congress had a son or daughter fighting in the Gulf War. The heads of British intelligence are now appointed in their late forties. The average age of journalists on TV and in newspapers has fallen in the past ten years as proprietors struggle to reach the younger audience everyone needs to win the fierce battle for circulation and ratings.

This means that many of the political leaders and most of their advisers have no concept of the political and personal consequences of warfare. There have been no world wars to devastate families, none like those that so scarred our parents and grandparents. Thus, there is no true conception of the real horrors of war. In the media, there are now very few reporters and editors who have covered conventional conflicts. In fifteen years of reporting on wars, revolutions and terrorism around the world, I have only covered two conflicts that might be considered conventional: the Iran-Iraq War and the Gulf War against Saddam Hussein. The balance was made up by a large number of smaller wars, low-intensity conflicts and acts of terror.

It is hardly surprising, then, that the current generation that is leading the media and government has a very limited vision of war and the capability of the armed forces. It is a view formed by each individual's experience, which is confined almost entirely to television, movie images and, to some extent, what they have read in the newspapers. This is a small world, where action is concentrated on the human drama, the big picture writ small so that it is understandable to the average person. It is a world where attention spans are short.

For example, the way news is reported has undergone a striking transformation in the past twenty-five years. The average length of a presidential sound bite on a network evening news broadcast fell from forty-two seconds in 1968 to less than ten seconds in 1988. The three major networks gave to George Bush and Michael Dukakis one fourth of the airtime they gave to Richard Nixon and Hubert Humphrey. In the last election and since the inauguration of President Clinton, there has been even less time allotted to the attempts by the President and his staff to communicate directly with the people. Gone are the days when the networks were suitably deferential when the White House called. Instead, a cult of personality has evolved, built around the anchorman and -woman who help present, refine and make the news. The reason for this is that Presidents are, by and large, rather boring and do little for the ratings, while a network's own staff can present material in a way that suits them and enhances the ratings. Or, as Dan Rather put it in a speech:

> They've got us putting more fuzz and wuzz on the air, cop show stuff, so as to compete not with other news programs but with entertainment programs—including those posing as news programs—for dead bodies, mayhem and lurid tales. We have allowed this great instrument, this resource, this weapon for good to be squandered and cheapened. The best among us hang our head in embarrassment, even shame. We should all be ashamed of what we have and have not done, measured against what we could do . . . ashamed of many of the things we have allowed our craft, our profession, our life's work to become.

This mea culpa by one of the very people who have replaced Presidents with punditry is well founded. As the television medium has trivialized the news, so newspapers have to seek ways of presenting their information in a lively and exciting way to retain their audience. That has meant not just a narrowing of the focus but a concentration on the trivial, the marginal and the irrelevant in the search for excitement. In war-fighting terms, this means that sound bites have replaced sense. There is no longer a search for an understanding of the bigger picture— the "strategic vision," if you like. Instead, what matters is the here and now, and modern communication means that what happens over there is presented here and now to the millions of Americans who form the body politic.

The reporting of conflict has changed as the media has evolved so that the camera and the laptop bring home the single image of casualties, civilian or military. The new media and political elite demand that these casualties be counted in single figures to be acceptable. For example, it was striking in both the Falklands and the Gulf War that casualties were extraordinarily light. Even so, every death was analyzed and agonized over and investigated in an attempt to find unrealistic certainties in the chaos of war. It is that drive to minimize loss of life that is going to be a primary factor of decision making in the foreseeable future.

Of course, reducing casualties is a laudable goal. But death and injury are unfortunate consequences of committing military forces to a conflict. Soldiers are trained to kill people, and yet there seems to be a broad view that crisis management today can somehow be handled without loss of life.

The media have played a large part in developing this view. The media always demand excellence in others; in terms of crisis management this translates as a successful resolution with minimum cost to "our" side. In the past ten years, the way the media form opinion has changed dramatically. CNN is everywhere, and where CNN goes, all the other media outlets swiftly follow. Censorship today is virtually impossible with backpack satellite broadcast systems and telephones that allow reporters to file their copy from anywhere in the world.

While competition among different media remains superficially fierce, much of it is artificial, with an increasing number of stories covered by correspondents who are paid for their name or their face rather than for the content of what they can deliver. The fact that so many major cities now have but one newspaper has reduced competition, producing a kind of corporate arrogance that places much less value on the old-fashioned scoop and more on establishing a reputation as a "newspaper of record." This in turn has made for a different kind of reporting where CNN and the wire services begin to set the agenda and are followed by the newspapers and magazines whose editors and writers take their ideas and their leads from those who fed at the news trough first.

News has an immediacy that drives the political process in ways that can be very unhealthy, particularly when so many of the decision makers have no experience of the world about which they are making decisions of life and death. However, as much as both the military and politicians would wish it otherwise, that is the new reality of the media covering crises today, and it will only get worse as the Internet becomes the preferred medium of expression and knowledge gathering.

This is not just a one-dimensional view of the media as news gatherers. Others who are part of the violence also have a much better understanding of how the media operates. I remember covering the takeover of the American embassy in Teheran in 1979 by militant students. Many of those holding the Americans hostage had been educated at American universities and had a good understanding of the vulnerabilities of that society. The daily press conferences were carefully scheduled for the American networks so that interviews could be obtained, fed to the satellite and reach the New York anchors in time for the evening news. It was skillful propaganda that proved very effective.

But the world has moved a long way from those early days. In the same way that the Pentagon routinely establishes a sophisticated public affairs operation for every military deployment, so protagonists understand the value of media attention. A good example of this came in a *New York Times* report of February 6, 1995, which described the activities of the Chechen rebels fighting the Russians.

> The Chechen military commander, General Aslan Maskhadov, sitting serenely in his headquarters here [Novogroznensky, Chechnya], a recent model Motorola walkie-talkie in one breast pocket. . . . They gave interviews and then they gave out their satellite phone numbers to reporters and left. . . . When Russia's most wanted men were done talking to the journalists, politicians and relatives of the hostages, General Maskhadov made a couple of brief telephone calls on his portable phone. . . .

That use of modern communications has been matched by the proliferation of different systems that give people all over the world access to the same kind of information at almost exactly the same time. A bomb goes off in London and the fact of its detonation is flashed around the world by many different wire services and television stations. Information about that explosion is exchanged among journalists on the spot and by others in different cities. For example, when the IRA cease-fire broke down in February 1995 with the detonation of a bomb in London, I reported the story from Washington using contacts I have in the intelligence community and governments in America and Britain. Because I have some knowledge of the subject, I in turn was contacted by several wire services in a number of different countries, the major news magazines in America and television stations around the world. A single source thus becomes a resource for a vast number of news outlets.

That trend is exaggerated by the developments on the Internet. According to the Associated Press, the numbers of North American newspapers offering information through online computer services tripled in

1995 to 175 and doubled again in 1996. Some newspapers have established sites on the World Wide Web; others offer stories and photographs through commercial online services; and a few others have launched their own stand-alone online services. And this is just the beginning. Every major newspaper group has plans to provide a complete download of its newspaper or magazine to portable laptop computers, which will interface with video, and to provide databases that will be accessible at a very low fee to any customer interested in getting more information on a given subject that is reported that day.

What that means today is that media coverage is highly selective and driven not necessarily by the importance of a story, but by the cost of covering it, or even by something as simple as who happens to be in the area at the time. The column inches and television time devoted to Somalia versus Liberia, Angola or Burundi bear no relation at all to the scale of the tragedy unfolding in any of those countries. Strategically and economically, Angola is a more important country than Somalia; and the human tragedy in Burundi is arguably worse than in Somalia. Nevertheless, the media herd went to Somalia because it captured the imagination of editors, it made good and colorful copy, it was accessible and—perhaps most important of all—everyone else was there. The sheep instinct that drives so much of the media prevailed.

This selective media focus, which tends to be extraordinarily intense and of limited duration, has the effect of driving policy. That in turn has led the media to believe that it has not only a duty to report the news but also the power to influence events. As the British Prime Minister Stanley Baldwin said of the press barons of his time: "What they aim at is power and power without responsibility—the prerogative of the harlot throughout the ages."

A century ago, a single incident that was deemed to impinge on national sovereignty would provoke an immediate and violent act of retribution. When General Gordon was killed in Khartoum, the British dispatched a punitive expedition that years later punished the perpetrators of the act. Today, when the body of a single American is dragged through the streets of Mogadishu, the American government reverses its foreign policy and begins a withdrawal from the country. This momentous change in the way foreign policy is handled bodes ill for the future.

With an attention span so short and a worldview so limited, it is difficult to conceive how a consistent policy for crisis management can be developed by the world's leading democracies. Is it conceivable that the world's only remaining superpower would deploy forces to an area like Northern Ireland to keep the peace in the way the British did thirty years ago? Is it conceivable that the American administration would send any troops to a similar environment? Under the present circumstances, the answer is no.

War, peacekeeping, and crisis management have never been about consensus and about opinion polls. Those factors may have had influence, but they have never been paramount. The successful prosecution of any military operation is about leadership and a strength of resolve that allows principle and conviction to ride over the often ill-formed media criticism and the snapshot reporting.

These developments will all have a critical impact on the future of war fighting in the Information Age. The proliferation of information sources has been matched by a reduction in the number of sources supplying raw information. The proliferation of methods of controlling information flow has been more than matched by the development of systems that allow information gatherers to communicate at will. It is quite clear that the days of effective censorship designed to impact public opinion in support of a government's policy are over. No longer can governments control the flow of information, however much they might wish to do so. It would be foolhardy indeed for any government to plan for the control of information in the event of tension or conflict. On the contrary, planning must be devoted not to controlling by the suppression of news, but controlling by the manipulation of information.

Yet it is one of the paradoxes of the Information Age that military planners and political leaders have more need now than ever before to try to maintain public support for political and military action. It is no longer the case that public support for national endeavors will be granted automatically. On the contrary, gaining public support for any kind of military action will be very hard. As Bosnia demonstrated, winning over Congress is tough and winning over the public at large is tougher still. Yet, taking action is only the first step on what will almost always be a very tough road. Casualties in any deployment of forces are inevitable and, as we have seen, there is simply no public, political or media awareness of what warfare is about and therefore little tolerance of any cost, let alone the costs that are inevitable in a full-scale conflict. The challenge then is to try to create an environment that at the same time bolsters public support at home while undermining public support in the enemy camp.

In an ideal world, military and political leaders would enlist the assistance of the media to propagate a view that would help the national interest. In reality, such a bargain is almost impossible to sustain, especially when the media are so diverse and modern technology makes a mockery of national borders. To those familiar with the concept of cyberspace and comfortable operating there, the very idea of nationalism is alien.

The most comprehensive review of the relationship between the media and the military was conducted by the Freedom Forum First Amendment Center in 1995. Among its findings, which were based on a poll of about 1,000 military officers and 350 members of the media, were:

- Fifty-five percent of military officers polled believed that the media should be allowed to report whatever it wants from the battlefield, without censorship, as long as guidelines developed jointly between the military and the news media are honored.

- Only 2 percent of the officers and 18 percent of the media thought reporters should be free to report "anything they want, with no restrictions."

- More than 80 percent of the officers said the news media was "just as necessary to maintaining US freedom" as the military.

While all that suggests a healthy mutual respect between the military and the media, there are still enough grounds for disagreement and conflict:

- Seventy percent of officers and 74 percent of the media agreed that "few members of the media are knowledgeable about national defense."

- Sixty percent of the military officers, but only 8 percent of the media, agreed "military leaders should be allowed to use the news media to deceive the enemy, thereby deceiving the American public."

- Sixty-four percent of the officers, compared to 17 percent of the media, believed the news media harmed the war effort in Vietnam.

- Ninety-one percent of the officers, compared to 30 percent of the media, thought the media is more interested in increasing readership or viewership than in telling the public what it needs to know.

Valuable as this study is, it reflects an old order that no longer exists and a debate that is largely irrelevant to the future of warfare. It is a fact accepted by every public affairs officer in the military that most reporters are extraordinarily ignorant about the subjects they cover. It is an accepted fact among the journalists that whatever the military might say in reply to a survey, it will fight to suppress the news once casualties mount and criticism begins in wartime. But even given those prejudices, in the Information Age it matters little what the American media or military think about each other. Equally, for the Pentagon and the media to earnestly debate about which reporters have access to which pools and when and exactly how much information is to be released is irrelevant. In the real world in a real conflict, there will be thousands of reporters on the ground from every country with media. Censor one reporter from Washington, then his competitor from Tokyo will get the information, file it with his editors and it will be published within hours and be back in Washington. That is how the Information Age works.

During the Gulf War, there was an extensive pooling system for all war correspondents that was designed to control the information flow. Yet my newspaper provided some of the best coverage of the war operating almost entirely outside the pool system and by deliberately avoiding all the press conferences and formal briefings. For every reporter controlled by the pool system there were four or five roaming around the Middle East and Western capitals trawling for information that was readily accessible. And, of course, information supplied to one reporter is almost immediately available to every reporter.

The reality is that information flow in the Information Age cannot be controlled. What can and should be done is to design a new architecture that uses cyberspace and the information revolution to help prosecute warfare. This need not involve lying—a short-term solution to a tactical problem that has few long-term strategic benefits. What it can involve is the manipulation of information so that the focus and flow of data are controlled. During the Gulf War both the CIA and Britain's MI6 had an active campaign of disinformation that specifically targeted the Arab press so that articles appeared that were damaging to Saddam Hussein and the Iraqi cause. But in England and America, the intelligence agencies were more careful not to attempt to manipulate or deceive on their home turf.

There are techniques and tactics now available that allow allied forces to take some control of an enemy's broadcast media and to play a significant part in controlling the print media as well. That is a valuable asset that moves what used to be propaganda into an important part of the strategic and tactical equation. In the future, every force commander should have available an array of such tools so that he can plan knowing that at a minimum the enemy's morale is going to be very fragile. This is where television and radio fall into the realm of psyops (psychological operations). The technology is readily available for taking over an enemy's TV signal and broadcasting a completely different message, or manipulating the enemy's own images through the use of morphing and other techniques to change the meaning of, for instance, a broadcast by Saddam Hussein. This is an extrapolation of the technology that allows psyops units to record enemy military radio signals and digitally alter them, then rebroadcasting them immediately, sowing confusion in the ranks. The downside of mixing up enemy transmissions is that they have a very limited utility. They can be used once or twice, then the enemy catches on and the game is up. Like a lie, such deception has a limited shelf life. Much better would be to tell the unvarnished truth, which has no sell-by date.

The prime exponent of using television to bend an enemy to one's will is Chuck de Caro, a frantically energetic producer who came to the medium via the Air Force Academy and the Green Berets. Having worked as a war correspondent for CNN in Central America, he has the ability to see the military-media relationship from both sides. It prompted him to formulate a new doctrine of information warfare called Soft War, about which he lectures regularly at the National Defense University. Taking no account of how many stars sit on the shoulders of his students' uniforms, de Caro blasts them with a lecture entitled "Sats, Lies and Video Rape," in which he tries to ram home the importance of a military that understands the power of television, and knows how to use it. Concurrently with his lecturing duties, de Caro runs a company called Aerobureau, the only airborne, multipurpose flying TV studio in the world. It consists of a Lockheed Electra ("The Amazin' Lady") equipped with a fully functional TV studio, complete with camera-bearing UAVs and with satellite uplinks for getting pictures out. Aero-bureau is designed for use by broadcast networks to report from war zones or scenes of natural disaster. So committed is de Caro to the notion of television as a weapon of warfare that he has created a variant of the Amazin' Lady for the military to beam positive television to regions where populations are held in thrall by monopolistic messages of hate.

In the summer of 1997, de Caro was invited to Bosnia by NATO's Office of the High Representative as a result of his reputation for inno-vative "out of the box" thinking he had established at the National Defense University. Moreover, a former student was at the time the general commanding the American forces in the north of Bosnia. Both organizations were frustrated at the ability of the accused war criminal Radovan Karadzic to maintain an iron grip over the Bosnian Serb popu-lation through a mixture of fear and propaganda, and wondered if there was a Soft War angle to breaking the impasse. De Caro's response was typically robust. He saw a situation in which the OHR/SFOR's military, political and media elements were hopelessly uncoordinated, a mish-mash of lines of communication crossing different organizations, armies and cultures, ranged against a tight, cohesively organized foe who held the strings of military, political and media power tightly in one hand. With every nuance of NATO's political decision making broadcast to the Serbs in advance by CNN, and with no coordinated message being beamed in support of NATO's drive for peace and stability, "a cabal of fifth rate Balkan bozos . . . thumb their noses at the US and NATO," as de Caro puts it in his Soft War lecture.

De Caro's prescription for overcoming this stranglehold that Karadzic has over the reins of power runs counter to the established military thinking, which sees an enemy and wonders how to beat him. To the military mind, winning such a standoff is done by either removing the key players on the enemy's side—in this instance by arresting suspected war criminals—or by comprehensively defeating him on the field of battle. In the volatile Balkans, neither option is attractive. De Caro proposes to ignore Karadzic and make him irrelevant by speaking directly to his followers. He sees the challenge with the same mind-set as a network television programmer. The "enemy" is SRT, Serbian Radio and Television, whose output, controlled by Karadzic, is well below the standards of what American or most other European viewers would accept as professional. "Just above junior college campus TV level," is how de Caro puts it.

The propaganda put out on SRT is crude, according to de Caro, who says you might see film clips of Nazi invaders during the Second World War intercut with pictures of the SFOR troops, with the commentator calling them the "SSFOR." Such interspersion of images has the subtlety of an axe, but it works. "It's lousy TV," de Caro says, "so you don't counterpropagandize, you counterprogram." This means broadcasting the very latest American TV shows with slickly produced local messages that offer an image of peace, stability and prosperity. "It's a ratings game," de Caro declares, "and I would win it hands down."

He bases his philosophy on the belief that the inhabitants of the former Yugoslavia are susceptible to the most attractive (some would say corrupting!) influence of them all—consumerism. He believes the sophisticated, prosperity-promising images he could beam across the shattered republics would turn people away from their narrow hatreds and instead focus their attention on the possibility that if they gave up killing each other they might end up with a decent life. In the process, the warlords like Karadzic would see their grip on the citizens slipping away. Ultimately they become irrelevant, marginalized by *Baywatch*.

Soft War is a concept that strikes some sparks in the heart of the military, but has yet to catch fire. It finds more support on the fringes, in areas where civilian and military life overlap, notably the National Guard. De Caro has a firm supporter in Brig. Gen. Bruce Lawlor of the Vermont National Guard, who has created an IW "red team" to fight war games with the military establishment in order to demonstrate the power of information warfare. It has not been easy. "Information has the potential to make military folk irrelevant," Lawlor says. "If through the use of information you can make a country do what you want it to, or stop it from doing what you don't want it to, then you don't need military force. This is truly revolutionary."

Yet, sustaining a successful hearts-and-minds campaign at home or in the homes of allied nations is a much tougher prospect than swaying citizens of potential enemies. In part, this can be achieved by open briefing but that will be unrealistic once the shooting really starts and the fragile coalition among the media, the politicians and the military that allowed the deployment in the first place begins to fragment. As the British discovered in the Falklands, the press in its drive for information and news and the military in its need for secrecy and the maintenance of morale are natural and often deadly competitors.

Too little attention has been paid to using the opportunities presented by cyberspace. Instead, the focus has been on how to control the information flow, something that every professional reporter knows is impossible today. The arrival of electronic media means that there can be huge force multiplication by inserting the right information in the right part of the network. For example, a video feed, one that accurately portrays a government's position, made accessible to every TV station and to every newspaper on the Internet, might be of more value than a conventional press briefing. Indeed the evidence for the wisdom of this is before the eyes of the military and political leadership. John Leo, the noted columnist for *U.S. News & World Report*, made this critical observation of the power of television:

> Now that we televise our wars, the images will very likely have more to do with building and sustaining support for any war than will the actual news of what's going on. This is what TV has done to our politics, and I don't see any reason why it shouldn't happen to our wars. A White House official once phoned Lesley Stahl [CBS News] to thank her for a report on President Reagan that she considered devastating. The White House understood, as she did not, that the report's images were so strongly pro-Reagan that the words didn't matter.
>
> For instance that first film of the air war [in the Gulf], showing a "smart" bomb seeking out and destroying a Baghdad installation probably settled the issue of collateral damage once and for all. No later findings of inaccuracy could ever have erased that powerful image of precision bombing and the emotional support it brought to the war. Facts now have to play catch up with the images—and rarely win.

But instead of embracing the new technology and the opportunities it brings, conventional thinking about war fighting and the manipulation of public opinion continues to predominate. This is dangerous traditionalism that makes no allowance either for the changing perception of wars among the media and political elite or the proliferation of information sources that are impossible to control.

However, as we look around the world today, it is difficult to find the leadership qualities that successful crisis management demands. With this lack of resolve, it is hard to see just what future peacekeeping will be like. The Pentagon and all the other defense ministries around the world have been war gaming the almost infinite number of scenarios that the current unstable world can produce. There is no doubt that the military can change its tactics and train its people. But what is it that will persuade this new generation of leaders in the media and politics to understand that peace has a price? I fear there are not enough politicians with the courage to pay the price, or enough members of the media who respect the bold decisions that may cost lives. Instead, there is a drive for quick, easy solutions to complex problems, and if those easy solutions do not work, then there appears to be no will to find the real answers.

Morality and Megabytes

I N her 1996 statement that the search for the truth about America's vulnerability to information warfare was "the new Manhattan Project," Deputy Attorney General Jamie Gorelick demonstrated she was on one hand being sensible in trying to stir some urgency into the debate but on the other she was well wide of the mark. Compared with that great and dreadful enterprise, which taxed the souls of those charged with its completion, IW is a subject that has hardly touched the military, scientific or public consciousness for the moral issues it raises.

The Manhattan Project was the name given to the development by the United States (with some input from the British) of the first atomic bombs. For as long as it lasted, the Manhattan Project was a top secret, military-controlled operation. But both before and after the Manhattan Project the principle of using nuclear energy for warlike purposes was the subject of continuous intellectual debate.

No such claim can be made about IW.

For despite Gorelick's attempt to ascribe real importance to the President's Commission on Critical Infrastructure Protection, its activities, although carried out in the public eye, appear almost to be taking place in an intellectual vacuum. To the public, the Internet is a playground or a productivity tool. It is not the means by which their plane might be pulled from the sky or their electric power supply switched off. In keeping with the commercial tenor of the times, the most dangerous purpose to which most see the Internet being put is the draining of their bank account or the theft of their credit card number by some malevolent hacker.

Before the Second World War, an age when science was a marvelous new frontier that captured the imagination of the public at large, atomic physics was the subject for breathless debate that accorded it a romantic, almost science fiction, status. Ever since the British physicist Ernest

Rutherford had discovered the nucleus of the atom, the press and public had been debating how radioactive energy could be produced effectively and harnessed for the greater good. But scientists at the University of Gottingen in Germany and at Cambridge University in England, the two powerhouses of nuclear science, had a more abstract view of what they were doing. Their science was pure, and even when the Nazis began to cast their chill pall over Europe, few of the scientists pondered the cataclysmic possibilities of their work.

Before World War II many of the top brains fled Germany because they were Jewish. Britain was their natural home and they carried on their work there. But two Hungarians who fled to America in 1933 did understand the Janus nature of nuclear science. In 1939 Leo Szilard and Eugene Wigner were among those who believed that Adolf Hitler had at his disposal the means that would allow him eventually to make an atom bomb. They urged the U.S. government, at one stage enlisting Albert Einstein to plead their case, to launch its own bomb program. Shortly thereafter in Britain, Austrian physicist Otto Frisch and his German colleague Rudolf Peierls came to the same conclusion and made similar representations to the British government.

But even as the scientists turned their work toward the creation of a new weapon, the moral debate about what they were doing continued among them. Some, like Max Born, refused to use their skills for war making. Others like Szilard went ahead anyway, but feared the conse- quences. In 1939, the French scientist Frederic Joliot-Curie proved the theory that would make the creation of the bomb possible. In America, Szilard and the Italian Enrico Fermi reproduced the experiment. Szilard would later recall his somber feeling.

> Everyone was ready. All we had to do was to turn a switch, lean back and watch the screen of a television tube. If flashes of light appeared on the screen it would mean that large-scale liberation of atomic energy was just around the corner. We just turned the switch and saw the flashes. We watched for a little while and then we went home. That night there was very little doubt in my mind that the world was headed for grief.

He was right, in that the Joliot-Curie experiment triggered the race for the bomb. As the bomb neared reality, the scientists started to ponder the impact their work would have on the future of the world. In 1944, the Danish scientist Niels Bohr, who had fled to Britain and then Amer- ica via Sweden in 1943, implored President Franklin Roosevelt to share the secret of the bomb with America's ally, the Soviet Union. This, he believed, was the only way to guarantee a rational agreement when peace finally came to exert international control over atomic power. Roosevelt thought about it, and appeared sympathetic. But under the

1943 Quebec Agreement with the British, he was unable to take any action without first consulting Winston Churchill, so Bohr was dispatched to talk to the British Prime Minister. He received a cold reception. The secrecy-obsessed Churchill found him "woolly-headed," and dismissed his entreaty. He even questioned whether Bohr was not in fact a Soviet spy.

Bohr was not alone in his fears. The scientists at the Manhattan Project's Chicago Labs sent a report to the government urging that America go public with its knowledge and place the bomb into the care of an international agency. Further, they said that rather than drop the bomb on Japan, a neutral demonstration of the bomb's power would be enough to shock the Japanese into surrender. This was the culmination of all the moral fears that had been creeping into some of the scientists' souls as they came closer to their goal, when the science no longer looked quite so pure and its consequences seemed ready to change the world. The report never made it beyond the military and civilian leadership of the Manhattan Project, Gen. Leslie Groves and Dr. Robert Oppenheimer, who rejected the suggestions.

It was not just the scientists who agonized over the morality of what was being done. The diplomat to whom fell the job of seeking Churchill's approval to use the bomb (as required by the Quebec Agreement) was John Winant, the U.S. ambassador to London. He was so morally troubled by the bomb that after it was dropped there were fears for his sanity. An observer recalled, "Winant was beside himself. He paced back and forth wringing his hands."

The public had yet to absorb the moral terror that some of those close to the bomb had felt. In the early days after Nagasaki, atomic power resumed its place as the wondrous new energy source of the future, predictions for its use becoming ever more hyperbolic. "Heat will be so plentiful that it will even be used to melt snow as it falls," it was written. And not by a science fiction fantasist, nor an overexcited journalist, but the president of Chicago University.

But soon came the knowledge that American ownership of this technology was by no means exclusive and that an ally-become-enemy, the Soviet Union, possessed the power to devastate America. The chill of the Cold War shrouded mankind's consciousness, and atomic power became in people's minds a terrible force capable of ending their world.

The words "morality" and "warfare" sit uncomfortably together. Indeed it is on the point of morality that the justification for waging war stands or falls. Before World War I and the horrors that the mass media permitted the world to witness, revulsion over the effects of war was confined to those whose religion and philosophy rebelled against destruction and slaughter. Such people have stood for humanity's conscience through the ages. But they stood in such a tiny minority that their views were irrelevant in the face of the notion that Might Is Right.

That notion became further aggrandized by the enlistment of the Almighty into the cause, often on both sides simultaneously. It was a wonderful way to encourage the troops and to charge any resulting destruction and death to God's account. For if He is with us, then our cause is a noble one.

Even in Desert Storm, the name of God was used by political and military leaders as coalition troops were sent to fight Saddam Hussein, as if somehow the good guys had a monopoly on His attention. No matter that in Iraq troops were exhorted by their commanders in the name of Allah (and no doubt in the name of Saddam Hussein as well). While it is clear that Saddam suffered a serious defeat, does the same hold true for Allah? Of course not.

In the latter part of the twentieth century, the God on Our Side philosophy has been adapted into the concept of the Just War, which allows an enlightened society to wage war against manifestly malevolent enemies. Adolf Hitler is the outstanding example of such an enemy, which is why it was no accident that George Bush painted Saddam as a latter-day Hitler as he set about building support for the Gulf War coalition. Cynics pointed out at the time, and have continued to do so since, that the cause of ousting Saddam from Kuwait might have appeared a little less "just," and therefore less worth accomplishing, if overriding national interests of the United States and other members of the international coalition had not been so threatened. In fairness, the very fact that so many varied nations did join the coalition spoke for the potency of the argument that Saddam was guilty of an egregious breach of international law.

But the fact that the coalition forces were ordered to rid Kuwait of Iraqi forces did not mean they had carte blanche to do so at any price. In the 1990s, a Just War is one that is waged with care not to spread casualties beyond military targets. To do anything else would be to lessen an enlightened nation's sense of its own moral worth. Of course to an absolutist all war is evil and compromise of that can not be countenanced. For the main part, the rest of society has broadly accepted, within very strict bounds, that there is justification for an aggressive war to right a terrible wrong, or a limited action taken to defend those incapable of defending themselves against genocide.

Fundamental to that moral argument is the effect war has on civilian populations. Such specific concern for civilians is new. Wholesale slaughter, looting and raping of noncombatants have been the hallmarks of warfare throughout history. In the days of city-states, long sieges were designed to bring an opponent to his knees by mass starvation of the besieged populace. When the walls fell, the people inside were put to the sword. Tales of ruthless slaughter would flourish and discourage others from resisting the aggressor's might.

Over the centuries the emergence of professional armies, using cavalry and eventually firearms, took the fighting away from the cities and onto the battlefield, the location of which was determined by a mixture of tactics and convenience. Civilians still suffered badly, either through scorched-earth policies that destroyed the agricultural base of the nation or from the depredations of a conquering invader. But the slaughter gradually receded from the civil sector. The last war in which civilians were virtually unaffected by the direct application of military force was the First World War of 1914–18. Apart from the dropping of some bombs on London by zeppelins, and the immediate impact on the people of France and Flanders of having the main part of the war fought over their land, it was mainly a military affair.

The airplane ended that, first in the Spanish Civil War when the Luftwaffe, acting on behalf of General Franco's Nationalist forces, refined the art of aerial bombardment on cities like Guernica and then in the Second World War when the mass bombing of civilian targets was used to force nations to their knees. The air war against Britain's cities and factories failed to subdue the nation, but the carpet bombing of German civilian and industrial targets by British and American air forces did hasten the end of the war, at a terrible price in civilian life. The process by which war was being removed from immediately impacting civilians was thrust brutally in reverse, inevitably coarsening attitudes to the sanctity of life and turning the fate of civilians into part of the war-making process. Nicholas Fotion, the noted military ethicist at Emory University in Atlanta, saw in World War II the complete negation of anything that might be regarded as just in warfare.

> You're supposed to tell the difference between the soldiers and their babies. [But in World War II] they talked about "de-housing" the people and trying to ruin their morale. There's an interesting notion that we talk about in philosophy called the slippery slope. You talk about 10,000 killed in a night and 15,000 doesn't sound too bad. You get to accept all kinds of things after a while. People got used to bombing civilians.

The slippery slope ended with the most awesome demonstration of power ever witnessed, namely the dropping of the atomic bombs on Hiroshima and Nagasaki in August 1945. Fotion and others noted the irony that it was not the number of people killed on those two occasions —the death toll in Nagasaki was lower than in Tokyo following an incendiary bomb attack from 334 B-29 bombers on the night of March

9, 1945—but the speed and manner of their destruction that so shocked and awed the world, and finally brought it to its senses.

The dropping of the bombs awakened a new awareness of mankind's vulnerability. Never again could citizens of any land feel complete security. As the shattered cities of Europe and Asia started to rebuild, they did so under the looming shadow of a weapon that threatened to level everything, for good. Air power had demonstrated that ordinary people hundreds of miles from the front line were vulnerable to attack. The A-bomb had demonstrated how the world might end. It is little wonder that the existence of such a weapon should focus the general public's mind on the morality and ethics of war.

Postwar political and military thinking was shaped by the public's fear and revulsion over nuclear weapons. As this public consciousness made itself increasingly felt in the late 1950s and onward, the concept of using such weapons in anger became less of an issue than using them as levers to gain diplomatic and political ascendancy. This attitude reached its ultimate expression in the 1980s when President Ronald Reagan, with unusual clarity for a politician, sank billions of dollars into building up the nuclear arsenal purely as a means of spending the Soviet Union into bankruptcy. The Cold War did not last long after that.

While it appeared to the public that the American war machine was being geared to a philosophy of mass destruction—indeed, the end of the world—military planners were quietly designing weapons that once again started to draw the focus of damage away from the indiscriminate obliteration of cities and populations toward ever more precisely defined targets. The results of that effort astonished the world's TV viewing audience when minicameras on the noses of laser-guided bombs demonstrated the pinpoint accuracy that modern military technology made possible. This was the turning point in the way the public thought about weapons. Not only had the threat of nuclear war, and the threat of MAD —Mutually Assured Destruction—receded, but America had proved that even in the middle of cities military targets could be found and destroyed with little or no harm to civilians, euphemistically known as "collateral damage" in military circles. American power could be wielded with justice and discretion. According to foreign policy expert A. J. Bacevich, "This new military revolution—one might call this sanitary war—would enable Americans henceforth to satisfy their yearning to believe themselves virtuous even as they exercised commanding influence across the globe."

The sanitary war concept was no accident. The American military had started a thoroughgoing examination of morality and ethics as part of the development of new means of waging war, and from the mid-1970s classes in morality and ethics had become fixtures on the schedules of the academies of the main fighting services. "The most important thing a cadet can do to train for military leadership is to concern himself with his own moral character," believes Col. Malham Wakin of the U.S. Air Force Academy. Yet the public had not paid attention to this growing influence in military life. A. J. Bacevich, executive director of the Foreign Policy Institute at the Paul H. Nitze School of Advanced International Studies at Johns Hopkins, believes the sea change in public conscious-ness toward the morality and ethics of warfare only occurred in the aftermath of the Gulf War. And he immediately warned of its dangers. For not only had the Gulf War demonstrated that a just war could be fought with precision, but it also gave rise to the illusion that wars could be fought at very little human cost to the United States. Gen. Charles "Chuck" Horner, the architect of the allied air war in the Gulf, wrote that this was a clear aim of the allied leadership.

> We gave casualty avoidance priority over military effectiveness because it was the morally correct thing to do. The American people have demonstrated unbelievable tolerance at the losses of sons and daugh-ters in battle when they believe in the cause, but no President or general can overestimate the speed at which that patience will dis-appear if they are perceived to be spending lives foolishly. Public sensitivity to casualties can dominate our political and military decision-making in a crisis.

This policy, based as it was on a noble aim, led to a realignment in public perception about the cost of war that Bacevich perceived as unfortunate.

> The expectation that Desert Storm has endowed the United States with the capacity to dominate world events without soiling itself in the process only sets Americans up for bewildering and painful disappoint-ments.

Arguably Somalia was the first such disappointment. Bacevich con-tends that the burden this places on the military is even more difficult and complex.

> As a result, tolerance for inaccuracy or even human error diminishes. Soldiers in the field may find themselves hard-pressed to satisfy de-mands for virtually no-fault performance—especially in a media-saturated theater of operations.

In their reaction to the crisis in Bosnia, political and military leaders demonstrated their awareness of the unintended cul-de-sac down which the moral argument had been driven. Far from feeling empowered by their dominance, the Americans appeared hamstrung by fears of U.S. casualties and the limitations imposed on their military technology by a difficult physical environment. This pervasive nervousness went hand in hand with America's struggle to understand its role in the post–Cold War period, and left the door open for extremists who cared little for the cost of human life when compared with their ultimate aim to inflict severe damage. In his bitter satire "How We Lost the High-Tech War of the Future," Charles Dunlap writes as a revolutionary leader from a presumably Middle Eastern group or nation which turned modern views on war morality into a weapon against America.

> Our strategy was to make warfare so psychologically costly that the Americans lost their will to win. To do so we freed ourselves from the decadent West's notions of legal and moral restraint. . . . We would rather be feared than respected.

IW offers new opportunities, and pitfalls, in the search for a form of warfare that permits national or international goals to be achieved while at the same time satisfying an increasingly rigid code of moral conduct. As Harry Stonecipher, President and CEO of McDonnell Douglas, told an audience at MIT:

> Through the emergence of precision-guided, stand-off missiles, combined with the common desire of advanced nations to hold casualties to a minimum, we may arrive at a less destructive form of combat, in which the objective is not to overwhelm the enemy or to destroy his country but merely to remove his ability to launch or carry out an attack.

Stonecipher said the downside of this scenario was, as Charles Dunlap also pointed out, that not everyone would sign on to the same rulebook as the so-called advanced nations. He said increasingly sophisticated weapons technology would eventually fall into the hands of disaffected nations, groups or even individuals and that as a result the challenges facing political and military leaders would become harder rather than easier. It would be human qualities more than anything else that would determine our collective fate. "We should remember that weakness—whether moral or physical—will always invite attack," he said. Stonecipher would find a ready audience for that view among elements of the U.S. military who fear that soldiers wielding computer mice will lose sight of the objective of warfare, namely victory. As Maj. Ralph Peters, a contributor to *Parameters,* the journal of the Army War College, wrote:

> We will fight information warfare, but we will fight with infantry. Our potential national weakness will be the failure to maintain the moral and raw physical courage to thrust that bayonet into an enemy's heart.

Even if the bayonet is shaped like a computer mouse. The challenge to some military officers is in determining where war begins and ends. This is because the very practicality of the element of IW that involves computer attacks on an enemy's infrastructure lies in surprise. It becomes immeasurably harder to lay electronic siege to an enemy if he knows the attack is on its way. Yet that implies launching an attack before a formal declaration of war, or in the absence of such a declaration at least some notification of war or deadline beyond which action will be taken.

What does the Geneva Convention say about going on and cutting cellular phones in Northern Africa and can we do that? Is putting a virus into a computer considered an act of war? There aren't a lot of treaties or rules and these are questions we should be asking.

Those questions were posed by Lt. Col. Charles Arneson, a staff officer at TRADOC, the Army command tasked with guiding warriors through the dilemmas posed by the new era in war fighting. So while IW may appear to be a splendid means of constricting an enemy's capacity for fighting by, for instance, attacking his electrical power grid as happened in the early hours of the Gulf War, a more objective view might equate such an action with terrorism, as A. J. Bacevich warns.

To the world beyond our borders, it may appear the Americans are asserting a double standard, denouncing as reprehensible the bomb placed in a parking garage (to which the United States may be particularly vulnerable), while deeming the disabling of an urban electrical grid by remote missile attack (which the United States is uniquely equipped to launch) to be altogether acceptable.

In other words, one's view of what is just and moral in war might not derive from any absolute, universally accepted ethics. Also, given that IW attacks aimed at infrastructure inevitably affect civilians, to what extent is IW a retrograde step in the development of a moral form of warfare? In moral terms, the precision-guided weapon that can take out a radar station may be seen by historians as the peak of ethical weapons development, for any loss of life in such an attack falls within accepted norms of warfare. An IW attack that disrupts a computer network, shutting down a power grid or a telecommunications system, may conversely cost the life of a baby in an incubator, or a hospital patient in intensive care. The debate has only just begun into these new ethical questions and possible contradictions posed by a form of warfare that at first blush appears to offer a relatively bloodless means of fighting.

There is another side to the advent of IW and its impact on world peace. For all that protesters screamed their abuse at the weapons of the nuclear age, the fact remains that the very existence of Mutually Assured Destruction was sufficient guarantee that the peace would be kept. It is an irony that the fear of nuclear attack with which many people lived was actually a price they paid for the guarantee that it would never happen. MAD was the ultimate deterrent that rendered the quest for first-strike capability meaningless. It is a further irony that it was the inability of either the United States or the Soviet Union to fight a war without risking complete destruction that kept the peace. For all their world-destroying strengths, each was fundamentally weak, because victory could not be assured. Again, as Ronald Reagan demonstrated, the only possible path to victory was economic, not military.

And so the Cold War ended, unleashing not a golden era of peace but a Pandora's box of regional and ethnic conflicts. With the Soviet Union gone, and Russia now a tentative ally, the possibility of MAD has receded, and thus the ultimate deterrence value of nuclear weapons has correspondingly been diluted. Ironically, citizens now feel safer in their beds at night, while the threat to their peace and security has actually increased. Terrorism, organized crime and renegade states all pose threats that were kept under the lid in the "bad old days" of MAD.

So is IW a force for stability in these wild times or not? If deterring war is a moral act, as most people probably agree, then a force that increases the likelihood of war must by definition be immoral. Military thinkers and philosophers are only just coming to terms with the manner in which IW has changed the ethical and legal landscape. We have already considered the possibility that IW, as a method of warfare that can directly impact civilian populations and degrade their infrastructure, can be seen as a retrograde step. The attitude of the U.S. Army on this point is quite clear. In the seminal document *FM 100-6 Information Operations,* issued by the Department of the Army in August 1996, we are told that IW:

- Contributes to defusing crises.

- Reduces the period of confrontation and enhances the impact of informational, diplomatic, economic and military efforts.

- Forestalls or eliminates the need to employ combat forces.

The document adds, "The ability of the U.S. government to influence the perceptions and decision making of others greatly impacts the effectiveness of deterrence, power projection, and other strategic concepts." It is a persuasive argument, and easy to buy. But is it right?

The issue has received little attention, given the comparative novelty of IW. As might be expected, proponents of IW as the coming wave of warfare affirm IW's deterrent capability simply because it allows the dominant IW power (America, of course) to so disrupt an enemy's networks of command, control, communications, computers and intelligence (C4I) that nobody would attempt to engage. One of the foremost American analysts of deterrence, Peter D. Feaver of the Political Science Department at Duke University, agrees with this broad thesis, in as much as "IW should allow us to influence with unprecedented precision the enemy's decision-making capabilities," and adds, "to the extent that IW works as promised, it does appear to promise a 'silver bullet' approach to offering a deterrent threat." But Feaver takes his analysis to forensic levels, and builds a strong argument that there is a yawning gap between what IW as a technology might promise and what the practitioners of IW can deliver.

> Here the problem is not with the technology *per se*—it might work as promised—but rather the way organizations are likely to handle the new technology. As in the early days of the nuclear age, IW capabilities are shrouded in multiple layers of secrecy. The IW arena is among the most highly compartmentalized in the entire US defense establishment. The right hand quite simply does not know what the left hand *can do,* let alone what it in fact is doing.

Feaver illustrates this with a scenario in which a friendly IW operation has "captured" an enemy information node and is using it to put out information designed to advance the friendly cause. Friendly sensors outside the IW unit that carried out this operation pick up that information and report it up the friendly chain of command. The possibility for information contamination is obvious. The friendly side is collecting intelligence believed to originate from the enemy, but which was in fact planted by other friendly units. Feaver says it might be common sense to assume that at some point in the chain there is someone with sufficient authority, competence and oversight to know what is fact and what is fiction. But he points out that life does not always obey the rules of common sense, citing the case of CIA mole Aldrich Ames as an example of what he calls "information fratricide" or "blowback." Thanks to Ames, the Soviets turned many American agents in place, and used them to feed false information back to the CIA. The CIA actually *knew* some of this intelligence was corrupt, but in order not to alert Ames to its counterintelligence operation against him, the information was treated as accurate. It was then passed up the chain of command *without anyone being told it was corrupt.* And thus the U.S. political and military leadership's attitudes to the Soviet Union were being shaped by false information. Feaver says, "The IW capability that allows for information exploitation will almost certainly be more sensitive and closely held [by the government] than the intelligence capability that might collect on the fruits of that information exploitation. It strains credulity that the US government can efficiently manage this problem in every case."

The impact of such contamination of the information process on deterrence is severe, Feaver suggests.

> If blowback confuses the IW-waging state, then that state's ability to interpret the target's signals—and hence the target's ability to signal back—is degraded. For deterrence and compellence to work, it is necessary that the target signal his response to the threat, whether to comply or not to comply. If the coercing state is manipulating the target's information environment, it is that much harder for the target state to signal a response.

Add to this the potential for IW to sow seeds of doubt in the enemy's chain of command and decision-making process and you remove one of the factors that analysts saw as crucial in maintaining deterrence during the Cold War, namely a tight, clear connection between the enemy's political and military operations. Then add in the perception IW creates of a dominant force possessing first-strike capability, which the United States undoubtedly has, and one can see imbalance, confusion and fear, the very factors that dilute deterrence. Thus Feaver makes a plausible case for IW, on balance, working against the interests of stability. If America looked set to shut down a potential enemy's networks as a first strike, it is not hard to imagine an enemy's fallback strategy that would include a first-strike use of terrorist weapons on the U.S. homeland. Unlike nuclear weapons, an IW capability does not invite retaliation in kind.

The debate over the deterrent value of IW is a new one, and will mature in the coming years as analysts gather more solid evidence and experience on which to base their theories. But it seems credible to suggest that IW does have the intrinsic potential for causing confusion in both friendly and hostile camps.

There is another layer of confusion to lay over this theoretical scenario, one that blurs traditional perceptions of sovereignty and national boundaries. The authors of the Rand Corporation analysis *Strategic Information Warfare: A New Face of War* warn that virtually any national delineation will become less clear in the global informational environment, be it "geographical, bureaucratic, jurisdictional or even conceptual," and they cite as an example of how this is already happening in other fields the emergence of global financial and monetary markets, which dilute the power of national governments to influence policy that directly affects their own people. The same might also be said of the spread of global corporations whose only loyalty is to the bottom line and their shareholders. The Rand study maintains that this blurring of boundaries which results from the spread of global networks will make identifying sources of attack very difficult, leading to a weakening of cooperation between domestic law enforcement agencies and national security agencies, all of whom will, in time-honored fashion, fight tooth and nail for control.

For instance, what happens when a cyberattack hits an American city's water supply system? A hacker competent enough to do that will also take pains to cover his tracks. So the list of suspects could include a domestic terrorist from a militia group, a domestic or foreign crime group seeking new ways to extort money, a terrorist group operating from abroad, or a third country seeking to discomfit and embarrass the United States. Those familiar with the politics of government will be able to imagine the feeding frenzy as various three-letter agencies vie to be the lead investigator. It would not be a pretty sight, and illustrates the new uncertainties IW can bring to a system of government invented way before the computer age and showing no signs of adapting to the new times with any great speed.

The sovereignty issue also promises to make cooperation between nation states difficult, since electrons recognize no national borders. The attack on the Air Force's Rome Labs was a classic case, in which the hacker's route stretched from London, Latvia, Canada, the United States and into Korea's Atomic Research Institute, whose files were plundered. But which Korea? This posed a potentially critical issue of sovereignty, for if it had been North Korea's Atomic Research Institute that had been breached by Datastream Cowboy, the United States might have had a real problem on its hands. "The concern was that the North Koreans would think the transfer [of data by the hacker] was an intrusion by the U.S. Air Force, which could be perceived as an act of war," the official report into the incident related. Of course it transpired that the Atomic Research Institute involved was in South Korea, which properly regarded the intrusion as a criminal, rather than a hostile, act. Investigators chasing hackers, malicious or otherwise, across national boundaries generally rely on the common cause of law enforcement everywhere that sees the downside of letting hackers roam around the world's databases. Police forces tend to cooperate with each other rather than obstruct, but such an assumption is not a strong enough basis for international policy.

In the turbulent post–Cold War period, which initially seemed to promise so much in terms of a New World Order, trying to reach some kind of international consensus on how to deal with all of these issues will not be easy. The United Nations is seeking to redeem itself in the eyes of critics who see it as inept and corrupt, both politically and economically. The spotty record of U.N. peacekeeping operations, which have not been helped by America's own difficulties in understanding what role it should play, has not encouraged the world community to believe that the U.N. is the correct body in which to hold the necessary debates about our changing world, where borders are being rendered irrelevant by electronic commerce and yet where ethnic and nationalistic savagery is being committed on as large a scale as has ever been witnessed. In the absence of an emerging consensus for the future of international law in the electronic age, the voice of those who believe the liberating nature of information will provide answers should be heard, for the power of information to create the paradigm we are searching for has already been witnessed. It was television that fueled the collapse of communism. "The world sees you!" protesters in Prague chanted at the riot police who in another age might have brutally suppressed them. But the world really did see them, through the cameras of CNN.

A global village will have global customs. Denying people human rights or democratic freedoms no longer means denying them an abstraction they have never experienced, but violating the established customs of the village. It hardly matters that only a minority of the world's people enjoy such freedoms or the prosperity that goes with them; these are now the benchmarks. More and more people around the globe are demanding more say in their own destiny. Once people are convinced this is possible, an enormous burden of proof falls on those who would deny them.

Conclusion

W A R has changed forever. The end of the Cold War and the collapse of communism have completely altered the strategic playing field. The West's common enemy has gone, weakening and ending traditional alliances and leaving a confused world picture. The United States straddles the globe, the most powerful nation ever seen, with an array of weapons and an economic vitality that produces a stream of innovation no nation can come close to matching.

And yet that awesome power has few anchors. Back in the days of the Cold War it was relatively simple to know the enemy. Communism was a convenient and real foe which united the NATO nations under American leadership. Across the world, the lines were sharply delineated between communism and capitalism, and many of the political and economic rivalries that produced conflict were a subset of that clash of culture and belief. Today, the Warsaw Pact has disintegrated along with the Soviet Union, and while NATO continues as a group of like-minded nations, it is struggling to find an identity and a mission.

The collapse of a superpower rivalry that had lasted forty-five years coincided with another revolution. The Information Age has brought with it computers, microchips, processing power and all the other parts of a language that nobody knew a decade ago and which today is just a normal part of routine communication. The practical effect of this revolution is evident in all walks of life in developed societies from the car to air traffic control and the Internet. The computer has made possible opportunities and things that would have been unthinkable a few years ago. As microchips have gotten smaller and computing power has grown exponentially, so the influence of the Information Age has begun to be felt at every level of society.

What sets this revolution apart is the pace and the breadth of its effects. Only ten years ago, the world of the Internet was largely unknown and yet today it is everywhere. A new generation is being brought up in a virtual world where paper is old-fashioned and online there is a new world of work and play. These changes have begun to undermine the very fabric of society, the delicate bonds that helped hold together towns, cities and countries. The fragile controls that nations have over their people have begun to disintegrate as the concepts of country, loyalty and national identity have eroded in the face of an assault which offers the promise of a different world where no such inconvenient restraints exist.

Several centuries ago the Crusaders traveled thousands of miles to fight the infidel. In the European Reformation people died for their faith and defending that faith became part of America's raison d'etre. Today, while faith remains a powerful mobilizing force in some parts of the world, it is inconceivable that developed nations would find religion a rallying cry for war.

Patriotism, which allowed politicians to send their men to fight and die for their country, is becoming an old-fashioned concept with little resonance among a generation that has loyalty to cyberspace but little loyalty to a world where sacrifice, honor and duty are the price that might have to be paid for membership in a society with standards, morals and principles. Patriotism has for centuries been enough to persuade people to lay down their lives. People talked of the "nobility" of sacrifice. "*Dulce et decorum est pro patria mori*—it is sweet and fitting to die for your country." Heroism and sacrifice are the hallmarks of soldiers through the centuries, from Sparta to Somalia. In ancient times men fought because they had to; their city or their state had to be defended from almost certain destruction. In the modern professional services, men and women join because the pay is good, the training excellent and among some people, service to your country still matters. But the men who died in the dusty streets of Mogadishu did so for their friends whom they vowed they would never abandon on the field of battle. Their country had no compelling reason to ask them to die, and the politicians who watched them die, courtesy of television, found in their very public deaths a reason to change their policy. It was a demonstration of just how weak political will has become in the face of public apathy and fear of media criticism.

In a compelling and controversial analysis of the American condition as the next millennium beckons, former CIA analyst Michael Vlahos maintains that America in the late 1990s is in the same mental state as France was in the years prior to 1871. Then, France was the unchallenged leader in new weapons technology. Its army was the envy of the world for its professionalism and its skill at arms. With no major wars to fight, and no armies close to being strong enough to mount a challenge, it created a new role for its military, capable of fighting small regional conflicts in Europe and even far-flung Mexico. France was proud, strong and preeminent. And about to be crushed.

In 1871, at Sedan, the great French army was humiliated by the forces of the German federation, and the French nation suffered a blow to its national pride that crippled its power, and most importantly its own sense of empowerment. It was a blow from which the French have never recovered.

Vlahos wrote:

> France did not lose in 1871 because it was stupid or unlucky. Certainly, its war technologies and combat experience were superior to Germany's. France lost in 1871 because a revolution had swept Europe, and that revolution changed war. The French military and the French ruling elite had not kept up. They had missed the revolution in warmaking.

The revolution they had missed was the mushrooming growth of the railroad in late-nineteenth-century Europe. The Prussian military genius Gen. Helmuth von Moltke understood how to harness the rail system to war fighting and was able to deliver troops to fight battles with a degree of surprise and logistics flexibility the French could not match. Moltke was the general whose strategy and tactics gave rise to the saying "war by time table."

Vlahos believes that America is in the same state as France was over 100 years ago.

> So we await our Sedan. To dismiss this prospect, we must dismiss the possibility of a future foe, an equal challenger with an evil intent. To dismiss this possibility, we must dismiss the transformative power of Big Change and assert that we can keep control not only of our world system, but what happens to it, forever. But no one can do this.

To adapt Vlahos's idea, for the European railroad read cyberspace. The Big Change he writes of is the explosive transformation of society that is being wrought by the elimination of barriers between people across the world by the Internet. Everything about what is taken for granted today as "daily life" will change, from commercial transactions, working habits, commuting, to the very notion of employment.

> This change will mean dislocation, anxiety, fear of the future. It is happening right now, but it will get much worse. Many will do very well in this new world, as they are doing now. But many will not. . . .
> As millions of Americans find themselves in the fearful, rollicking slide of Big Change, stripped of old meaning, they will demand new

meaning: What is my status in society? How do I belong? What is my worth?

And the inability of the entrenched, self-preserving political and military elites to answer those questions, or even to understand why they are being asked, will render them irrelevant. They will be replaced.

So who will be tomorrow's German Federation, tomorrow's Moltke who makes an astounding end run around the mighty, unbeatable American power? A revitalized Russia? An economically and militarily powerful China? Iran? Perhaps one of these, but they still exist in the envelope of today's thinking. For thinkers like Vlahos, the challenge is finding the true dimensions and deciding exactly who will wield power in cyberspace —indeed what exactly power will mean in a place where much reality is virtual and traditional concepts of territory are obsolete.

The way these questions are answered tends to be restrained by what we know today. The answers naturally fall in the direction of traditional threats such as Islamic terrorists, ethnic minorities. But with national boundaries collapsing as the nation state becomes increasingly irrelevant in the Information Age, the real answers probably lie beyond our current imagining.

Vlahos's Big Change concept produces new alliances, new formations that lie beyond anybody's current imagination. This is by no means a preposterous concept. The Internet is proving to be a fertile ground for people with identical interests from all across the globe to meet and pursue their objectives in a space that has no physical existence, no national boundaries, no police or armies; it is simply cyberspace.

When you log on to the Internet, enter the URL www.mtnforum.com. That will get you to the home page of the Mountain Forum, an entirely peaceful and peace-seeking community of people from all over the world who have one common interest—they all live on mountains. Their goals are to further the interests of mountain-dwelling people by keeping national decision makers in their various countries informed of the issues confronting mountain people and their environment.

In the winter of 1996, futurist Erin Whitney-Smith put together a conference on the Internet that brought together mountain people from all over the world. She found that men and women in the Urals, the Rockies and the Alps in common hated the people in the valley, who they felt exploited them; were independent; disliked governments that interfered with that independence; and cared deeply about the environment. That group has now become a virtual community and shares experiences, successes and failures in ways that would have been unimaginable a few years ago. Mountain people have bonded through the virtual identity they have developed on the Web. The question that should concern government and law enforcement agencies around the world is that there are thousands of such groups spawning each year. Each group brings together like-minded people who arguably share more common interests and a greater loyalty to each other than they do to the country where they were born.

Moore's Law states that computing power doubles every eighteen months, a pace that will determine the speed of the information revolution and the revolution that is becoming known as information warfare. Virtual reality will become actual reality. For people whose lives become increasingly less defined by the place they work, because now they work online; by their colleagues, because they never see them anymore; by trips to the store, because they shop online; and by their sense of national identity, because they find they have more in common with people they meet 7,500 miles away than they do people just around the corner, cyberspace becomes their reality. It is where life, art, commerce and perhaps love will be conducted. And it is where information warfare will rule.

War, as it is defined now that the superpower rivalry is over between the United States and the former Soviet Union, poses some unique challenges. There is now no prospect of Soviet hordes pouring through Germany's Fulda Gap to fight on the West German plain with NATO forces. The Warsaw Pact is a spent force and today's Russian military is incapable of fighting even a guerrilla war against a band of desperadoes like the Chechen rebels. In that conflict where appalling generalship by the Russians resulted in unnecessary casualties, the Russians were reduced to asking their diplomats in New York to see if they could buy some cellular telephones that would match the Chechen technology. There is no prospect for the Russians of fighting a conventional war against a modern army without facing certain defeat.

But that certainty has failed to produce the stability that everyone expected. Instead, one stable quadrant has been replaced by many other instabilities, some of them as dangerous as anything experienced in the Cold War. The developed world is now in a state of permanent war with organized crime, drug barons, terrorists and economic spies. Those are wars that already require the full resources of the developed nations' intelligence communities and, in some cases, their armed forces. But that is only the beginning. Across the world ancient ethnic tensions that once were held in check by the overarching geopolitical imperative of the superpower rivalry are now causing conflicts in countries as diverse as Somalia and Bosnia. Such tensions are certain to get worse in the years ahead.

Yet, the nations with the most resources and the most sophisticated weapons have shown themselves remarkably reluctant to get directly involved in these problems. Part of this is institutional inertia, part is a new introspection which leads to people demanding that their politicians focus on home rather than overseas.

But this reluctance also reflects a new reality in the political, military and media elites. There are no leaders in the world today who have direct experience of conflict as it has been understood for centuries. Instead, conflict is seen through the narrow prism of the television or movie screen or through the pages of newspapers that are in turn a reflection of their reporters' inexperience and zero tolerance for failure.

Information warfare promises real solutions to the challenges of the post–Cold War world. Not only might it be possible to take on the drug barons and terrorists with the new tools that are either available or being developed, but there are wars that might be won as well. Revolutionary new weapons will be available, such as sensor grass, ant spies, surveillance dust and exoskeletal uniforms that make supermen of mortal soldiers. Standoff, the ability to fight a war so that the enemy is within range of your precision weapons but you are beyond his, will be the fundamental strategy. Inside networks, nations that pursue IW will create ever more sophisticated viruses and worms that will attack enemy information systems with the same degree of precision and devastation that we normally associate with missile technology.

The difficulty in dealing with all these startling new developments and the challenges they pose is in the sheer pace of the change that is producing them. The government, including the defense and intelligence communities, are entrenched bureaucracies highly resistant to change. While some people may personally be capable of agile response, the framework that surrounds them prevents it.

It was therefore no surprise when the Quadrennial Defense Review was unveiled by Defense Secretary William Cohen in June 1997, that despite all the signs of innovation in parts of the armed forces, the military establishment fell back on the certitudes of the Cold War. They had an opportunity to spring forward, got halfway out of their chairs . . . then sat back exhausted. Like the politicians who think computers are something secretaries use for writing letters, the military establishment is falling hopelessly behind the information curve.

All the innovation and energy are now in industry. Young people fresh out of college are eschewing the idea of joining big companies and instead are starting their own in their garage. That was the model Bill Gates and his friends set with Microsoft and it is happening still. Indeed, Microsoft itself is becoming the establishment and its heels are being nipped by aggressive, agile young businesses.

This wondrous economic vitality has placed power beyond our wildest dreams in the hands of individuals sitting in their homes. Through the culture and through the simple expedient of spending time online, citizens are shaping their own ability to use, for good and evil, the power of information warfare. They are carving out new homesteads in cyberspace. These individuals instinctively understand, as they join global chat groups swapping ideas and philosophies with no heed to national boundaries, what the establishment does not, namely that the old way of life, and the old way of war, has gone forever.

A startling example of how out of touch the establishment has become with what is happening in the real world came in an astonishing attack on the Internet by two of Washington's top journalists, the husband and wife team of Steven and Cokie Roberts. In a newspaper column, they deplored the fact that ordinary people were now using computers to communicate and deliberate on issues, creating the prospect of an electronic democracy that "makes our blood run cold . . . no more deliberation, no more consideration of an issue over a long period of time, no more balancing of regional and ethnic interests, no more protection of minority views."

It was a classic case of people inside the castle walls ridiculing any pretension by those outside to know what is good for them, despite the American public constantly demonstrating it was a great deal wiser than politicians and pollsters ever gave it credit for. The fact that the great deliberative democracy the Robertses were defending so bravely is demonstrably corrupt did not seem to concern them. A few reforms here or there, they argued, would fix it. The fact is they saw the cozy world that looked after them so well being inexorably eclipsed.

The Robertses would no doubt disagree with this conclusion and the conclusions of others who heaped scorn on their attack. Similarly, America's military leaders would deny they are stuck in some Cold War mind-set. They would point to the billions of dollars being spent on IW technologies and on initiatives like the Air Force Information Warfare Center and the Navy's Fleet Information Warfare Center (AFIWC and FIWC). What they do not grasp is that IW is as much about attitude as about ammunition. A twenty-two-year-old hacker has a better understanding of what cyberspace and its challenges represent than a fifty-five-year-old four-star general. This is why the best hope the U.S. military could have is to listen to its younger officers, like thirty-three-year-old Navy Lt. Ross Mitchell, whose Ring of Fire concept will radically change naval warfare. Or Charles Dunlap, a colonel in the U.S. Air Force, who, like Mitchell and many others in the middle and lower ranks, demonstrates an ability to think laterally about issues.

In January 1996, Dunlap wrote an article for the *Weekly Standard* magazine that offered an astounding, and chilling, portrait of a conflict yet to come. His premise was simple, yet it had the immediate and fundamental resonance of truth. He said that all the high-tech weaponry and IW cyberpower could not prevent America from being beaten by an enemy who has a strong will, cunning honed by the necessity of survival, a warrior breed which fights battles every generation and whose society is hardened against suffering by deprivation. Dunlap wrote:

The following is the transcript of a secret address delivered by the Holy Leader to the Supreme War Council late in the year 2007.

In the name of The One Above, I offer greetings to my fellow warriors! Today, with His grace, I speak of our great victory over our most evil enemy, America. A little more than 10 years ago experts thought that what became known as the Revolution in Military Affairs would leave developing nations like ours incapable of opposing a high-tech power like the United States. With the help of The One Above, we proved them wrong. They were guilty, as those who defy the sayings of the divine usually are, of idolatry—though in this case they did not worship graven images, but the silicon chip. As though a speck of sand could defeat the will of The One Above.

Dunlap was not saying that the development of IW capabilities, non-lethal weapons, and smart and brilliant weapons was wrong, just that they are only part of the picture. He was warning, correctly, that rushing headlong into a gleaming, technology-driven future without any guiding philosophy or leadership renders the giant vulnerable.

Throughout history war has never been about silver bullets. There are no magic solutions to the problems confronting the world on the brink of a new century. Conventional conflicts as well as the struggles against organized crime and terrorism will require the usual mix of raw courage and high technology. But there are two important differences about the information revolution: the speed at which it is unfolding, and the power it is delivering into the hands of ordinary men and women around the world. These will pose unique challenges for the future. As the revolution takes shape, the countries with the greatest access to the technology will have to move very fast if they are going to keep pace and not cede control of their own destiny to others who are more creative, flexible and agile.

Traditional government is ill-equipped to deal with the speed of this revolution. Bureaucracies by their very nature abhor change and yet the pace of this revolution demands almost constant change. While the tools of information warfare do offer promise, the politicians and the lawyers who guide them have been slow to understand the potential. Guidance for the military and the intelligence communities has been almost non-existent and there has been no public debate about the perils and possibilities of IW. This lack of leadership has produced a vacuum which has been filled by rivalry between different armed forces and different intelligence agencies. The squandering of public resources in an inefficient system is a familiar story. But in an environment where real change is measured not in decades but months, such incompetence casually disregards national security needs and could cause America and its allies to lose a vital technological edge in the Information Age.

This revolution also requires the political and military leadership to understand the purpose and consequences of war and the risks that attach to any military action. On recent evidence, none of those attributes are present to any degree, and across the world a risk-averse approach to warfare in all its forms has seeped into the corridors of power. That in turn will lead to an increasing dependence on IW as the perfect solution for fighting wars with no risk of casualties and at relatively low financial cost. But that is to seek the very silver bullet that history has shown does not exist. As David proved against Goliath, strength can be beaten. America today looks uncomfortably like Goliath, arrogant in its power, armed to the teeth, ignorant of its weakness.

Glossary

AFIWC   Air Force Information Warfare Center

AFOSI   Air Force Office of Special Investigations

AIA   Air Intelligence Agency

Appliqué   A wireless intranet for the battlefield

Arsenal Ship   Floating robot operated by remote control that can carry 500 cruise missiles.

Artificial Intelligence   A broad range of applications that exhibit human intelligence and behavior, including robots, expert systems, voice recognition, natural and foreign language processing. It also implies the ability to learn or adapt through experience.

ASIM   Automated Security Incident Measurement

ASTAMIDS   Airborne Standoff Mine Detection System

ATO   Air Tasking Order

AWACS   Airborne Warning and Control System

AWE   Advanced Warfighting Exercise

BDA   Bomb Damage Assessment

BIES   Battlecube Information Exchange System

Biological weapon   Microorganism that causes disease in man, plants, or animals, or causes the deterioration of matériel.

BW   Biological Weapons; Biological Warfare

C2   Command and Control

C2W   Command and Control Warfare. The integrated use of operations security (OPSEC), military deception, psychological operations (psyops), electronic warfare (EW) and physical destruction, mutually supported by intelligence, to deny information to, influence, degrade or destroy adversary C2 capabilities, while protecting friendly C2 capabilities against such actions. Command and Control Warfare applies across the operational continuum and all levels of conflict. C2W is both offensive and defensive: 1) Counter-CC2—to prevent effective C2 of adversary forces by denying information to, influencing, degrading or destroying the adversary C2 system; 2) C2-Protection—to maintain effective command and control of own forces by turning to friendly advantage or negating adversary efforts to deny information to, influence, degrade or destroy the friendly C2 system.

C3I   Command, Control, Communications and Intelligence

C4I   Command, Control, Communications, Computers and Intelligence

CAFMS   Computer Assisted Force Management System

CENTCOM   U.S. Central Command

CERT   Computer Emergency Response Team

CI   Counterintelligence

CIA   Central Intelligence Agency

CIC   Central Intelligence Center

CINC   Commander in Chief

CNN War   Influence of the independent media on a particular real-world perception of an actual event (or anticipation of same).

CNO   Chief of Naval Operations

COMPUSEC   Computer Security

COMSEC   Communications Security. Measures and controls taken to deny unauthorized persons information derived from telecommunications and ensure the authenticity of such telecommunications. Communications security includes cryptosecurity, transmission security and physical security of COMSEC material.

Copernicus   The code name under which the Navy plans to reformulate its command and control structures in response to the realization that information is a weapon. Through Copernicus, war fighters will get the information that they need to make tactical decisions. The architecture of Copernicus was designed by Vice Adm. Jerry O. Tuttle.

Critical Infrastructures   Infrastructures that are deemed to be so vital that their incapacity or destruction would have a debilitating regional or national impact. They include at least seven categories: telecommunications; electrical power systems; gas and oil; banking and finance; transportation; water supply systems; continuity of government and government operations. Emergency services (including medical, police, and fire and rescue services) might also be considered critical infrastructures.

CTAPS   Contingency Tactical Air Control Automated Planning System

Cyber   The prefix is from the Greek root *kybernan*, meaning to steer or govern, and a related work, *kybernetes*, meaning pilot, governor or helmsman.

Cyberwar   Conducting, and preparing to conduct military operations according to information-related principles.

DARPA   Defense Advanced Research Projects Agency

DCI   Director of Central Intelligence

DES   Data Encryption Standard

DIA   Defense Intelligence Agency

Digital Data Warfare   Malicious computer code covertly introduced into one or more specific computer systems or networks, by an attacker to meet military, political, economic or personal objectives.

DII   Defense Information Infrastructure. The DII encompasses information transfer and processing resources, including information and data storage, manipulation, retrieval and display. More specifically, the DII is the interconnected systems of computers, communications, data, applications, security, people, training and other support structure, serving the Defense Department's local and worldwide information needs. The DII 1) connects the DOD mission support, command and control, and intelligence computers and users through voice, data, imagery, video and multimedia services, and 2) provides information processing and value-added services to subscribers over the DISN. Unique user data, information and user applications are not considered part of the DII.

DISA   Defense Information Systems Administration. Military organization charged with responsibility to provide information systems support to fighting units.

DISN   Defense Information Systems Network. A subelement of the DII, the DISN is the Defense Department's consolidated worldwide enterprise-level telecommunications infrastructure that provides the end-to information transfer network for supporting military operations. It is transparent to its users, facilitates the management of information resources and is responsive to national security and defense needs under all conditions in the most efficient manner.

DISSP   Defense-Wide Information Systems Security Program

DIW   Defensive Information Warfare

DOD   Department of Defense

DOE   Department of Energy

DRA   Defense Research Agencies

DSCS   Defense Satellite Communications System

EA   Electronic Attack. That division of electronic warfare involving the use of electromagnetic or directed energy to attack personnel, facilities or equipment with the intent of degrading, neutralizing or destroying enemy combat capability. EA includes: 1) actions taken to prevent or reduce an enemy's effective use of the electromagnetic spectrum, such as jamming and electromagnetic deception, and 2) employment of weapons that use either electromagnetic or directed energy as their primary destructive mechanism (lasers, radio frequency weapons, particle beams). See EW.

Elint   Electronic intelligence

EMP   Electromagnetic Pulse. The electromagnetic radiation from a nuclear explosion caused by Compton-recoil electrons and photoelectrons from photons scattered in the materials of the nuclear device or in a surrounding medium. The resulting electric and magnetic fields may couple with electrical/electronic systems to produce damaging current and voltage surges. May also be caused by nonnuclear means. Also called EM.

Encrypt   To convert plain text into unintelligible forms by means of a cryptosystem. (Note: The term encrypt covers the meaning of encipher and encode.)

EUCOM   European Command

EW   Electronic Warfare. 1) Any military action involving the use of electromagnetic and directed energy to control the electromagnetic spectrum or to attack the enemy. The three major subdivisions within electronic warfare are: electronic attack, electronic protection and electronic warfare support. 2) Military action involving, (1) the use of electromagnetic or directed energy to attack an enemy's combat capability, (2) protection of friendly combat capability against undesirable effects of friendly or enemy employment of electronic warfare or, (3) surveillance of the electromagnetic spectrum for immediate threat recognition in support of electronic warfare operations and other tactical actions such as threat avoidance, targeting and horning.

FBI   Federal Bureau of Investigation

FIWC   Fleet Information Warfare Center

Flying Dutchman   A Trojan horse that erases all traces of itself after performing its mission. This is a common feature of Trojan horses that helps defeat subsequent investigation.

FORCE XXI   The U.S. Army's blueprint for the army of the twenty-first century.

Front-Door Coupling   Accessing a target computer system by using media for which the system is designed.

GCHQ   Government Communication Headquarters

GII   Global Information Infrastructure. Includes the information systems of all countries, international and multinational organizations and multi-international commercial communications services.

Global Information Environment   Non—Department of Defense information systems (media, government agencies, nongovernmental organizations, international organizations, foreign governments and industry) which collect, process and disseminate information about operations. These organizations create perceptions and opinions through the particular bias they place on information in the dissemination phase. Their representation of the facts may be tailored to promote a particular agenda. These systems largely operate autonomously and are not subject to direct control. They significantly impact military decision making and execution.

GPS   Global Positioning System

Hacker   1) A person who enjoys exploring the details of programmable systems and how to stretch their capabilities, as opposed to most users, who prefer to learn only the minimum necessary; 2) Unauthorized user who attempts or gains access to an information system.

HARM   High-Speed Anti-Radiation Missile

HERF   High Energy Radio Frequency. As in HERF gun: a device that can disrupt the normal operation of digital equipment such as computers and navigational equipment by directing HERF emissions at them.

Humint   Human intelligence. Information derived from human observations and experiences.

IC   Intelligence Community

IFF   Identify Friend or Foe

Information   Knowledge such as facts, data or opinions, including numerical, graphic or narrative forms, whether oral or maintained in any medium.

Information Assurance   The availability of services and information integrity.

INSCOM   Intelligence and Security Command

IO   Information Operations

IT   Information Technology

IW   Information Warfare. 1) Aggressive use of information means to achieve national objectives. 2) The sequence of actions undertaken by all sides in a conflict to destroy, degrade and exploit the information systems of their adversaries. Conversely, information warfare also comprises all the actions aimed at protecting information systems against hostile attempts at destruction, degradation, and exploitation. Information warfare actions take place in all phases of conflict evolution; peace, crisis, escalation, war, deescalation and post-conflict periods.

J-2   Intelligence

JCS   Joint Chiefs of Staff

JIVA   Joint Intelligence Virtual Architecture

JSIWC   Joint Services Information Warfare Command

LAN   Local Area Network

Land Warrior   An integrated intelligence-gathering and sensor suite for the soldier that includes night vision, laser rangefinding and video links with headquarters.

MAD   Mutually Assured Destruction

MADS   Mini-ATO Distribution System

Malicious Computer Code   Any computer code that is on a system without the consent of the owner.

MARV   Miniature Autonomous Robotic Vehicle

Maskirovka   Russian Radioelectronic Combat term meaning concealment through deception.

MEMS   Micro Electro-Mechanical System

Milnet   The network of military computers.

MITI   Ministry of International Trade and Industry (Japan)

NBC   Nuclear, Biological, Chemical

NCA   National Command Authority

NDU   National Defense University

Netwar   1) Information-related conflict at a grand level between nations or societies. 2) Trying to disrupt, damage or modify what a target population knows or thinks it knows about itself and the world around it.

NII   National Information Infrastructure. The NII is a system of high-speed telecommunications networks, databases and advance computer systems that will make electronic information widely available and accessible. The NII is being designed, built, owned, operated and used by the private sector. In addition, the government is a significant user of the NII. The NII includes the Internet, the public switched network, and cable, wireless and satellite communications. It includes public and private networks. As these networks become more interconnected, individuals, organizations and governments will use the NII to engage in multimedia communications, buy and sell goods electronically, share information holdings and receive government services and benefits.

NIWA   Naval Information Warfare Activity

NLW   Nonlethal Weapons

NN-EMP   Nonnuclear Electromagnetic Pulse

NOC   Nonofficial Cover

NRO   National Reconnaissance Office

NSA   National Security Agency

NSC   National Security Council

NSDD   National Security Decision Directive
NWC   Naval War College
OIS   Office of Information Security
OIW   Offensive Information Warfare
OODA   Orient, Observe, Decide, Act
OPSEC   Operations Security. OPSEC is a process of identifying critical information and subsequently analyzing friendly actions attendant to military operations and other activities to: identify those actions that can be observed by adversary intelligence systems; determine indicators adversary intelligence systems might obtain that could be interpreted or pieced together to derive critical information in time to be useful to adversaries; select and execute measures that eliminate or reduce to an acceptable level the vulnerabilities of friendly actions to adversary exploitation.
PGP   Pretty Good Privacy
PIN   Personal Identification Number
PKCS   Public Key Cryptography Standards
PSN   Public Switched Network
Psyops Psychological Operations. Psyops are operations planned to convey selected information and indicators to foreign audiences to influence their emotions, motives, objective reasoning and, ultimately, the behavior of foreign governments, organizations, groups and individuals. The purpose of psyops is to induce or reinforce foreign attitudes and behavior favorable to the originator's objectives. Psyops are a vital part of the broad range of U.S. political, military, economic and informational activities. When properly employed, psyops can lower the morale and reduce the efficiency of enemy forces and could create dissension and disaffection within their ranks.
R&D   Research and Development
RMA   Revolution in Military Affairs. The realization by the military that information and information technologies must be considered as a weapon in achieving national objectives via military activity.
SAM   Surface-to-Air Missile
SC-21   Surface Combatant for the 21st Century
SEALs   Seal Air Land (Navy special operations)
Signit Signals Intelligence. The interception and analysis of electromagnetic signals.
SIS   Secret Intelligence Service
Sonata Future U.S. naval warfare.
Spiders Software that traverses the World Wide Web to locate new files and examine and record the content of the files.
TACC   Tactical Air Control Center
TEMPEST Military code name for activities related to van Eck monitoring, and technology to defend against such monitoring.
TMD   Theater Missile Defense
TOT   Time on Target
TRADOC U.S. Army Training and Doctrine Command
Trapdoor A hidden software mechanism triggered to circumvent system security measures. This can be a legitimate programming technique that allows a

developer to bypass lengthy log-on routines or access source code directly. Its existence, if known by unauthorized persons, however, can be the source of a significant security breach.

Trashing   Hacker term for physically searching garbage for useful information about the target site, such as manuals, telephone number passwords, proprietary information, internal memos and so forth.

Trojan horse   Malicious computer code that is located within a desirable block of code (e.g., an application program or operating system software). To be a Trojan horse, the presence of the code must be unknown and it must perform an act that is not expected by the owner of the system.

UAV   Unmanned Aerial Vehicle

USIA   United States Information Agency

van Eck Monitoring   Monitoring the activity of a computer or other electronic equipment by detecting low levels of electromagnetic emissions from the device. Named after Dr. Wim van Eck, who published on the topic in 1985.

VDU   Visual Display Unit

WBOM   War by Other Means

Notes

Introduction
Page
13 *"It's all part of information warfare"*: Interview, March 12, 1996.
16 *"First is the continuing"*: Statement before the Senate Select Intelligence Committee, February 5, 1997.
17 *Making full use of today's*: John Arquilla and David Ronfeldt, "Information, Power and Grand Strategy," draft paper July 1995, p. 19.

Chapter 1: War in the Infosphere
Page
23 *For two decades now*: Roger C. Molander et al., *Strategic Information Warfare: A New Face of War* (NDRI/Rand, Santa Monica, Calif., 1996).
24 *At any one time*: David Schneider et al., "Power Plays—Economic Implications of the Power Shortage in China," *The China Business Review,* November-December 1993, quoted in Patrick L. Clawson, ed., *Energy and National Security in the 21st Century* (Institute for National Strategic Studies, October 1995).
24 *But the country also needed*: Ibid.
29 *Naval air forces had in-flight refueling capability*: Andrew F. Krepinevich Jr., *The Conflict Environment of 2016: A Scenario-Based Approach* (Center for Strategic and Budgetary Assessments, Washington, D.C., 1996).
33 *The jewel in the hydroelectric crown*: Clawson, *Energy and National Security.*

Chapter 2: A Desert Myth
Page
36 *At the outset, his intentions*: Colin Powell, *My American Journey* (New York: Random House, 1995), pp. 459–62.
36 *"Wars should be the politics"*: Ibid., p. 148.
36 *Be sure to win political*: U.S. News & World Report, April 14, 1997, p. 9.
37 *During a routine war-fighting exercise*: Rick Atkinson, *Crusade* (New York: Houghton Mifflin, 1993).

38 *Immediately after the invasion of Kuwait:* Interview with British intelligence official, September 10, 1996; the staff of *U.S. News & World Report*, *Triumph Without Victory* (New York: Times Books, 1992), pp. 224–25.

39 *"It was a very frustrating":* Interview, May 8, 1997.

39 *Yet it became swiftly apparent:* Alan D. Campen et al., eds., *The First Information War* (Fairfax, Va.: AFCEA International Press, 1992).

40 *"Much of what they did":* Ibid., p. xi.

41 *"In the first 90 days":* Vincent Kiernan, "Cooper Lifts Veil of Secrecy to Applaud DSP," *Space News,* April 1–7, 1991.

41 *The 40,000 strong British force:* The Sunday Times, January 27, 1991.

41 *"The air war was an operation":* Gen. Sir Peter de la Billiere, *Storm Command: A Personal Account of the Gulf War* (London: HarperCollins, 1992), p. 205.

42 *Victory often shrouds doubts and failures:* John Paul Hyde, Johann W. Pfeiffer and Toby C. Logan, "CAFMS Goes to War," in Campen, *The First Information War.*

43 *It was the secret batch:* Atkinson, *Crusade.*

44 *Following the first wave of Stealth and Tomahawk attacks:* Ibid.

46 *"The Iraqis were good":* Robert S. Hopkins III, "Ears of the Storm," in Campen, *The First Information War.*

46 *A second amphibious rehearsal:* The staff of *U.S. News & World Report, Triumph Without Victory,* p. 170.

46 *"We told no lies":* Interview, September 21, 1996.

47 *Two other teams successfully carried out:* Atkinson, *Crusade,* pp. 370–71, 386–91.

47 *"Three members of Bravo Two Zero":* Gen. Sir Peter de la Billiere, *Looking for Trouble* (London: HarperCollins, 1994), p. 411.

47 *That story was eventually told:* De la Billiere, *Storm Command,* pp. 235–49; Andy McNab, *Bravo Two Zero* (London: Bantam, 1993); Chris Ryan, *The One That Got Away* (London: Century, 1995).

48 *"If we tried that":* Interview, December 12, 1996.

48 *They removed a stretch:* De la Billiere, *Storm Command,* pp. 235–49; McNab, *Bravo Two Zero.*

50 *It is also testament to the degree:* Atkinson, *Crusade,* pp. 336–37.

Chapter 3: The Challenge of the Chip
Page

52 *location that has come to be called cyberspace:* The term cyberspace is used in a multitude of ways these days, from meaning anything to do with computers and the Internet to, in the military sense, that part of the battlefield which requires the use of information technology–based weapons. The actual word "cyberspace" was coined by the futurist thinker and writer William Gibson in his novel *Neuromancer* (New York: Ace Science Fiction, 1984). Gibson used it to describe the shadow world of data and artificial intelligence where a new breed of criminal mischief maker could play havoc by breaking into information systems and stealing data. Little wonder that this is one of the seminal works in the hacker/cracker community.

55 *"The Information Age has dawned"*: Gen. Colin L. Powell, "Information-Age Warriors," *Byte,* July 1992, p. 370.

55 *Absolutely critical in this process was a concept paper:* Gen. Glenn Otis, Coleman Research Corp., and Dr. Peter Cherry, Vector Research Inc., *Information Campaigns* (Vector Research Inc., Ann Arbor, Mich., 1991).

56 *In fact, the Soviets discussed the issue first:* Interview with Fred Giessler, NDU, September 5, 1997.

56 *They can be summed up:* Andrew Krepinevich, "Cavalry to Computer," *The National Interest,* Fall 1994.

57 *Now it commanded less than 1 percent:* Oliver Morton, "The Software Revolution: The Information Advantage," *The Economist,* June 10, 1995.

57 *"A downsized force":* Powell, "Information-Age Warriors," p. 370.

57 *The day of the million-dollar toilet seat:* New equipment destined solely for the armed services necessarily has a limited production run, so when the total cost of designing and building a new plane is divided by the number of planes built, grotesque prices for prosaic items emerge, giving critics and comedians plenty of material to use against the defense establishment. The million-dollar toilet seat was just such an item.

57 *computer hacker:* A hacker is someone who uses computers and telephone lines to break into other people's computers for the fun and mischief derived therefrom. The ultimate intent is neither malicious nor particularly defined. The process is what matters. To a cracker, a name for a malicious hacker, the process is the means toward a definite end, which could range from data theft to manipulating a third-party computer to do something different from what its owner wants.

57 *"actions taken to achieve":* Office of the Chairman of the Joint Chiefs of Staff, *Joint Doctrine for Command and Control Warfare,* quoted in John I. Alger, "Declaring Information War," *International Defense Review* (Jane's Information Group, London, July 1996).

58 *C2W was now defined:* Office of the Chairman of the Joint Chiefs of Staff, *Memorandum of Policy Number 30,* quoted in Alger, "Declaring Information War."

58 *"Coming to grips":* Martin C. Libicki, *What Is Information Warfare?* (Washington, D.C.: National Defense University, August 1995).

Chapter 4: The Gang That Couldn't Shoot Straight
Page

61 *Twelve aircraft:* Kenneth Allard, *Somalia Operations: Lessons Learned* (Washington, D.C.: National Defense University, 1995).

61 *By the time the TV crews arrived:* Ibid.

62 *"the major air and sea ports":* Ibid.

63 *Thus did the world see:* James Adams, "The Role of the Media," in *Cyberwar: Security, Strategy and Conflict in the Information Age,* Alan Campen et al., eds. (Fairfax, Va.: AFCEA International Press, May 1996), p. 107.

63 *More food and supplies:* Allard, *Somalia Operations.*

63 *But the U.N. action:* Ibid.

64 *On June 5, 1993:* Rick Atkinson, *Washington Post,* December 8, 1993.

64 *He strongly denied responsibility:* Patrick J. Sloyan, *Newsday,* December 5, 1993.

64 *"We didn't plan to kill him":* Ibid.

64 *"We cannot have a situation":* Ibid.

65 *Caustic Brimstone evolved into Gothic Serpent:* Rick Atkinson, *Washington Post,* January 30, 1994.

65 *The British politely refused:* Sloyan, *Newsday,* December 5, 1993.

65 *The SAS assessment:* Interview with SAS source, July 5, 1996.

66 *"We had to do something":* *Newsday,* December 5, 1993.

66 *"We were going to set":* Ibid.

66 *In Mogadishu, a CIA cell:* Atkinson, *Washington Post,* January 30, 1994.

66 *The chief Somali agent:* Ibid.

67 *"They never came close":* Patrick J. Sloyan, *Newsday,* December 6, 1993.

67 *"We look like the gang":* Ibid.

67 *These were the tactics:* Ibid.

67 *Carter stressed to Clinton:* Ibid.

68 *It was clear to Hoar:* Ibid.

68 *"That's really Indian country.":* Atkinson, *Washington Post,* January 30, 1994.

68 *The target was actually a block away:* Ibid.

68 *"We're ready to get out of Dodge":* Atkinson, *Washington Post,* January 30, 1994.

70 *"What's wrong?":* Douglas Waller, *Washington Post,* May 15, 1994.

70 *Gordon collected them:* Citation for the award of the Medal of Honor to M. Sgt. Gary I. Gordon and Sfc. Randall D. Shugart.

71 *"It's not part of our mindset":* Rick Atkinson, *Washington Post,* January 31, 1994.

72 *"By Saturday and Sunday":* Interview, May 23, 1997.

72 *"The thing that haunts me":* *Atlanta Journal and Constitution,* March 29, 1994.

73 *"He didn't know what to say":* Patrick J. Sloyan, *Newsday,* December 8, 1993.

73 *"I don't want to end up":* Ibid.

73 *Others deny that:* Ibid.

73 *"The blame for my son's death":* James Adams, *The Sunday Times,* May 29, 1994. The source for this story, which was avoided by mainstream American newspapers and TV, was a senior White House official who was present at the confrontation.

74 *He claimed to be surprised and angry:* Art Pine, *Los Angeles Times,* May 15, 1994.

74 *given that policy was shifting:* Michael R. Gordon, *New York Times,* May 13, 1994.

74 *The families later said:* Associated Press, May 14, 1994.

74 *"Cardinal rules were violated":* *Newsday,* December 8, 1993.

74 *"were betrayed by an administration":* *USA Today,* May 19, 1994.

Chapter 5: Riding the Tiger
Page

76 *The commission's report of February 28, 1994:* Joint Security Commission, *Redefining Security,* February 28, 1994.

78 *Close to 70 million calls:* Ibid.

80 *"The way the Pentagon":* Interview, September 6, 1996.

Chapter 6: Trial by Strength
Page

81 *The island had a decades-long reputation: Encarta 96,* Microsoft Corp.

81 *There were rumors that:* Interviews with CIA sources, 1996.

82 *The rush became a flood: Time,* July 18, 1994.

82 *"I would give them": New York Times,* May 27, 1992.

82 *The President was warned:* Interview with National Security Council official, September 19, 1997.

82 *"Those who leave Haiti":* Associated Press, January 14, 1993.

83 *To compensate perhaps for the mixed signals: New York Times,* September 16, 1994.

83 *"They are slowly turning":* Associated Press, July 13, 1994.

83 *"Your time is up.": New York Times,* September 16, 1994.

84 *One step up from boosting:* Bill Gertz, *Washington Times,* September 13, 1994.

84 *The leaflets read:* Ibid.

84 *Although Carter did not know it:* Interview with Pentagon official, September 5, 1997.

85 *On Sunday afternoon:* Judy Keen, *USA Today,* September 21, 1994.

85 *Carter, in an interview with CNN: Showdown in Haiti,* CNN, September 19, 1994.

86 *"Nobody has ever done that":* Interview, March 12, 1995.

86 *"Intelligence is a dirty word":* Mark Urban, *UK Eyes Alpha: The Inside Story of British Intelligence* (London: Faber and Faber, 1996), p. 214.

87 *Whatever nuggets they could extract: U.S. News & World Report,* April 15, 1996, p. 47.

87 *Within six months of the new Serbian government:* Interview with senior American intelligence official, April 1996.

88 *"Since these were French":* Ibid.

88 *"Every time you transmit": Washington Times,* January 22, 1996.

88 *A brigade-sized unit: Washington Post,* January 19, 1996.

88 *"see a thousand meters": Defense Daily,* January 14, 1997.

89 *"This is gold": New York Times,* January 19, 1996. Quoted on condition of anonymity.

89 *"We called the embassy":* Maj. Gen. Michael V. Hayden, *Integrating and Conducting Info Operations: Structuring the AIA Vision,* Air Intelligence Agency internal document.

89 *At the same time:* Interview with National Security Council official, September 19, 1997; interview with British intelligence source, January 12, 1997.

90 *The signal was piped onto the Internet: Washington Post,* January 2, 1997.

90 *"Would the station run": Washington Post National Weekly Edition,* June 9, 1997.

90 *It proved to be a major blow: Washington Post,* October 8, 1997.

91 *The report said that:* International Institute for Strategic Studies, *Strategic Survey 1996/97,* London 1997.

91 *The report went over:* General Accounting Office, *United Nations: Limitations in Leading Missions Requiring Force to Restore Peace* (Washington, D.C., March 27, 1997).

Chapter 7: The March of the Revolutionaries
Page

93 *"Know the enemy and know yourself":* Sun Tzu, *The Art of War,* ed., James Clavell (New York: Delacorte, 1983), p. 18.

93 *Martin Van Creveld:* Carl von Clausewitz, *On War,* eds., Michael Howard and Peter Paret (Princeton: Princeton University Press, 1984).

94 *Guerrilla warfare also falls outside:* Oliver Morton, "The Software Revolution: The Information Advantage," *The Economist,* June 10, 1995.

94 *It did not come to blows:* "Blood Feud," *Fortune,* April 14, 1997.

94 *Thomas Homer-Dixon:* Thomas Homer-Dixon, "On the Threshold: Environmental Changes as Causes of Acute Conflict," in *International Security,* Fall 1991.

95 *"I am not quoting Clausewitz":* Gen. Frederick M. Franks, Jr., "Winning the Information War," Association of the U.S. Army Symposium, Orlando, Florida, May 15, 1994.

95 *"What is my mission?":* Ibid.

96 *"stand in the way":* Ibid.

96 *"The ability to move information rapidly":* TRADOC 525-5, Department of Defense, Washington, D.C., August 1994.

97 *Michael L. Brown, a former Army officer:* Michael L. Brown, *The Revolution in Military Affairs: The Information Dimension* (Fairfax, Va.: AFCEA International Press, 1996).

98 *Their point was a none too subtle:* Stacey Evers, "Stopping the Hacking of Cyber Information," *Jane's Defence Weekly,* April 10, 1996.

98 *"Technology is not simply":* Aerospace Daily, March 20, 1996, p. 425.

98 *"We have crossed":* Sonata, U.S. Navy, 1994.

99 *"the integrated use of operations security":* Copernicus . . . Forward, Department of Defense, Washington, D.C., June 1995.

99 *The* New World Vistas *document:* Air Force Scientific Advisory Board, *New World Vistas,* December 15, 1995.

100 *"The Force XXI Army":* Department of Defense, Washington, D.C., *Concept for Information Operations,* TRADOC Pamphlet 525-69, August 1, 1995.

100 *a doctrine for information warfare:* On December 10, 1996, the Pentagon officially switched from using the term Information Warfare to Information Operations. This is more than semantics; it represents a significant step

back from the concept of IW as a new and separate kind of war, instead of confirming IW as a subsection of conventional warfare. For the purposes of this book, I will continue to refer to IW because I think the Pentagon was wrong to relegate IW in this way. Clearly it is a much bigger issue than they are prepared to see.

100 *"This is an Army"*: Interview, December 12, 1996.

100 *"The focus of our current"*: Andrew Krepinevich, testimony to Senate Armed Services Committee hearings, 1997.

Chapter 8: From Double Tap to Double Click

Page

109 *"This is the first weapons system"*: Interview, November 11, 1996.

110 *"In future land battles"*: Ibid.

111 *The next stage will involve: The Times*, June 29, 1997.

112 *"If I know where you are"*: Interview, December 16, 1996.

112 *one general called "the greatest revolution"*: Quoted in interview with Neil Siegel, Director of Army Systems at TRW, Data Technologies Division.

113 *the 1st Brigade: USA Today*, March 6, 1997.

113 *The whiz kid programmer: C4I News*, Philips Business Information Inc., April 10, 1997.

113 *In all, in its role as EXFOR: Washington Post*, March 31, 1997.

113 *At that stage, it was apparent:* "Ready! Aim! Boot!," *U.S. News & World Report*, January 20, 1997.

114 *"They know where all our assets are"*: Ibid.

114 *But if the UAVs delivered the intelligence:* Ibid.

114 *"The first sergeant from Bravo Company"*: *USA Today*, March 6, 1997.

114 *"Applique is by far a definite keep"*: *Defense Daily*, March 28, 1997.

114 *His Applique set broke down:* E-mail from officer at Fort Irwin to military Usenet group, March 20, 1997.

115 *"It was a digital traffic jam"*: Ibid.

115 *"We're still in the crawl and walk phase"*: *Washington Post*, March 31, 1997.

115 *EXFOR in turn was judged:* Ibid.

115 *"immature and unreliable"*: *Army Times*, May 26, 1997.

115 *"an outstanding job"*: Statement by Philip E. Coyle III, Director, Office of Operational Test and Evaluation.

115 *"I've got 2,400 soldiers"*: *Washington Post*, March 31, 1997.

115 *"2,500 soldier crick in the neck"*: *Defense Daily*, March 31, 1977.

116 *"I've gone from double tap to double click"*: *Navy Times* (USMC Edition), March 24, 1997.

116 *He suggested his own plan: San Diego Union-Tribune*, March 17, 1997.

117 *"It was essentially a concept"*: Interview, March 1997.

117 *"This is exciting stuff"*: *San Diego Union-Tribune*, March 17, 1997.

117 *The same data, combined with intelligence:* In a real action the SPMAGTF command center would have been aboard a ship such as the USS *Coronado*. The duplication in command centers was due to the differing agenda of the Fleet Battle Alpha exercise.

117 *"This is going to allow us"*: Los Angeles Times, March 17, 1997.

118 *"I think the temptation to look inward"*: Inside the Pentagon, March 13, 1997.

118 *"They would get facts."*: Ibid.

119 *"They were surrogates to test the concept"*: Copley News Service, April 3, 1997.

119 *"Teamwork you can build."*: Los Angeles Times, March 17, 1997.

Chapter 9: Fly on the Wall: Weapons and Wasps
Page

123 *Ho's invention is capable*: Interview, December 13, 1996.

123 *"This is the first time"*: The Sunday Times, November 20, 1994.

124 *"We're sketching out the scenario"*: Interview, June 11, 1996.

124 *"When you approach technical people"*: Georgia Institute of Technology Research News, February 6, 1997.

124 *"It seems essential"*: Philadelphia Inquirer, March 27, 1997.

124 *To demonstrate that microsize is achievable*: Georgia Institute of Technology Research News, February 6, 1997.

125 *"We may have to learn"*: Ibid.

126 *"This is the first totally new"*: Washington Post, June 23, 1996.

126 *"We basically unplugged the nervous system"*: Virginian-Pilot, August 13, 1996.

126 *The computers replaced fifty sailors*: Ibid.

126 *For the purposes of the exercise*: Aerospace Daily, March 11, 1997.

126 *"The whole speed of battle"*: Defense News, April 14–20, 1997, p. 8.

127 *The Pentagon is thinking seriously*: Navy News and Undersea Technology, August 26, 1996.

127 *In a trenchant attack*: Retired Officer, November 1996.

128 *The Air Force has been scrambling*: Aviation Week and Space Technology, January 6, 1997.

128 *The system can be used*: Interviews with senior Northrop Grumman officials, December 1996.

129 *He says the market for them*: U.S. Commercial and Military Unmanned Aerial Vehicle Markets, Frost and Sullivan report, 1996.

129 *"The rapid, accurate intelligence"*: Defense News, August 12–18, 1996.

129 *In the civilian sector that could include*: Baker Spring, Heritage Foundation, in ibid.

129 *The Army is to place*: Defense Daily, August 12, 1996, p. 234.

130 *"We want to take the human"*: Ibid.

130 *These mine-eating bugs*: European Stars and Stripes, October 23, 1996.

130 *"The future of air combat"*: Defense News, August 12–18, 1996.

130 *"Far-out ideas"*: Speech to Electronic Industries Association meeting, October 17, 1996, quoted in Defense News, November 11–17, 1996, p. 4.

131 *The little machines, themselves disguised*: San Diego Union-Tribune, November 20, 1996.

131 *New generations of these triggers*: Business Week, April 26, 1993, p. 92.

131 *have brought the cost of activating*: New York Times, January 27, 1997.

132 *With intellectual and financial leadership*: Business Week, April 23, 1993, p. 92.

132 *A 1992 study by the Rand Corporation*: Keith Brendley and Randall Steeb, *Military Applications of Microelectromechanical Systems* (Santa Monica, Calif.: Rand Corp., 1992).

132 *"We clearly have much more"*: Interview, June 11, 1996.

133 *By adopting the same technology*: Business Week, April 26, 1993, p. 92.

133 *Many of these devices are not*: Descriptions of these devices, cost estimates and savings were derived from the report *MEMS—DOD Dual Use Technology Assessment*, Department of Defense, Washington, D.C., December 1995.

Chapter 10: Set Tennis Balls to Stun
Page

138 *With twenty-three ships of the U.N. Task Force*: Los Angeles Times, March 8, 1995.

139 *nonlethal weapons (NLW)*: The phrase "nonlethal weapons" is as enticing as it is potentially misleading. Enticing because to many it conjures up a romantic vision of the soldier removed from the deadly and brutal practice of war, that war can somehow be made bloodless. That just is not going to happen and no soldier or commander will ever permit the use of deadly force to be supplanted by nonlethal weapons. To do so would put the life of the soldier at unnecessary risk. The phrase is also misleading because few of the weapons described as nonlethal are actually guaranteed not to kill. Ask members of the British army who have fired rubber bullets in Northern Ireland. These weapons are supposed to quell riots with nonlethal force, but on several occasions rioters have died after being struck in the head by them. Alternatives for the phrase "nonlethal weapons" have therefore been tried out, including Less-Than-Lethal, Tunable Lethality, Less-Lethal, Limited Effects Technologies, Soft Kill and even Pre-Lethal. Despite this range of options and despite the stated shortcomings of "nonlethal" as a completely accurate descriptor, NLW remains the convention. The closest anyone has come to defining it officially was Charles Swett, the Assistant Secretary of Defense for Special Operations and Low Intensity Conflict, who described NLW as "Discriminating weapons that are explicitly designed and employed so as to incapacitate personnel or materiel while minimizing facilities and undesired damage to property and the environment" (Quoted in Xavier K. Maruyama, *Technologies in Support of International Peace Operations* (Monterey, Calif.: Naval Postgraduate School, 1996).).

139 *to take to Somalia*: Gary Anderson, "The Whys and Hows of Non-Lethal Military Force," Washington Times, November 30, 1995. Anderson is a Marine Corps officer who directed the Marine Corps Experimental Unit during the planning and implementation of Operation United Shield.

140 *"Even before this there was"*: Interview, December 3, 1996.

140 *The relative absence of the mobs*: Anderson, "The Whys and Hows."

140 *"It is generally accepted"*: Maj. Gen. Edward G. Anderson, testimony to

Senate Armed Services Subcommittee on Nonlethal Technologies, March 15, 1996.

141 *The list of items to be taken into Somalia*: Transcript by Federal News Service of Defense Department background briefing on nonlethal weapons, February 17, 1995.

141 *The Somali militia drove their vehicles*: Maruyama, *Technologies in Support of International Peace Operations.*

142 *the Laser Dazzler*: Scott Gourley, "Less Than Lethal Weapons," *Jane's Defence Weekly*, July 17, 1996.

142 *The bright red splash of light*: Ibid.

142 *"With non-lethal weapons"*: Gen. Tony Zinni, USMC *Proceedings*, U.S. Naval Institute, December 1996.

143 *"The first is never to put"*: Anderson, "The Whys and Hows."

144 *Would it not be desirable*: Interview, December 3, 1996.

144 *Sleep agents*: Maruyama, *Technologies in Support of International Peace Operations.*

144 *Strobe lights*: Harold Levie, Lawrence Livermore National Laboratory, "Disorienting Pulsed Light," presented at "Less-Than-Lethal Program Review and Technology Demonstration," sponsored by the National Institute of Justice, Rockville, Md., March 13, 1995.

144 *Dazzling lights and lasers*: Richard Kokoski, "Non-Lethal Weapons: A Case Study of New Technology Developments," in *SIPRI Yearbook, 1994* (Stockholm International Peace Research Institute, Oxford University Press, 1994).

144 *Liquid stun gun*: Maruyama, *Technologies in Support of International Peace Operations.* Maruyama refers to Product Description REM-31912 of the Jaycor Company, San Diego, Calif., January 31, 1995.

145 *Possibilities begin to open up*: Interview with Jon Becker of Aardvark Tactical Inc., Arcadia, Calif., October 1996. Aardvark Tactical is a regular supplier of NLW technologies to law enforcement and also helped in the planning and equipping of Operation United Shield.

145 *Acoustic guns capable of shooting*: Ibid.

145 *"You can make someone's guts shake"*: Ibid.

146 *Los Alamos National Laboratory has proposed*: White Papers for Non-Lethal Initiatives, DOD Programs Office, Los Alamos National Laboratory, April 1994.

146 *"We saw a ten hertz"*: *Defense Electronics*, March 1994, p. 12.

146 *"The work was really outstanding"*: *U.S. News & World Report*, July 7, 1997.

147 *The spray contains metals*: Maruyama, *Technologies in Support of International Peace Operations.*

147 *Superacids also achieve*: Thomas A. O'Donnell, *Superacids and Acidic Metals as Inorganic Chemical Reaction Media* (New York: VCH Publishers, 1992).

147 *Superacids can also eat rubber*: White Papers for Non-Lethal Initiatives, DOD Programs Office, Los Alamos National Laboratory, April 1994.

147 *Weapons firing very fine dust*: Richard L. Garwin, "New Applications of Non-Lethal and Less-Than-Lethal Technology" at the American Assembly

Book Conference on U.S. Intervention in the Post–Cold War World: New Challenges and New Resources, April 7–10, 1994, Arden House, Harriman, N.Y.

147 *Superlubricants known as "slickums"*: Donald L. Schmidt et al., "Water-Based Non-Stick Hydrophobic Coatings," *Nature*, Vol. 368, March 3, 1994.

147 *"Stickums" have the opposite effect*: Maruyama, *Technologies in Support of International Peace Operations*.

148 *Surprisingly, given the relatively benign nature*: Ibid. Maruyama cites several works as relevant: Steven Aftergood, "The Soft-Kill Fallacy," *Bulletin of the Atomic Scientists*, September-October 1994; Kokoski, "Non-Lethal Weapons"; Garwin, "New Applications of Non-Lethal and Less-Than-Lethal Technology."

148 *Anti-personnel NLW may also*: The Biological Weapons Convention, the Chemical Weapons Convention, the Geneva Protocol and the Certain Conventional Weapons Convention, also known as the Inhumane Weapons Convention. As cited by Maruyama, ibid.

148 *This has obliged the Clinton administration*: Letter from President Bill Clinton to the Senate, June 23, 1994.

148 *For instance, the International Red Cross*: International Committee of the Red Cross, *Blinding Weapons: Reports of the Meeting of Experts Convened by the International Committee of the Red Cross on Battlefield Laser Weapons, 1989–1991*, Geneva, 1993.

148 *"People had a lot of trouble"*: Interview with Jon Becker, October 1996.

149 *As one horrific incident*: Incident reported in "Killing Them Softly: New Electronic Warfare Applications," *Journal of Electronic Defense*, which cited "Case Study 1—The Forrestal Incident," Compliance Engineering 1993.

149 *If a nuclear EMP was generally*: Ibid.

150 *Scientists testing the creation*: D. A. Fulghum, "EMP Weapons Lead Race for Non-Lethal Technology," *Aviation Week*, May 24, 1993.

150 *"As far as we knew"*: Interview, April 20, 1996.

151 *"If validated, Beer Cans"*: Interview, January 15, 1997.

151 *"What we seem to have here"*: Ibid.

151 *"Dark Tangent"*: Forbes ASAP, June 3, 1996, p. 96.

151 *As Xavier Maruyama argues*: Maruyama, *Technologies in Support of International Peace Operations*.

151 *the 1977 Geneva Protocol stipulates*: Article I-57.2.a.ii, Geneva Protocol I.

152 *This technology is called*: Maruyama, *Technologies in Support of International Peace Operations*.

153 *On one terrible occasion in Phoenix*: Interview with senior Justice Department official, December 1996.

154 *All were agreed that committing forces*: Described to the author by the class, June 25, 1997.

Chapter 11: The Wrong Hands

Page

156 *Petersen was later arrested:* Los Angeles Times, November 28, 1995.

157 *"The Feds turned Petersen":* Information Week, March 13, 1995, p. 12.

157 *"What a loser":* Los Angeles Times, July 31, 1994.

157 *Either way, he siphoned $150,000:* Los Angeles Times, November 28, 1995.

157 *"[Using Petersen] was about as dangerous":* Information Week, March 13, 1995, p. 12.

157 *"Very dangerous":* Phrack, 1993, quoted in Los Angeles Times, July 31, 1994.

157 *"Most hackers would have sold":* Ibid.

158 *"To be physically inside":* New Orleans Times-Picayune, September 15, 1996.

158 *Hippie leader Abbie Hoffman:* Richard Power, Current and Future Danger: A CSI Primer on Computer Crime and Information Warfare (San Francisco: Computer Security Institute. 1996).

158 *He discovered that a toy whistle:* Ibid.

160 *"Knowledgeable hackers can provide":* Ibid.

160 *The hackers had been hacked:* Interview with intelligence source, July 1, 1996.

160 *He was picked up in 1987:* Intelligence Newsletter, October 12, 1995.

161 *By 1995 he had become fully legitimate:* Ibid.

161 *"Whoever you are hiring":* Power, Current and Future Danger.

161 *"If you have a billing dispute":* Ibid., p. 13.

162 *"The concept of Blacknet is real":* Information Week, August 28, 1995, p. 30.

162 *"We're concerned about the overall issue":* Information Week, August 28, 1995, p. 30.

162 *Inevitably, turf wars were fought:* Michelle Slatalla and Joshua Quittner, Masters of Deception: The Gang That Ruled Cyberspace (New York: HarperCollins, 1995). Indispensable reading for those wishing to achieve a deeper understanding about the hacker culture and the hacker gang wars.

164 *Their phone line was taken over:* Phoenix Gazette, December 12, 1994.

165 *The Maryland Police and Correctional Training Commission:* U.S. News & World Report, April 21, 1997, p. 25.

165 *"The Internet has quadrupled":* New York Times Magazine, February 25, 1996.

165 *The Cyber Nazi Group:* CNG Web site.

165 *Entry-level hacking was much easier:* "Computer Hacker Information Available on the Internet," GAO testimony to Permanent Subcommittee on Investigations, Committee on Governmental Affairs, June 5, 1996.

165 *"This is the sector":* NBC News, April 18, 1996.

166 *He claimed it contravened:* U.S. News & World Report, April 21, 1997, p. 25.

167 *"Are people learning":* Reuters, April 17, 1996.

167 *"This scenario of a nuclear"*: Keynote address to the Conference on Terrorism, Weapons of Mass Destruction, and U.S. Strategy, the University of Georgia, Athens, Georgia, April 28, 1997.
167 *Welfare fraud allegedly netted*: Associated Press, May 24, 1997.
168 *Within days, the respected teacher*: Ibid.
168 *Needless to say, the university*: Associated Press, May 26, 1997.
168 *It was, as the program title indicated*: Steve Emerson, producer of the PBS documentary "Jihad in America," quoted by the Associated Press, May 24, 1997.
169 *The former Bulgarian spy agency*: Scott Charney, Computer Crime Unit, Department of Justice, *Computer Crime*, January 24, 1996.
169 *Cyberterrorists share access codes*: *Washington Times*, May 14, 1995.
169 *Anyone sending him an e-mail*: *New York Times*, May 5, 1997.
169 Radikal *was a serious worry*: *New York Times*, June 6, 1997.
169 *"A CyberTerrorist will remotely"*: Barry C. Collin, "The Future of CyberTerrorism: Where the Physical and Virtual Worlds Converge," remarks to Eleventh Annual International Symposium on Criminal Justice Issues, Office of International Criminal Justice, the University of Illinois at Chicago.

Chapter 12: The Back Door's Open
Page
173 *In 1991, a farmer*: *The Economist*, January 13, 1996, p. 77.
173 *In a startling investigation*: Neal Stephenson, *Wired*, December 1996; Internet posting by Peter Leyden, features editor, *Wired*, April 16, 1997.
173 *The third incident happened*: *San Francisco Chronicle*, May 14, 1997.
174 *An attack of this magnitude*: *USA Today*, July 2, 1996.
174 *These were the alleged masterminds*: *St. Petersburg Press*, October 3, 1995.
175 *According to FBI testimony*: *USA Today*, July 2, 1996.
175 *"There's a huge disincentive"*: Ibid.
175 *An expert on banking*: Interview, November 19, 1996.
175 *"It appears that the Soviets"*: Ibid.
175 *In 1994 the New York Stock Exchange*: Agence France-Presse, July 18, 1996.
176 *What jolted the administration*: Molander et al., *Strategic Information Warfare*.
178 *Throughout the government*: *Security in Cyberspace*, Minority Staff statement, U.S. Senate Permanent Subcommittee on Investigations, Hearings on Security in Cyberspace, June 5, 1996, p. 19.
178 *"I have seen many people"*: Interview, September 13, 1996.
179 *The staff found this vague attitude*: *Security in Cyberspace*, p. 19.
179 *The Inspector General stated*: Ibid., p. 22.
179 *The extent to which the government*: *Orange County Register*, May 13, 1995.
179 *One agency assembled*: *Security in Cyberspace*, pp. 22, 27.
179 *"Don't wait for the intelligence community"*: Interview, September 13, 1996.

179 *This despite the fact*: Testimony before Congress, June 25, 1996.
180 *"the President shall submit"*: Kyl Amendment to Intelligence Authorization Bill for FY1997 (Sec. 1053).
180 *"Usually they can just pull"*: *Security in Cyberspace*, p. 28.
180 *"To date, Congress has not"*: House Committee on National Security, report on H.R. 3230, the National Defense Authorization Act for FY1997.
181 *"Vulnerability is all over"*: *Computerworld*, July 22, 1996, p. 29.
181 *"Many of them recognize"*: *USA Today*, November 21, 1996.
182 *"The objective of warfare"*: Report of the Defense Science Board Task Force on Information Warfare, January 1997.
182 *"Thus does war follow"*: Martin Libicki, *Defending Cyberspace and Other Metaphors* (Washington, D.C.: National Defense University, February 1997), p. 3.
182 *"There is a need"*: Report of the Defense Science Board Task Force on Information Warfare, January 1997.
182 *"We couldn't find"*: Reuters, June 19, 1997.
183 *"We are entering"*: Interview, September 26, 1997.
183 *Critical infrastructures underpin*: Critical Foundations, Protecting America's Infrastructures, The Report of the President's Commission on Critical Infrastructure Protection, Washington, D.C., October 1997, p. 24.
186 *"For example, there"*: Interview, November 15, 1997.
186 *"DoD is very interested"*: Statement by John Hamre, Deputy Secretary of Defense, before the Senate Committee on Judiciary, Subcommittee on Technology, Terrorism and Government Information, November 5, 1997.
187 *Eligible Receiver*: Details of the exercise were obtained in interviews conducted in January, February and March, 1998.
188 *"Eligible Receiver was a real shock"*: Interview with the author, January 6, 1998.
188 *In April 1997, Defense Secretary William Cohen*: William H. Cohen, speech to Conference on Terrorism, Weapons of Mass Destruction, and U.S. Strategy, the University of Georgia, Athens, Georgia, April 28, 1997.
188 *"We wondered what would happen"*: Howard Whetzel and Kenneth Allard, "Internet Insecurity May Prove Deadly," *Wall Street Journal*, May 14, 1997.
188 *This is what their terrorist*: Ibid.
189 *"If our democratic freedoms"*: Ibid.
189 *"Technology, combined with"*: Lt. Gen. Patrick Hughes, "Global Threats and Challenges: The Decades Ahead," Statement for the Senate Select Committee on Intelligence, January 28, 1998.

Chapter 13: Venimus, Vidimus, Dolavimus
            (We Came, We Saw, We Hacked)
Page
193 *The startled computer team*: Appendix to Staff Statement, U.S. Senate Permanent Subcommittee on Investigations, May 1996.
194 *The notorious Legion of Doom*: Ibid.

197 *It was the unfamiliarity*: Washington Post, March 30, 1996.

198 *This was the first court-authorized*: CHIPS, July 1996, p. 8.

198 *For two months they watched*: Sworn affidavit of Special Agent Peter Garza, Naval Criminal Investigative Service, March 1996.

198 *"I've infiltrated the U.S. Navy"*: The Dominion (Wellington, New Zealand), April 18, 1996.

199 *"We are using"*: Department of Justice news release, March 29, 1996.

199 *"These Yankees don't have"*: Washington Post, March 30, 1996.

199 *"attacks on Defense computer systems"*: General Accounting Office, Computer Attacks at Department of Defense Pose Increasing Risks, May 1996, p. 4.

200 *"attackers have obtained and corrupted"*: Ibid.

200 *One security company*: Interview with intelligence official, November 17, 1996.

200 *"Defense's policies on information security"*: General Accounting Office, Computer Attacks, p. 34.

201 *I was the first journalist*: The visit took place on April 8, 1997.

202 *Commanders will try to deploy*: Air Force Magazine, March 1997, p. 20.

202 *Planting viruses and worms*: Air Force Week, March 31, 1997, p. 7.

202 *"These viruses would lie dormant"*: Ibid.

202 *"Information is either a place"*: Interview, April 8, 1997.

203 *"AFIWC has 1,000 people."*: Ibid.

203 *"We are at war every day"*: Lt. Col. David Srulowitz, AFCERT Web page, April 1996.

204 *"The technology was not difficult"*: Interview, April 8, 1997.

205 *"I am not an intelligence officer"*: Interview, Kelly AFB, April 8, 1997.

205 *The document* Cornerstones of Information Warfare: U.S. Air Force, Cornerstones of Information Warfare.

206 *"On March 25, 1996"*: Interview, April 8, 1997.

207 *"We haven't decided"*: Interview, October 28, 1996.

208 *"We're here for the protection"*: Interview, October 2, 1996.

208 *"computers, software and monitoring"*: Interview, October 2, 1996.

209 *"More than anything right now"*: Interview with Fred Giessler, NDU, June 1996.

209 *"We are losing the initiative"*: Interview, April 8, 1997.

209 *"At present there is no"*: Interview, October 2, 1996.

210 *"The strategic landscape"*: Defense News, June 18, 1995, p. 1.

210 *"Since our name has been out there"*: Interview, December 1996.

210 *"We are anxious to find out"*: Interview.

212 *French intelligence has hired former hackers*: Sunday Telegraph, March 23, 1997.

Chapter 14: Big Ears and Noddies
Page

213 *But of all the brilliant successes*: Baltimore Sun series on the NSA, December 3–15, 1995.

214 *The NSA wanted Hagelin:* Ibid. The *Baltimore Sun* quoted several sources for this story, including a biography of William Friedman written by Ronald Clark in 1977, and James Bamford's 1982 classic on the NSA, *The Puzzle Palace.*

214 *It was an intelligence gold mine:* Ibid.

215 *"We sent the S-boxes":* Bruce Schneier, *Applied Cryptography* (New York: John Wiley, 1996), p. 301.

216 *Regarding the DES:* M. S. Conn letter to Joe Abernathy, NSA, ser: Q43-111-92, June 10, 1992.

217 *"The common person needs":* Radio interview with Phil Zimmerman by Russell D. Hoffman on *High Tech Today,* WALE, Providence, R.I.

218 *As well as granting Jim Bidzos: Communications Week,* June 10, 1996, p. 8.

218 *He said it would cost $50,000:* Matt Blaze, e-mail to Cypherpunks Group, "Distributed DES Crack," July 1996.

218 *M. J. Wiener of Carleton University:* M. J. Wiener, *Efficient DES Key Search* (Ottawa: School of Computer Science, Carleton University, 1994).

218 *This was the "Chinese Lottery" system:* J. J. Quisquater and Y. G. Desmedt, "Chinese Lottery as an Exhaustive Code Breaking Machine," *Computer,* Vol. 24, No. 11, November 1991, pp. 102–9.

219 *Cracking any other message: PC,* October 25, 1996, p. 29.

219 *"If we do not address this matter":* Interview, May 21, 1997.

220 *Under severe pressure: Fortune,* November 11, 1996, p. 11.

220 *"If they had said":* Interview, October 31, 1996.

221 *Yet the prisons are full:* Address to National Information Systems Security Conference, Baltimore, October 1996.

221 *They paid the price:* Interview with intelligence source, October 1996.

221 *"Most evildoers are like us":* Interview, October 1996.

222 *"After a plane bombing":* Interview, October 24, 1996.

Chapter 15: Puzzles and Mysteries
Page

225 *So much so that she dropped them:* Interview with Foreign Office official, June 15, 1996.

226 *"a toddler soccer game": Security in Cyberspace.*

226 *"The Intelligence Community":* Consortium for the Study of Intelligence. *The Future of U.S. Intelligence,* April 1996.

227 *CIA NOCs in Colombia: Time,* February 20, 1995, p. 30.

228 *At least being an organization: The Guardian,* May 24, 1997.

229 *"Hello, my name is":* Ibid.

Chapter 16: The New Arms Race
Page

233 *"At the end of the 20th century":* Georgiy Smolyan, Vitaliy Tsygichko and Dimitriy Chereshekin, *Konfident,* November-December 1996, pp. 19–21.

234 *"We used to have great drinking sessions":* Interview, March 3, 1997.

235 *The campaign was remarkably successful*: Christopher Andrew and Oleg Gordievsky, *KGB: The Inside Story* (London: Hodder & Stoughton, 1990), pp. 384–85.

235 *For years after the story*: Alvin Snyder, *Warriors of Disinformation* (New York: Arcade, 1995), pp. 104–5.

235 *For instance what the Russians call*: Neil Munro, *The Quick and the Dead* (New York: St. Martin's, 1991), p. 124.

235 *"Reflexive Control is understood"*: Quoted in "Reflexive Control: A Subsystem of Information War," by Fred Giessler, presented to Information-based Warfare course, National Defense University, July 21, 1993.

236 *"[Reflexive Control is] conveying"*: Ibid.

236 *As the new headquarters*: Filipp Bobkov, *KGB and Power* (Moscow: Veesan MP Publishing House, 1995), pp. 82–84; Pete Earley, *Confessions of a Spy: The Real Story of Aldrich Ames* (London: Hodder & Stoughton, 1997), p. 117.

236 *"The CIA's greatest coup."*: *The Sunday Times*, February 4, 1996.

237 *The Russian federal agency*: *The White Book of the Russian Special Services* (Moscow: Obozrevatel Publishing House, 1995), p. 225.

237 *"I agree that we ourselves"*: *Pravda*, December 21, 1996.

239 *A year later, as the information revolution*: Interview, March 4, 1997.

239 *"Now we use all Western equipment"*: Ibid.

239 *"Until the introduction of certifiable means"*: *Segodnia*, September 8, 1995.

239 *"There are those who wish to possess"*: *Obshchaya Gazeta*, May 22–28, 1997, pp. 1–3.

241 *"Along with the maintenance"*: Tass, July 18, 1996.

241 *"the technical level"*: *Journal of Slavic Military Studies*, December 1996, p. 8.

242 *In the Russian context*: *Intelligence Newsletter*, Indigo Publications, Paris, France, 1996.

242 *"The American concept"*: Major M. Boytsov, *Moskoy Sbornik*, No. 10, 1995, pp. 69–73.

243 *Among the weapons that have been developed*: James Adams, *The New Spies* (London: Pimlico, 1995), pp. 270–83.

243 *"We believe that Yeltsin"*: Interview, March 4, 1997.

243 *"the United States is consistently . . ."* Georgiy Smolyan, Vitaliy Tsygichko and Dimitriy Chereshekin (of the Systems Analysis Institute at the Russian Academy of Science). "Perspective: A Weapon That May Be More Dangerous Than a Nuclear Weapon: The Realities of Information Warfare," *Nezavisimoye Voyennoye Obozreniye*, Moscow, November 18, 1995, pp. 1–2.

244 *"We would like an international agreement"*: Interview, March 5, 1997.

Chapter 17: A Mole in the Oval Office
Page

245 *Between 1989 and 1996*: *American Spectator*, December 1996.

245 *Riady changed his name*: *International Herald Tribune*, May 29, 1997.

245 *"an espionage coup"*: Interview, May 22, 1997.

246 *"a very, very big discount"*: *South China Morning Post*, November 7, 1992.

247 *Riady bragged to friends: New York Times,* November 5, 1996.

247 *She made the call: Newsweek,* March 24, 1997, p. 36.

247 *On the guest list: New York Times,* March 10, 1997.

248 *In yet another example: Washington Post,* March 9, 1997.

248 *CIA agents occupy: Baltimore Sun,* November 1, 1996.

249 *"He was at the Chinese embassy": New York Daily News,* June 10, 1997.

249 *"extremely serious and dangerous": Washington Times,* July 1, 1997.

249 *"Was he a spy?":* Interview, June 27, 1997.

249 *And despite the scandal:* Interview, April 22, 1997.

249 *Both are concerned about revealing:* Interview, April 22, 1997.

249 *The final report of: Washington Post,* February 7, 1998.

250 *"There are indications that": New York Times,* February 16, 1998.

250 *At the time of writing:* Interview with the author, February 26, 1998.

251 *"The sanguinary type of war":* Wei Jincheng, "New Form of People's War," *Jiefangjun Bao,* June 25, 1996.

251 *Prodigy signed a joint venture:* Reuters, April 29, 1997.

251 *"The Chinese understand":* Interview with the author, September 12, 1996.

252 *It is doubtful however that circuitry:* Xue Lianfong and Wei Yuejiang, "Digitized Forces Killer Has Come into Being," *Jiefangjun Bao,* April 30, 1996.

252 *"When we engage in war":* Maj. Gen. Wang Pufeng, "Meeting the Challenge of Information Warfare," *Zhongguo Junshi Kexue,* February 20, 1995.

253 *The Chinese even tried: Los Angeles Times,* February 9, 1997.

253 *Film companies and television stations: The Sunday Times,* December 22, 1996. This author has direct experience of this from three Hollywood projects where China was rejected as a part of the plot because of fears of pressure from Beijing.

254 *"Much of the American foreign policy": Washington Post,* March 16, 1994.

254 *Other former high officers:* John B. Judis, "Chinatown: How China Bought the Establishment," *New Republic,* March 10, 1997, p. 17.

254 *Clinton actually did de-couple: Los Angeles Times,* May 27, 1994.

254 *One such was China Yuchai International:* Kenneth Timmerman, "While America Sleeps," *American Spectator,* June 1997, p. 36.

255 *The Pentagon objected to the sale: Newsweek,* March 24, 1997, p. 38.

255 *As soon as they landed in China:* Ibid.

255 *Aronson lost $8.5 million: Insight,* March 24, 1997, p. 8.

255 *"Those who believe":* Zhu Xiaoli and Zhao Xiaozhu. *America, Russia and the Revolution in Military Affairs,* FBI's translation.

255 *"The disposition of the Chinese": Defense News,* June 9–15, 1997, p. 1.

256 *"Anybody who makes the case": Defense News,* June 9, 1997.

256 *"They want to do what":* Ibid.

256 *It is no accident: Chicago Tribune,* February 18, 1996.

256 *"If you know the enemy":* Sun Tzu, *The Art of War,* p. 18.

257 *"hostile forces and reactionary, organizations":* Lo Ping, "Unstable Regions Set Forth," quoted in Nicholas Eftimiades, *Chinese Intelligence Operations* (Annapolis: Naval Institute Press, 1994), p. 13.

Chapter 18: It's the Economy, Stupid
Page
259 *One of the most sclerotic nations*: Paul M. Joyal, *Industrial Espionage Today and Information Wars of Tomorrow* (Washington, D.C.: Integer Security, 1996).
259 *According to Pierre Marion*: *International Herald Tribune*, September 14, 1991. Marion appeared on the NBC program *Expose* on September 13, 1991.
259 *"the economic KGB"*: Bill Gertz, *Washington Times*, February 9, 1992.
259 *The agents were sent home*: *Chemical Marketing Reporter*, July 31, 1995, p. 7.
260 *a plain brown envelope to the CIA*: *The Sunday Times*, April 11, 1993.
260 *to the Knight-Ridder newspaper chain*: *Washington Post*, April 27, 1993.
260 *This memo*: *Washington Times*, June 4, 1993.
260 *"They just lied"*: Interview, April 29, 1993.
260 *"This executive was introduced"*: *Washington Times*, June 4, 1993.
260 *The CIA had sent out*: *Washington Post*, April 27, 1993.
260 *Such was American annoyance*: *Houston Chronicle*, May 5, 1993.
261 *She took him to a series of meetings*: *Washington Post*, February 26, 1995.
261 *The operation was rolled up*: Ibid.
261 *It was a total triumph*: *Los Angeles Times*, October 11, 1995.
262 *They found out what had really happened*: *Washington Post*, February 26, 1995.
262 *compounded by CIA intercepts*: Ibid.
262 *It was set up in the late 1970s*: Oleg Kalugin, remarks at National Information Systems Security Conference, Baltimore, October 1996.
263 *"I believe it is necessary"*: *Interfax*, Moscow, June 26, 1996.
263 *"It's an ominous sign"*: Testimony to Joint Hearing of Senate Intelligence and Judiciary Committees, February 28, 1996.
263 *"Don't look at Russia"*: Oleg Kalugin, remarks at NISSC, Baltimore, October 1996.
263 *Country B, for example*: *Washington Times*, February 22, 1996.
263 *"highly sensitive information"*: *Washington Post*, May 7, 1997.
264 *"The ambassador wants me"*: *Los Angeles Times*, May 18, 1997.
264 *"The story is absolutely baseless"*: *The Sunday Times*, May 11, 1997.
264 *"We are dealing with reality here"*: Interview, July 2, 1997.
264 *"I've read a two inch-thick"*: Interview, May 9, 1997.
265 *The missing equipment had malfunctioned*: *International Herald Tribune*, August 17, 1996.
265 *"There was video"*: *The Sunday Times*, May 11, 1997.
266 *"[The Japanese are] creatures"*: *The Sunday Times*, June 16, 1991.
266 *So it directed all those energies*: Richard Deacon, *Kempei Tai: A History of the Japanese Secret Service* (New York: Beaufort Books, 1983).
267 *In 1984, CIA Director William Casey*: Bob Woodward, *Veil: The Secret Wars of the CIA 1981–1987* (New York: Simon & Schuster, 1987).
267 *So crushing was Hitachi's defeat*: *Nihon Keizai Shimbun*, October 11, 1993.

268 *"We have approximately 800":* Joint Hearing of the Senate Intelligence
Committee and the Terrorism, Technology and Government Information
Subcommittee of the Senate Judiciary Committee, February 28, 1996.

268 *"The visit to the island":* New York Times, May 22, 1995.

269 *He was sent to jail:* International Herald Tribune, July 11, 1996.

269 *In 1995 the computers:* The Sunday Times, August 4, 1996.

269 *"Every two or three years":* Washington Post, June 27, 1993.

270 *There are a number of reasons:* Interview, January 14, 1992.

270 *"People who do it":* Speech to the National Information Systems Security
Conference, Baltimore, Maryland, October, 1996.

Chapter 19: Every Picture Tells a Story
Page

272 *The battle may have been lost:* Financial Times, December 24, 1994.

273 *When it was finally shut down:* Interpress Service, August 4, 1994.

273 *" 'That is your mission' ":* Interview, June 26, 1997.

274 *"These broadcasts were largely responsible":* R. A. Dallaire and B. Poulin,
"UNAMIR—Mission to Rwanda," *Joint Force Quarterly,* Spring 1995,
p. 66.

276 *The story that eventually ran:* The Sunday Times, October 1, 1995.

276 *It was frustrating because video images:* The Times, May 31, 1997.

277 *"We are very concerned":* Interview, August 2, 1996.

278 *In other words, it was an accident:* Los Angeles Times, February 5, 1996.

280 *In the last election:* New Republic, June 21, 1993, p. 30.

280 *"They've got us putting more fuzz":* Boston Globe, October 7, 1993.

282 *According to the Associated Press:* Associated Press, February 6, 1996.

283 *"What they aim at is power":* Financial Times, July 6, 1995.

284 *The most comprehensive review:* Tennessean, September 22, 1995.

287 *for the military to beam positive television:* The military actually possesses
aircraft specifically designed to broadcast TV and radio, or to jam enemy
broadcasts. They are variants on the Lockheed C-130, called EC-130E, or
Commando Solo. But it is one thing to own the technology, quite another
to make it work. "These guys aren't TV producers!" Chuck de Caro says
with no small element of scorn in his voice. "What do they know about
putting out decent TV?" This may be slightly unfair, for in situations such
as Haiti where there is no other way to get a necessary message across, the
six Commando Solo planes provide valuable service. In that case, they
broadcast speeches by the returning President Aristide.

287 *"a cabal of fifth rate Balkan bozos":* Chuck de Caro, SOFT WAR: Sats, Lies
and Video Rape, Aerobureau Corp., 1996.

288 *"Just above junior college":* Interview, November 13, 1996.

288 *"It's lousy TV":* Interview, September 17, 1997.

288 *"Information has the potential":* Interview, November 7, 1996.

289 *"Now that we televise our wars":* John Leo, "Lessons from a Sanitized War,"
U.S. News & World Report, March 18, 1991, p. 26.

Chapter 20: Morality and Megabytes
Page
291 *Before the Second World War:* The author is indebted to *The Nuclear Age* by John Newhouse (London: Michael Joseph, 1989) and *The Nuclear Barons* by Peter Pringle and James Spigelman (London: Michael Joseph, 1981) for background on the atomic debate.
292 *"Everyone was ready.":* Pringle and Spigelman, *The Nuclear Barons.*
293 *He even questioned:* Newhouse, *The Nuclear Age.*
293 *"Winant was beside himself.":* Ibid.
293 *"Heat will be so plentiful":* Ibid.
295 *"You're supposed to tell the difference":* United Press International, August 4, 1995.
295 *Fotion and others noted:* Ibid.
296 *"This new military revolution":* A. J. Bacevich, "Morality and High Technology," *The National Interest,* Fall 1996, p. 37.
296 *"The most important thing a cadet":* Christian Science Monitor, March 1, 1985.
297 *"We gave casualty avoidance":* Bacevich, "Morality and High Technology," p. 37.
298 *"Our strategy was to make warfare":* Charles Dunlap, "How We Lost the High-Tech War of the Future." *The Weekly Standard,* January 29, 1996, p. 22.
298 *"Through the emergence of precision-guided":* Harry C. Stonecipher, "At What Price Peace? Freedom and Security in the 21st Century," speech to MIT Club, Boston, Mass., November 8, 1995.
298 *"We will fight information warfare":* Maj. Ralph Peters, U.S. Army War College, *Parameters,* quoted in *News and Record* (Greensboro, N.C.), July 6, 1997.
299 *"What does the Geneva Convention":* Defense News, December 24, 1995.
299 *"To the world beyond our borders":* Bacevich, "Morality and High Technology," p. 37.
300 *In the seminal document:* Department of the Army, *FM 100-6 Information Operations,* August 1996, p. 2.
301 *"IW should allow us":* "Deterrence in the Information Warfare Age," Peter D. Feaver, paper submitted to the International Studies Association, Toronto, Ontario, March 18–22, 1997, p. 20.
301 *"Here the problem":* Ibid., p. 25.
301 *"The IW capability that allows":* Ibid., p. 27.
302 *"If blowback confuses":* Ibid., p. 29.
302 *Rand Corporation analysis:* Molander et al., *Strategic Information Warfare,* p. 19.
303 *"The concern was that the North Koreans":* Peter Grier, "At War with Sweepers, Sniffers, Trapdoors and Worms," *Air Force Magazine,* March 1997, p. 20.
304 *"The world sees you!":* Walter B. Wriston, "Bits, Bytes and Diplomacy," *Foreign Affairs,* September-October 1997, p. 172.
304 *"A global village will have":* Ibid.

Conclusion
Page
306 *In a compelling and controversial analysis:* Michael Vlahos, "Byte City,"
    Washington Quarterly, Spring 1997, p. 43. "Byte City" was based on the
    chapter by Michael Vlahos in *The Information Revolution and National
    Security: Dimensions and Directions,* ed., Stuart J. D. Schwartzstein (Wash-
    ington, D.C.: CSIS, 1996).
309 *Moore's Law states that computing: Business Week,* June 23, 1997, p. 120.
    Moore's Law is named for Gordon Moore, who in 1965 was head of
    research at Fairchild Semiconductor Corp. His prediction was that the
    number of transistors on a microchip would double each year, from the
    original four in 1961 to the thousands found today. He changed his predic-
    tion in 1971 to a rate of doubling every two years. The actual rate of
    increase since 1965 has been every eighteen months. Moore went on to
    create Intel, the world's preeminent maker of computer microprocessors.
311 *"makes our blood run cold":* Steven and Cokie Roberts, *Salt Lake Tribune,*
    April 5, 1997.
311 *Dunlap wrote an article:* Dunlap, "How We Lost the High-Tech War of
    2007: A Warning from the Future," p. 22.

Bibliography

Adams, James. *The New Spies*. London: Pimlico, 1995.

Alberts, Dr. David S., and Dr. Richard E. Hayes. *Command Arrangements for Peace Operations*. Washington, D.C.: National Defense University, 1995.

Alberts, Dr. David S., and Daniel S. Papp. *Information Age Anthology*. Parts 1–4. Washington, D.C.: National Defense University, 1997.

Allard, Kenneth. *Somalia Operations: Lessons Learned*. Washington, D.C.: National Defense University, 1995.

Anderbert, Bengt, Maj. Gen., and Dr. Myron L. Wolbarsht. *Laser Weapons*. New York: Plenum, 1992.

Andrew, Christopher, and Oleg Gordievsky. *KGB: The Inside Story*. London: Hodder & Stoughton, 1990.

Atkinson, Rick. *Crusade*. New York: Houghton Mifflin, 1993.

Barnaby, Frank. *The Automated Battlefield*. Oxford: University Press, 1987.

Benedikt, Michael. *Cyberspace*. Cambridge, Mass.: MIT Press, 1994.

Bernstein, Richard, and Ross H. Munro. *The Coming Conflict with China*. New York: Knopf, 1997.

bin Khaled, HRH Gen. Sultan. *Desert Warrior*. New York: HarperCollins, 1995.

Bobkov, Filipp. *KGB and Power*. Moscow: Veesan MP Publishing House, 1995.

Brendley, Keith, and Randall Steeb. *Military Applications of Microelectromechanical Systems*. Santa Monica, Calif.: Rand Corp., 1992.

Brockman, John. *Digerati*. San Francisco: HardWired, 1996.

Brown, Ben, and David Shukman. *All Necessary Means*. London: BBC Books, 1991.

Brown, Michael L. *The Revolution in Military Affairs: The Information Dimension*. Fairfax, Va.: AFCEA International Press, 1996.

Campen, Alan D., et al., eds. *The First Information War*. Fairfax, Va.: AFCEA International Press, 1992.

Chesneaux, Jean. *Secret Societies in China*. London: Heinemann Educational Books, 1971.

Clawson, Patrick L., ed. *Energy and National Security in the 21st Century*. Washington, D.C.: Institute for National Strategic Studies, October, 1995.

Codevilla, Angelo. *Informing Statecraft*. New York: Free Press, 1992.

Cordesman, Anthony H., and Abraham R. Wagner. *The Lessons of Modern War*. Boulder: Westview, 1990.

Davis, Paul K. *New Challenges for Defense Planning*. Santa Monica, Calif.: Rand Corp., 1989.

Deacon, Richard. *Kempei Tai: A History of the Japanese Secret Service*. New York: Beaufort, 1983.

de la Billiere, Gen. Sir Peter. *Looking for Trouble*. London: HarperCollins, 1994.

———. *Storm Command*. London: HarperCollins, 1992.

Dewar, Col. Michael. *The Art of Deception in Warfare*. Newton Abbot: David and Charles Military Books, 1989.

Earley, Pete. *Confessions of a Spy*. London: Hodder & Stoughton, 1997.

Eftimiades, Nicholas. *Chinese Intelligence Operations*. Annapolis: Naval Institute Press, 1994.

Epstein, Edward Jay. *Deception*. London: Whallen, 1989.

Faligot, Roger, and Pascal Krop. *La Piscine*. Oxford: Basil Blackwell, 1989.

Faligot, Roger, and Remi Kauffer. *The Chinese Secret Service*. London: Headline Books, 1989.

Fialka, John J. *War by Other Means*. New York: Norton, 1997.

Freedman, Lawrence, and Efraim Karsh. *The Gulf Conflict, 1990–1991*. London: Faber and Faber, 1993.

Friedman, George, and Meredith Friedman. *The Future of War*. New York: Crown, 1996.

Godson, Roy, Ernest R. May and Gary Schmitt. *U.S. Intelligence at the Crossroads*. Washington, D.C.: National Strategy Information Center, 1995.

Hansen, James H. *Japanese Intelligence*. Washington, D.C.: NIBC Press, 1996.

Homer-Dixon, Thomas. "On the Threshold: Environmental Changes as Causes of Acute Conflict." International Security, Fall 1991.

Kahn, David. *The Code Breakers*. New York: Scribner, 1996.

Knight, Amy. *Spies Without Cloak*. Princeton: University Press, 1996.

Krepinevich, Andrew F. *The Bottom-Up Review: An Assessment*. Washington, D.C.: Defense Budget Project, February 1994.

———. *The Conflict Environment of 2016: A Scenario-Based Approach*. Washington, D.C.: Center for Strategic and Budgetary Assessments, 1996.

Larson, Eric V. *Casualties and Consensus*. Santa Monica, Calif.: Rand Corp., 1996.

Levathes, Louise. *When China Ruled the Seas*. New York: Simon & Schuster, 1994.

Libicki, Martin. *Defending Cyberspace and Other Metaphors*. Washington, D.C.: National Defense University, February 1997.

———. *What Is Information Warfare?* Washington, D.C.: National Defense University, August 1995.

Littman, Jonathan. *The Fugitive Game*. New York: Little, Brown, 1996.

Marenches, Count de, and David A. Andelman. *The Fourth World War*. New York: Morrow, 1992.

Maruyama, Xavier K. *Technologies in Support of International Peace Operations.* Monterey, Calif.: Naval Postgraduate School, 1996.

McCarthy, Shaun Paul. *The Function of Intelligence in Crisis Management.* St. Andrews: University of St. Andrews, 1996.

McNab, Andy. *Bravo Two Zero.* London: Bantam, 1993.

Molander, Roger C., et al. *Strategic Information Warfare: A New Face of War.* Santa Monica, Calif.: NDRI/Rand, 1996.

Munro, Neil. *The Quick and the Dead.* New York: St. Martin's, 1991.

Newhouse, John. *The Nuclear Age.* London: Michael Joseph, 1989.

O'Donnell, Thomas A. *Superacids and Acidic Metals as Inorganic Chemical Reaction Media.* New York: VCH Publishers, 1992.

Otis, Glenn, and Dr. Peter Cherry. *Information Campaigns.* Ann Arbor, Mich.: Vector Research Inc., 1991.

Paschall, Rod. *LIC 2010.* Washington, D.C.: Brassey's, 1990.

Porch, Douglas. *The French Secret Service.* New York: Farrar, Straus & Giroux, 1995.

Powell, Colin, and Joseph E. Persico. *My American Journey.* New York: Random House, 1995.

Pringle, Peter, and James Spigelman. *The Nuclear Barons.* London: Michael Joseph, 1981.

Richelson, Jeffrey T. *The U.S. Intelligence Community.* Boulder: Westview, 1995.

Romerstein, Herbert, and Stanislav Levchenko. *The KGB Against the "Main Enemy."* Lexington, Mass.: Lexington Books, 1989.

Ryan, Chris. *The One That Got Away.* London: Century, 1995.

Schneier, Bruce. *Applied Cryptography.* New York: John Wiley, 1996.

Schwartzstein, Stuart J. D. *The Information Revolution and National Security.* Washington, D.C.: Center for Strategic and International Studies, 1996.

Schweizer, Peter. *Friendly Spies.* New York: Atlantic Monthly Press, 1993.

Shaker, Steven M., and Alan R. Wise. *War Without Men.* Washington, D.C.: Pergamon, 1988.

Shimomura, Tsutomu. *Takedown.* New York: Hyperion, 1996.

Slatalla, Michelle, and Joshua Quittner. *Masters of Deception.* New York: HarperCollins, 1995.

Smith, Michael. *New Cloak, Old Dagger.* London: Victor Gollancz, 1996.

Snyder, Alvin. *Warriors of Disinformation.* New York: Arcade, 1995.

Stoll, Clifford. *The Cuckoo's Egg.* New York: Doubleday, 1989.

———. *Silicon Snake Oil.* New York: Doubleday, 1995.

Sun Tzu. *The Art of War.* New York: Delacorte, 1983.

Teicher, Howard, and Gail Radley Teicher. *Twin Pillars to Desert Storm.* New York: Morrow, 1993.

Urban, Mark. *UK Eyes Alpha: The Inside Story of British Intelligence.* London: Faber and Faber, 1996.

U.S. News & World Report, staff. *Triumph Without Victory.* New York: Times Books, 1992.

Van Creveld, Martin. *Technology and War.* Oxford, U.K.: Pergamon, 1991.

Watson, Bruce W., Bruce George, MP, Peter Tsouras, and B. L. Cyr. *Military Lessons of the Gulf War.* London: Greenhill, 1991.

Weinberger, Caspar, and Peter Schweitzer. *The Next War.* Washington, D.C.: Regnery, 1996.

Wiener, M. J. *Efficient DES Key Search.* Ottawa: School of Computer Science, Carleton University, 1994.

Winkler, Ira. *Corporate Espionage.* Rocklin, Calif.: Prima Publishing, 1997.

Woodward, Bob. *Veil: The Secret Wars of the CIA 1981–1987.* New York: Simon & Schuster, 1987.

Young, John Robert. *The Dragon's Teeth.* London: Century Hutchinson, 1987.

Other Publications:

C2 Warfare. Air Force Staff College Student Text, 1996.

Cryptography's Role in Securing the Information Society. National Research Council, 1996.

Glossary of INFOSEC and INFOSEC Related Terms. Vols. 1 and 2. Idaho State University, 1996.

Hacker Chronicles. CD-Rom produced and distributed by P-80 Systems.

IC21: Intelligence Community in the 21st Century. Permanent Select Committee on Intelligence, 1996.

Information Architecture for the Battlefield. Report of the Defense Science Board Summer Study Task Force. Washington, D.C.: Office of the Undersecretary of Defense, October 1994.

Information Guide. Department of Defense Institute, 1997.

Information Operations—FM 100-6. Headquarters, Department of the Army, August 1996.

InfoWarCon5: Conference Proceedings. NCSA, 1996.

In from the Cold. Twentieth Century Fund Task Force on the Future of U.S. Intelligence. New York: Twentieth Century Fund Press, 1996.

Intel 2000: A Quarterly Publication of Intelligence, Counterintelligence and Espionage. Vol. 1, No. 2, Fall 1996. Washington, D.C.

Intelligence and the New World Order. International Freedom Foundation. Buxtehude, 1992.

Interagency and Political-Military Dimensions of Peace Operations: Haiti—A Case Study. Washington, D.C.: National Defense University, 1995.

National Information Systems Security Conference Proceedings. Vols. 1 and 2. National Institute of Standards and Technology, 1996.

Overseas Chinese Business Networks in Asia. Australia: Department of Foreign Affairs and Trade, 1995.

Sun Tzu and Information Warfare. Washington, D.C.: National Defense University, 1997.

Terrorism with Chemical and Biological Weapons. Alexandria, Va.: Chemical and Biological Arms Control Institute, 1997.

Today's Cryptography, Frequently Asked Questions. RSA, 1993.

White Book of the Russian Special Services. Moscow: Obozrevatel Publishing House, 1995.

Index